18753

G. SCHWARZENBACH
Eidg. Technische Hochschule, Zurich, Switzerland
A. A. SMALES
United Kingdom Atomic Energy Authority, Harwell, England
EDWARD WICHERS
National Bureau of Standards, Washington, D.C.

ADVANCES IN ANALYTICAL CHEMISTRY AND INSTRUMENTATION

Volume 11

New Developments in Gas Chromatography

CONTRIBUTORS

JOHN R. CONDER, *Department of Chemical Engineering, University of Wales, University College of Swansea, Swansea, Wales*

JAMES E. GUILLET, *Department of Chemistry, University of Toronto, Toronto, Ontario*

STANLEY H. LANGER, *Department of Chemical Engineering, University of Wisconsin, Madison, Wisconsin*

DAVID A. LEATHARD, *Department of Chemistry and Biology, Sheffield, Polytechnic, Sheffield, England*

PETER F. MCCREA, *Research Department, The Foxboro Company, Foxboro, Massachusetts*

D. A. PATTERSON, *Home Office Central Research Establishment, Aldermaston, Reading, Berkshire, England*

JAMES E. PATTON, *Research Laboratories, Eastman Kodak, Rochester, New York*

C. A. WELLINGTON, *Department of Chemistry, University College of Swansea, Swansea, Wales*

New Developments in Gas Chromatography

Edited by
HOWARD PURNELL

Department of Chemistry
University College of Swansea

Swansea, Wales

AN INTERSCIENCE®PUBLICATION

JOHN WILEY & SONS
New York · London · Sydney · Toronto

Library of Congress Cataloging in Publication Data

Purnell, Howard, 1925—
  New developments in gas chromatography.

  (Advances in analytical chemistry and instrumentation.)

  1. Gas chromatography. I. Title. II. Series.

QD71.A2 vol. 11 [QD117.C5]   543'.008s [544'.926]   73-4772

ISBN 0-471-70241-2

Printed in the United States of America

10 9 8 7 6 5 4 3 2 1

INTRODUCTION TO THE SERIES

The scope and even the purpose of analytical chemistry is growing so amazingly that even the dedicated scientist with time on his hands cannot follow the significant developments which appear now in every increasing numbers. Analytical chemistry has its new and wider roles and these new developments must become everyday working knowledge and be translated into practice. At present a serious time lag still exists between evolution and practice. This new venture aims to bridge the hiatus by presenting a continuing series of volumes whose chapters deal not only with significant new developments in ideas and techniques, but also with critical evaluations and the present status of important, but more classical, methods and approaches. The chapters will be contributed by outstanding workers having intimate knowledge and experience with their subject.

It is the hope and belief that *Advances in Analytical Chemistry and Instrumentation* will offer a new medium for the exchange of ideas and will help assist effective, fruitful communication between the various disciplines of analytical chemistry.

Additionally, articles of this series will discuss developments in other fields, such as biology, physics, electronics, and mathematics, which are within the scope of analytical chemistry in this broader context. These would include, for example, applications of kinetics, isotopic tracers, computers, and modern physical tools to studies of complex molecules, short-lived species, ultraclean systems, continuous plant streams, and living organisms.

These volumes contain articles covering a variety of topics presented from the standpoint of the nonspecialist but retaining a scholarly level of treatment. Although a reasonably complete review of recent developments is given, a dry and terse cataloguing of the literature without description or evaluation is avoided. The scope of the *Advances* is flexible and broad, hoping to be of service to the modern analytical chemist whose profession each day demands broader perspectives and solution of problems with increased complexity. The periodical literature is inherently specialized and the appearance of suitable monographs takes place only after many years. Reviews are frequently directed to the specialist and often lack adequate description or evaluation. *Advances* hope to fill in the resulting need for critical comprehensive articles surveying various topics on a high level satisfying the specialist and nonspecialist alike. Comments and suggestions from readers are heartily welcome.

THE EDITORS

PREFACE

For those of us who have been involved in gas chromatography more or less since its inception, it is a chastening thought that our once-exciting infant has come of age. Each month, the journals of the world used to contain accounts of dramatic new developments in theory and technique. These days are far behind us now, and progress is more placid, with a staidness perhaps more befitting to a technique entered into adulthood. Nevertheless, it is salutory to think that, for the newest class of college graduates, gas chromatography always was.

Five years ago in an earlier volume in this series, I described my editorial approach as an attempt to find a balance between established areas of work deserving of review and topics that would be likely to develop. This volume is constructed on the same basis; it must be confessed, however, that it now proves difficult to visualize more than a few areas on the analytical side where great advance can be hoped for in the short term. The most striking growth in gas chromatographic application in recent years—and, it may be anticipated, for the immediate future—is on the nonanalytical applications side. For this reason, such topics take up a major part of this book.

As before, I have been fortunate in securing the services of authors who have both the authority and distinction to make the editor's work minimal. Again, what success this volume may enjoy is theirs.

HOWARD PURNELL

Swansea, Wales
January 1973

vii

CONTENTS

Forensic Applications of Gas Chromatography

D. A. PATTERSON, *Home Office Central Research Establishment, Aldermaston, Reading, Berkshire, England*

I. INTRODUCTION

A eulogy of the gas chromatographic technique would not be within the scope of this chapter; but it would also be improper if the impact that its development has made on forensic science were not adequately acknowledged. First used in forensic science in 1955(1), gas chromatography is now employed in more analyses, numerically, than any other technique. Moreover, it is doubtful whether any analytical procedure has applicability to such a wide range of problems.

1

In the United Kingdom, when the Road Safety Act (1967) brought British law into line with that of a number of other countries, it became an offense to drive with more than the prescribed level of alcohol in the blood. Gas chromatography was adopted then as the standard method of analysis, and by 1970 more than 30,000 samples were being analyzed annually. But apart from this routine application, the inherent sensitivity, specificity, and versatility of the technique make it an indispensable part of the analytical armory of the forensic scientist, and it finds applications in such diverse fields as toxicology, drug abuse, and arson investigation, and in the identification of paints and fibers.

A review such as this cannot be comprehensive, but it is hoped that by reference to the works cited here, an analyst interested in a particular aspect will be able, at least indirectly, to trace the material or method that is relevant to his requirements.

II.   CARBON MONOXIDE

Poisoning by carbon monoxide must have presented a problem for the toxicologist since man discovered fire, and a vast amount of literature has appeared on the subject. Generally, methods for its determination are divided into those for the gas itself and those for carboxyhemoglobin. Within these divisions, physical, chemical, and biochemical methods have been described and admirable reviews are available (2, 3).

In the majority of cases, the forensic scientist is required to measure the amount of carboxyhemoglobin in a blood sample; if the blood is fresh and uncontaminated, a spectrophotometric method such as that described by Kampen and Klouwen (4) is eminently satisfactory. During the process of putrefaction, however, products that interfere with spectrophotometric procedures are formed, and a method allowing the specific estimation of carbon monoxide in the sample is necessary. The use of gas chromatography, together with subsidiary determinations for hemoglobin and iron (to calculate the total hemoglobin content) gives more reliable results in these cases (3,5–7, 11).

Usually, carbon monoxide is liberated from the blood sample by addition of potassium ferricyanide plus a detergent, although the use of sulfuric, hydrochloric, or lactic acid has also been reported (5, 8). A mixture containing 2.5% potassium ferricyanide, Teepol L or Triton X-100, and buffered to pH 9, liberates all carbon monoxide from a 1-ml blood sample within 10 minutes (3). The liberated gas is then transferred by way of a gas loop and sampling valve to the column. Blackmore (3) used a stainless steel column, 5 ft long and packed with 60-80 mesh molecular sieve 5A from which the fines had been removed, whereas Goldbaum et al. (5, 11) used the same material in the form of 1/16-in pellets in a 2-m column which was run at 75 to 100°C, with helium as the carrier gas. Detection was by means of a katharometer and a typical chromatogram (3) is shown in Figure 1. Dominguez and his co-workers (9) and McCredie and Jose (7) have also described procedures using aluminosilicate sieves.

An alternative, indirect method of detection is provided by the incorporation of a nickel-catalyzed reduction stage; here a flame ionization detector serves to detect the resulting methane. This system is more sensitive and has been used (6, 10, 12) for accurate measurements of the carbon monoxide content of the blood of smokers. Less than 100 μl of blood suffices for the analysis.

For purposes of calibration, investigators commonly use both pure carbon monoxide gas and commercially available gas mixtures, as well as prepared mixtures of fully carboxylated and fully oxygenated blood

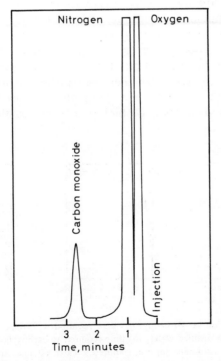

Fig. 1. Gas chromatogram (3) of a mixture of air and carbon monoxide from blood containing carboxyhemoglobin.
By courtesy of *Analyst*.

samples, peak height measurements being generally considered to give sufficient precision. Great care is necessary if mixed blood samples are to be used for calibration; for it has been shown (3) that blood stored at 20°C can lose 25% of its carbon monoxide combining power within a seven-week period. Treatment of fresh blood, with sodium dithionite, followed by carboxylation with pure carbon monoxide in a Lerquin apparatus and then flushing with oxygen-free nitrogen for 10 min, gives a 100% carboxylated sample. Figure 2 demonstrates the effect of sodium dithionite in achieving full carboxylation of fresh and stored blood.

III. VOLATILE SOLVENTS AND ANESTHETICS

The need to analyze for volatile low-molecular-weight compounds most commonly arises when deaths occur during general anesthesia or in connection with suicides involving the ingestion of household cleaners. Accidental deaths from "glue-sniffing" and from the injudicious use of cleaning solvents are not uncommon. Various methods for introducing these volatile compounds to the gas chromatograph have been recommended. Generally, a preliminary separation procedure, such as solvent extraction or steam distillation, is followed by injection; some workers, however, advocate direct injection of blood or urine samples (16). Bonnichsen and Maehly, in very comprehensive reports of poisonings by aromatic hydrocarbons (13) and chlorinated aliphatic hydrocarbons (14), favored preliminary steam distillation of biological samples, the steam distillate

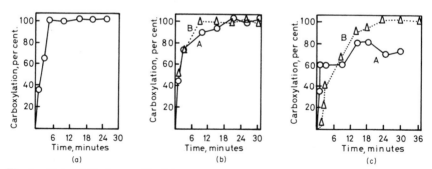

Fig. 2. Data of Blackmore (3) giving: (*a*) time required to carboxylate a sample of fresh blood diluted 1 : 4 with distilled water; (*b*) time required to carboxylate whole blood without *A* and with *B*, sodium dithionite; (*c*) time required to carboxylate a sample of blood stored at 37°C for 11 days and diluted with water before gassing: curve *A*, whole blood; curve *B*, whole blood and sodium dithionite.
By courtesy of *Analyst*.

being absorbed on to ammonium sulfate in a flask that is subsequently closed with a rubber cap. A headspace sample is then injected on to the column.

When the presence of chlorinated hydrocarbons is suspected, the very sensitive Fujiwara reaction can be used as a qualitative screening test prior to chromatography. Bonnichsen and Maehly (13) stressed the danger in relying too heavily on this test, however, and pointed out that a number of compounds give negative reactions, including hexachloroethane, ethyl chloride, monochloroethylene, *sym*-dichloroethylene, and monochloroethanol. Trichloroethanol, the metabolite of both trichloroethylene and chloral hydrate, gives a weak response.

Bonnichsen and Maehly used a 6-ft, 20% Apiezon column both for aliphatic hydrocarbons (125°C), and for aromatic hydrocarbons (190°C). A 2.5-m, 25% Carbowax 1540 column was used, in addition, for confirmation of the identity of the aliphatic chlorinated hydrocarbons. Retention data were given for benzene, toluene, *o*-, *m*-, and *p*-xylenes, indene, naphthalene, methylene chloride, chloroform, dichloroethane, carbon tetrachloride, trichloroethylene, and perchloroethylene. Analyses for carbon tetrachloride and methylene chloride in accidental poisoning cases have also been the subjects of monographs by other workers (17, 18), and Nelms and his co-workers (19) have demonstrated the presence of hydrocarbons in blood and tissue in two cases of gasoline poisoning. Hall and Hine (20) have reported two cases of trichloroethane intoxication.

Recently (14), a scheme of analysis based upon the gas chromatographic technique has been described for the identification of almost 600 drugs and metabolites in human viscera; of these, a large number are volatile, low-molecular-weight compounds. A solvent extraction procedure in the initial stages is followed by analysis with seven gas chromatographic systems. Confirmation of structure is by infrared spectrophotometry and by mass spectrometry.

Special attention has been paid (15, 16) to the identification of aerosol propellants. These are composed of hydrocarbons, chlorinated hydrocarbons, fluorinated hydrocarbons, and inert inorganic gases; mixed propellants are commonly encountered. In one study (15), a 22-ft column packed with 20% silicone grease on firebrick was used; the column, which was at ambient temperature, allowed separation of eight common components in 15 min. Carbon dioxide and nitrous oxide were not resolved under these conditions, however, and where either gas was present, it was necessary to use an alternative, 6-ft porous polymer (Poropak Q) column for identification.

Fluorinated hydrocarbons, often present as propellants in pressurized aerosols used in the treatment of asthma, have been implicated in sudden deaths of patients using them. Blood levels of the fluorocarbons in volunteers inhaling from placebo inhalers, have been measured (16), analysis being conducted with a 7-ft, 20% Carbowax 20M column at 70°C, in conjunction with an electron capture detector. Maximum blood concentrations of 1.7 μg/ml were observed. However, it is the author's opinion that whether the fluorocarbons are directly or indirectly causative of the cardiac crises in cases of this type has not been unequivocally established.

Of the volatile sedatives and general anesthetics, paraldehyde (21), ethchlorvynol (21, 22), ether (23), and halothane (24–26) have received most attention. In one report (26), in which a rapid, sensitive method for determination of halothane in blood was described, it was claimed that it was possible to monitor the halothane blood levels during the course of anesthesia. If this were indeed possible, the procedure could become a valuable adjunct to surgical techniques.

IV. ETHANOL

A number of comprehensive reviews of the early literature on alcohol analysis have appeared (27–30), although of necessity these described gas chromatographic methods only in passing. More recently, Walls and Brownlie (31) have dealt with alcohol and drugs in relation to driving and driving offenses, and this book, as an introduction and background to the subject, provides excellent reading.

The first report of the gas chromatographic separation of alcohols appeared in 1956 (32); but it was almost ten years before the method became generally accepted as the method of choice for the quantitative determination of ethanol. Like most other forensic science laboratories, the Government Laboratory in Stockholm, until 1962, had used a chemical as well as an enzymatic method for the determination of alcohol in blood and in urine; in some cases, large discrepancies had been noted between the results obtained by the two methods. For example, by use of a mixed polyethyleneglycol–polypropyleneglycol column in a "gas fractometer" equipped with a flame ionization detector and by direct injection of urine, the presence of methanol, acetone, and acetaldehyde, as well as ethanol, was detected in the samples; but the amounts recorded differed (33). Calculating from direct comparison of peak heights, it was found that in all cases the results obtained from the gas chromatographic determination of ethanol agreed more closely with those obtained from the enzymatic method than did those obtained by the chemical method. However, this technique was not applicable to blood samples and, from the forensic point of view, determination of the ethanol content of blood was considered to be more desirable, since this value was more likely to give a true reflection of the state of intoxication of the individual.

Subsequently, several approaches to the problem were adopted. These were: (a) direct injection of blood into the gas chromatographic column (34–36), (b) azeotropic distillation of the blood prior to chromatography (37), (c) preliminary separation of the ethanol by a solvent extraction procedure (38, 42), and (d) injection of head space samples (39–41). It was, however, the procedure described in 1966 by Curry, Walker, and Simpson (36) that first made it possible to perform the analysis with sufficient precision, accuracy, and speed that it became accepted for routine use in the United Kingdom. Furthermore, the procedure was applicable to small blood samples, in accordance with a later requirement of the Road Safety Act. The analysis was at 85°C with a 5-ft, 10% PEG 400 column connected to a flame ionization detector feeding an integrator. The blood (ca. 20 μl) was diluted with ten times its volume of aqueous n-propanol solution using a Griffin and George type 221 hemoglobin-type diluspence, the propanol serving as internal standard. Approximately 1 μl of the diluted blood sample was then injected directly into the column, and ratios of peak areas of ethanol to propanol were calculated. Calibration was achieved by using blood samples containing known amounts of ethanol and plotting the ratio of ethanol to propanol pulses from the integrator against the concentration of ethanol in the blood. The graph obtained was independent of the volume of blood injected; when 20 injections of the same solution were made (250 mg/100 ml), the error in the determined ethanol-to-propanol ratio had a standard deviation of less than 0.75, expressed as a percentage of the mean value. Chromatographic separation was complete in 4.5 min, and ether, acetaldehyde, acetone, methanol, and higher alcohols were all easily resolved from ethanol.

The foregoing procedure has been outlined in some detail because of its known excellence and also because it has been used widely without modification for more than five years now—a rare achievement in these days of rapid "progress." Figure 3 shows a typical chromatogram obtained using the procedure. If the method has a weakness, it is that isopropanol (which is sometimes used for swabbing the skin in the area from which blood is to be taken) is not completely resolved from ethanol by the PEG column. This difficulty is overcome in practice by duplication of the analysis using a column containing Poropak Q at 185°C when these two alcohols are easily resolved. Although the procedure described is based on direct injection of a diluted blood sample, contamination of the column is not a great problem. Some tailing of peaks occurs after about 1000 injections, but resolution can be restored by repacking the first few inches of the column. Even this necessity can be obviated, however, by choice of a

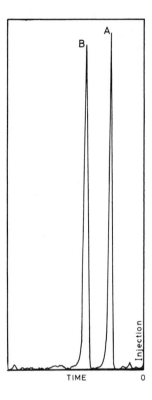

Fig. 3. Gas chromatogram of blood containing ethanol following procedure of Curry, Walker, and Simpson: curve *A*, ethanol; curve *B*, *n*-propanol.

gas chromatograph that allows the use of a glass insert near to the inlet end of the column. The blood is deposited within the insert, which can be replaced easily when necessary.

Adequate storage of blood samples pending analysis is of great importance, for it has been demonstrated (43) that many organisms commonly found in or on the body during life, or following post mortem contamination, are capable of producing significant quantities of ethanol from carbohydrate or tissue. In addition, of course, measures are necessary to prevent clotting of the samples, particularly if a direct injection procedure is to be used. In practice, stability and fluidity of the blood are readily achieved by addition of sodium fluoride and potassium oxalate, respectively, to the sample. In post mortem cases, bacterial formation of ethanol could well be underway by the time the blood sample is taken; indeed, it is very doubtful whether any interpretation at all can be placed on a finding of ethanol in a single blood sample from such cases. It is perhaps reassuring to note, however, that under laboratory conditions where ethanol is known to have been produced by bacteria, n-butyric acid and isobutyric acid have been consistently detected also (43). Since these are resolved from one another and from ethanol on the Poropak Q column and are not normally detected in fresh samples, a good indication of bacterial action can be obtained. As a result of his investigations into the bacterial production of ethyl alcohol, Blackmore (43) suggested the following very useful analytical sequence for the determination of ethanol in post mortem samples.

1.   All analyses should be undertaken by gas chromatography using two columns working in dissimilar operating conditions.

2.   Urine from an intact bladder is the fluid of choice, since ethanol is unlikely to be produced by its bacterial contamination.

3.   If urine is not available, blood sampling should be from left and right heart and one other peripheral source. The samples should be immediately preserved, and the presence of n-butyric acid or isobutyric acid should render the analysis suspect.

4.   In the absence of urine and blood, muscle should be taken from three peripheral sites and stored at 0°C before homogenization with distilled water and treatment in a similar manner to blood.

Current aims of those concerned with blood alcohol analyses are in the development of a fully automated procedure, although, so far as the author is aware, this has not yet been achieved. The development of the Perkin-Elmer Multifract F40 Gas Chromatograph represents a significant advance in this direction, however. Using this apparatus it is still necessary to perform the dilution of the blood manually, but thereafter, up to 30 samples can be analyzed without attention at the rate of 10 to 15 samples per hour. Accuracy and reproducibility of results are comparable to those obtained using the method of Curry, Walker, and Simpson.

The F40 comprises a thermostatted sampling turntable, which holds the samples (previously prepared for analysis) in vials closed with a rubber septum cap. When equilibrium has been reached, samples of vapor above the diluted blood are withdrawn and successively injected into the column by means of an electropneumatic device. Each stage of the analysis can be programmed in advance; automated baseline correction is incorporated, and the chart is marked during each analysis to allow identification of samples that have been analyzed.

In some areas of the world, evidence of breath alcohol levels is permissible in drunk driving cases, and at least one device based on gas chromatography of a breath sample is available (44). The subject can blow directly into the instrument and analysis is claimed to be rapid and accurate. The major disadvantage of such an approach, however, appears to lie in the difficulties involved in the efficient collection and storage, for further confirmatory analysis in the laboratory, of additional samples.

Finally, in connection with ethanol, inquiries into the illicit production of alcoholic beverages should be mentioned. Comparative analyses are usually required, and both inorganic and organic congeners can be investigated. Of the latter, acids, esters, and higher alcohols, which are co-products of the fermentation process, have been shown to be present in different beverages in varying proportions (45–47), and whiskies, gins, rums, and brandies have been characterized by gas chromatography of these compounds (46, 48).

V. DRUGS

The literature dealing with gas chromatographic separation and identification of drugs is now voluminous, and the forensic scientist has at least a potential use for a large proportion of this material. Fortunately, Gudzinowicz has comprehensively reviewed the subject to 1967, and other workers (50, 51) have abstracted methods which they consider to be particularly applicable in the forensic context. Of the several thousand drug preparations available, it is possible to analyze for a very large proportion by gas chromatography; but whether this is necessarily the technique of choice in the routine laboratories is another matter. Aspirin and lysergic acid diethylamide (LSD), for example, are substances for which gas chromatographic methods have been described; in practice, however, other methods are usually more convenient. In view of this, no attempt has been made here to give comprehensive coverage; rather, selected classes of substances have been chosen for which the gas chromatographic approach has a particular advantage.

Identification may be required of the pure drug itself, of the drug compounded into dosage form, or of the drug in a body fluid taken from a live patient or from the organs and body fluids taken post mortem. In difficult cases, the process of putrefaction may be underway or even well advanced while, to complicate the picture further, there is sometimes strong circumstantial evidence that a particular drug has been ingested, whereas in other cases a general "screen" for drugs is required. Usually, a preliminary solvent extraction procedure is necessary prior to chromatography, perhaps following a protein precipitation stage. Almost without exception, gas chromatography is only one of the several analytical techniques employed in the identification. For practical details of a reliable approach to the search for the unknown poison in human viscera, the reader is referred with confidence to the text book of Curry (50).

One important point that should be emphasized is that separate gas chromatographic columns must be reserved for the analysis of acidic and basic fractions from biological material, for an acid fraction would always contain nonvolatile acidic material and this having been injected onto a column, could result in salt formation on subsequent injection of a base. The converse also applies.

A. Acidic and Neutral Drugs

1. Barbiturates

The first attempt to identify barbiturates by gas chromatography was in 1960, when Janák (52) pyrolyzed their sodium salts and showed that the products presented characteristic pyrolysis patterns. This approach was of limited value, however, especially when more than one barbiturate was present or when the compounds were available only in the presence of impurities. Given this stimulus, however, within two years a large number of barbiturates had been separated on a range of columns of differing polarities. For example, Brochmann-Hanssen and Svensen (53, 54) used nonpolar Apiezon L and SE-30 phases, as well as the more polar polyesters and polyethylene glycols, Parker and Kirk (55) also used SE-30, and Van den Heuval et al. (56) had success with QF-1 and neopentyl glycol succinate. More recently, the use of Carbowax (57) and cyclohexanedimethanol succinate (58) have been reported, and elution characteristics for more than 20 barbiturates are recorded for each.

The author's principal experience has been with a 6-ft, 10% Apiezon-L column operated at 200 to 225°C, as recommended by Leach and Toseland (59). As Figure 4 indicates, excellent separation of commonly encountered barbiturates is achieved. For confirmation of identity, methyl (60–62) or trimethylsilyl derivatives (63) can be formed either on the column or prior to injection. Street, for example (63), reacted the barbiturates, dissolved in freshly distilled chloroform, with a 20% solution of bistrimethylsilylacetamide (BSA) in n-hexane to form the derivatives on the column. The resultant peaks were said to be more symmetrical than those of the unmodified compounds, and sensitivity of detection was increased. For methylation, one or other of dimethylsulfate, trimethylammonium hydroxide, and trimethylanilinium hydroxide is the favored reagent.

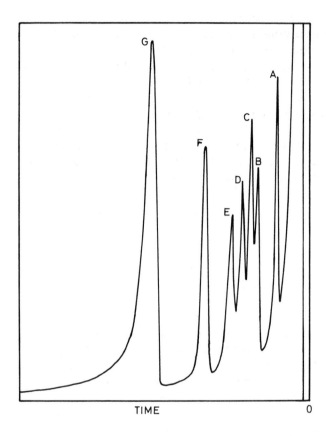

TIME   0

Fig. 4. Gas chromatogram of commonly encountered barbiturates. Analysis with Apiezon-L column at 215°C: curve *A*, barbitone; curve *B*, butobarbitone; curve *C*, amylobarbitone; curve *D*, pentobarbitone; curve *E*, quinalbarbitone; curve *F*, hexobarbitone; *G*, phenobarbitone.

Although most barbiturates disappear from the blood fairly rapidly, their elimination from the body as a whole takes place only over a period of several days. It is well established that phenobarbitone and amylobarbitone are hydroxylated in the body and that the hydroxy derivatives appear in the urine in both free and conjugated forms. Gas chromatographic analysis of urine samples collected following therapeutic dosage has clearly revealed the presence of *p*-hydroxyphenobarbitone (64) and of hydroxyamylobarbitone (65). For the latter, the presence of the metabolite in plasma has also been demonstrated; a 2% FFAP on Aeropak 30 column was used at 225°C, triphenylene serving as internal standard. Excretion of the metabolite was observed in the urine even after six days. The forensic toxicologist could measure the proportion of metabolite to unchanged drug excreted and, on the basis of the result, estimate the time that had elapsed since ingestion of the drug. This information could be very relevant in certain cases.

2. Diphenylhydantoin

Several cases have been reported of fatal poisoning (66–69), and adverse interaction of the anticonvulsant diphenylhydantoin with other drugs (70, 71). Methods involving ultraviolet spectrophotometry and thin-layer chromatography have been described, but gas chromatography is probably more specific. One procedure (72) for the unmodified drug uses a column of 5% DC-200 on Gas Chrom Q at 220°C but, more commonly, analysis involves derivative formation. Methylation can be achieved with diazomethane (73, 74) or with tetramethylammonium hydroxide (75), whereas a trimethylsilyl (TMS) derivative is readily formed with bistrimethylsilylacetamide (BSA) (76). The latter reagent also reacts with the p-hydroxyphenyl metabolite of diphenylhydantoin, allowing simultaneous determination of drug and metabolite. An internal standard is suggested in this case, and separation is best achieved on a 3% OV-17 column at 200°C.

Toxic blood levels range from 20 to 50 $\mu g/ml$; in cases of lethal overdose, levels of about 100 $\mu g/ml$ may be observed (77).

3. Glutethimide

Most workers who have been concerned with the gas chromatography of barbiturates have also recorded the retention characteristics of glutethimide under the same conditions. Although it is not a barbituric acid derivative, glutethimide has a similar pharmacological action to this class of drugs; moreover, it appears in the "acid fraction" resulting from an extraction of tissue or body fluid. The major metabolite of glutethimide is 2-phenyl glutarimide, and this too appears in the "acid fraction."

A number of methods for the gas chromatographic determination of glutethimide have been described (78–82), and the most recent one (82) is claimed to offer greater rapidity and sensitivity than others. Blood plasma (2 ml) is extracted with hexane and, after evaporation, di-n-butyl phthalate is added as internal standard. Chromatography is on a 5% SE-30 column at 195°C, and both glutethimide and the internal standard are eluted within 4 min. The whole analysis may be completed within 1 hr, the limit of measurement of glutethimide in plasma being 0.03 mg %. Since therapeutic levels range from 0.2 to 0.6 mg %, the sensitivity available is more than adequate. However, it would be of interest to know the retention time of 2-phenyl glutarimide under these conditions.

Grosjean and Noirfalise (83) have described a method for the separation of both glutethimide and its metabolite in which an 8-ft, 8% XF 1112 column at 195°C is used along with an internal standard for the purposes of quantitation. Separation occurs in 18 min.

4. Paracetamol

Paracetamol is readily available without prescription in the United Kingdom and self-poisoning has become a fairly common occurrence. Most known procedures for its determination in body fluids have been based on hydrolysis to p-aminophenol, followed by a coupling reaction to form an azo-dye, the latter compound then being estimated spectrophotometrically. Such approaches tend to be tedious, however, and a more rapid method is to be commended.

Prescott (83) has described a gas chromatographic method in which the extracted paracetamol is converted to its trimethylsilyl derivative which is chromatographed with a 10% OV-17 column at 200°C. Probably more convenient is the direct method of Grove (84) wherein plasma and urine are first saturated with ammonium sulfate and then extracted with ether; diphenylphthalate is added as internal standard and separation is achieved on a 2% FFAP column at 240°C. A typical analysis is shown in Figure 5. Both the internal standard and paracetamol elute within 11 min; aspirin, phenacetin, codeine, and orphenadrine, drugs that are commonly encountered in combination with paracetamol, do not interfere with the analysis. Phenacetin has a retention time of 1.5 min under these conditions, and the author suggests that the method may be applied to samples where phenacetin has been ingested. Plasma levels of paracetamol found following therapeutic and toxic doses of the drug are recorded.

5. Cannabis

Cannabis, although widely abused, is not a very toxic preparation and few deaths from its ingestion have been recorded; indeed, it is a matter for debate whether any of these have been fully authenticated. Scarce too, are reports of the detection of its components or metabolites in body fluids after ingestion and, similarly, there is considerable debate about whether this is possible. Yet next to alcohol, analysis for cannabis is required in more cases in forensic science laboratories throughout the world than for any other drug.

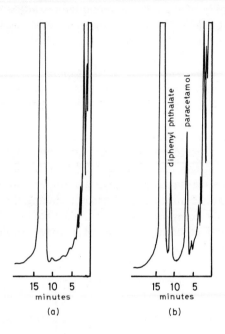

Fig. 5.   Results of Grove (84) giving: (*a*) chromatogram from 1 ml normal blank plasma with no added internal standard (diphenyl phthalate); (*b*) chromatogram from 1 ml of plasma from a volunteer taking paracetamol.
By courtesy of *Journal of Chromatography*.

In routine cases, where sufficient material is available, it is usual to carry out microscopical examination, spot tests, and thin-layer chromatography. But where the amount of substance is restricted (e.g., when pipe or cigarette residues are to be analyzed), gas chromatography can be of assistance.

It should be noted that cannabis is a crude plant preparation in which more than 80 components have been characterized, the proportions of these varying between samples according to the geographical source of the plant and the conditions of storage. Where charges of conspiracy or peddling are to be preferred, comparison of the composition of samples is likely to be required, and use of the gas chromatographic technique of analysis is almost essential.

Methods for gas chromatographic analysis date from 1961 when Kingston and Kirk (85) were able to detect at least six components with a 2% SE-30 column operated at 250°C. Subsequently, the use of Carbowax

20M (86), XE-60 (87), SE-52 (88), neopentylglycol adipate–trimer acid (89), OV-17 (90), and CDMS (91), were reported, and methyl, trimethylsilyl, trifluoroacetyl, and butyl derivatives of the compounds were prepared and used for analysis.

The author's experience has been principally with 1% CDMS on 80-100 mesh Diatomite CQ (91) and with the 2% OV-17 column described by Lerner (90). Both give excellent resolution of the major cannabinols and allow analysis of a sample in about 20 mins (see chromatogram of 1% CDMS, Figure 6). As an internal standard, Lerner has recommended (±)-methadone hydrochloride, but this elutes before cannabidiol and, in the author's opinion, the use of dibenzylphthalate is to be preferred. Dibenzylphthalate is advantageous in having a longer retention time than any of the major cannabis components, thereby reducing the probability of one of the more volatile components of cannabis having the same retention time as the internal standard (which, of course, would yield erroneous quantitative results).

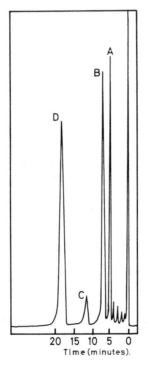

Fig. 6. Chromatogram from a typical sample of cannabis resin obtained with a column of 1% CDMS (see text): curve A, cannabidiol; curve B, Δ'-tetrahydrocannabinol; curve C, cannabinol; curve D, dibenzyl phthalate.

B. Basic Drugs

1. Amphetamines

Because of the widespread abuse of amphetamines methods for detecting and estimating this class of compounds have been widely investigated, and there is no doubt that gas chromatography is the most favored technique. The classical work in the field is due to Beckett and his coworkers (92,93), who used thin-layer chromatography and mass spectroscopy in conjunction with gas chromatography for identification of these and other stimulants in the urine of sportsmen.

In the method advocated by Beckett et al., urine is considered to be the most convenient biological sample. After it is made alkaline with sodium hydroxide, it is extracted first with diethylether and then with methylene dichloride. The two extracts are separately concentrated by evaporation at 40 and 55°C, respectively, and analyzed in turn on four columns at six temperatures, as follows: (a) 3 m, 5% Carbowax 6000/5% KOH at 155°C; (b) 1 m, 2% Carbowax 20M/5% KOH at both 140 and 180°C; (c) 2 m, 2.5% SE-30 at both 120 and 160°C; and (d) 2 m, 10% Apiezon L/10% KOH at 155°C. In all cases, the support is A/W DMCS-treated Chromosorb G (80-100 mesh).

Retention times were originally given for 42 compounds on all four columns (92); later (93), the work was extended to 74 more compounds, but using only column systems (b) and (c). Phenoxypropazine gave more than one major peak, and diethylpropion, methylphenidate, and pyrovalerone gave single major peaks with shoulders, an indication of decomposition. Generally, retention data were given for acetone, N-acetyl, and N-propionyl derivatives, where these could be formed; for amphetamine itself, data were presented for 12 different derivatives. Except for phentermine and chlorphentermine, where the possibility of steric hindrance in the molecule exists, all primary amines readily formed Schiff's bases. Ephedrine and pseudophedrine reacted with acetone at different rates, and this could be used as a distinguishing feature.

Toseland and Scott (94) have been similarly concerned with measurement and identification of amphetamine in body fluids. They also have described the formation of the N-acetyl derivative and made the useful, practical observation that the derivative is neutral and hence can be analyzed on a column otherwise used for the analysis of barbiturates. Their analyses were conducted with columns of 10% Apiezon at 180°C and 3% SE-30 at 120°C, the limit of detection being 60 ng. Figure 7a shows the separation of amphetamine and methylamphetamine, respectively, on an Apiezon/KOH column at 120°C; Figure 7b represents the separation of their N-acetyl derivatives under the same conditions but with the column

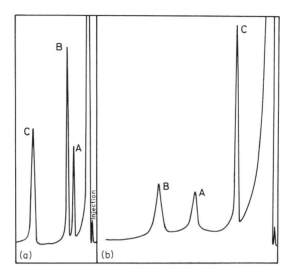

Fig. 7. (a) Analysis with Apiezon/KOH column at 120°C: curve A, amphetamine; curve B, methylamphetamine; curve C, nicotine. (b) Analysis with Apiezon/KOH column at 140°C of N-acetyl derivatives: curve A, amphetamine; curve B, methylamphetamine; curve C, nicotine. Nicotine is internal marker in each case.

temperature at 140°C. Nicotine is included in the latter case, as internal marker.

For maximum sensitivity of detection, it must be said, the formation of halogenated derivatives is necessary, together with the use of an electron-capture detector. Heptafluorobutyryl (92, 95) and trichoracetyl (96) derivatives of amphetamine have been described.

In rare cases it may be essential to differentiate between optical isomers. By reacting amphetamine with trifluoroacetyl-1-prolyl chloride, followed by chromatography on a 1% Carbowax 20M column at 185°C, a good separation is said to be achieved (97, 98).

2. Antihistamines

Although relatively nontoxic, the antihistamines are widely used, and many of them produce drowsiness as a side effect. Therefore, they are particularly likely to be implicated in driving cases. The first report of their gas chromatographic separation was by Fontan et al. (99), who used a 2% Carbowax 20M column at 190°C. With an injection port temperature of

230°C, it was demonstrated that the hydrochloride salts decomposed, so that the observed peaks in the chromatogram were due to the free bases. Most of the common members of the group eluted within 20 min, but triprolidine decomposed, giving two peaks, and no peak was observed with clemizole.

McDonald and Pflaum (100) used an injection port temperature of 256°C in separating 16 antihistamines on a column of 1% SE-30 and, similarly, noted some decomposition. Indeed, commonly encountered compounds such as chlorcyclizine, chlorpheniramine, cyclizine, and promethazine were all observed to give more than one chromatographic peak.

Of the other methods described (101–104), that of Jain and Kirk (105) is probably most effective. They were able to separate 22 antihistamines using a column of 1% CDMS on 100-120 mesh silanized Gas Chrom P. The drugs could be extracted from blood using an acetone–ether mixture and were chromatographed either at 160 or 190°C. The conditions also allow the separation of a number of other basic drugs that are likely to be present in samples, and this constitutes an additional advantage of the procedure.

3. Benzodiazepines

Chlordiazepodixe, oxazepam, diazepam, and nitrazepam are the principal members of the benzodiazepine group of compounds, although more than 20 derivatives of the 1,4-benzodiazepine nucleus have similar pharmacological properties. Gas chromatographic separation has been achieved with columns of OV-1 (106), OV-17 (107), and SE-30 (108). In one of the methods (106), the five most commonly encountered members of the group were very well separated with a column of 3% OV-1 at 245°C; the support was Gas Chrom Q (60-80 mesh), and the column was 2 m long. In preparation for analysis, the drugs were extracted at pH 7 using diethylether; the ether extracts were then evaporated to dryness, and the residue was dissolved in acetonitrile. The authors stated, without qualification, that ether containing not more than 0.00005% peroxide must be used from a bottle opened no more than one day previously.

4. Phenothiazines and Tricyclic Antidepressants

Important members of the phenothiazine and tricyclic antidepressant groups of drugs are chlorpromazine and trifluoperazine, which are derivatives of the phenothiazine nucleus, imipramine, and desiprimine (which are dibenzazepines), and amitriptyline and nortriptyline (which are substituted dibenzocycloheptenes). Metabolism studies on these classes of compounds provide fascinating reading, almost 30 metabolites of chlorpromazine alone having been detected in urine. In the majority of cases of acute toxicity, ultraviolet and infrared spectrophotometry and thin-layer chromatography suffice to solve the problem. When mixtures of the drugs have been ingested, however, it may be necessary to resort to gas chromatography. Leach (51) gave retention times relative to codeine, for most of the compounds in the series, for elution from a 2.5% SE-30 column at 225°C.

5. Methaqualone

The hypnotic methaqualone is widely prescribed in the United Kingdom, both alone and compounded with diphenhydramine; according to one report (109), it ranks after barbiturates and salicylates as the drug most often ingested in self-poisoning cases. Methaqualone has a characteristic ultraviolet absorption spectrum, and spectrophotometric methods for its analysis in plasma have been described (110,111). Unless the samples available are fresh, however, background interference may render such a determination inaccurate. Gas chromatographic methods for the separation of the drugs on columns of SE-30 (112,113) and Apiezon L (59) have been reported. To date, however, only the method of Berry (114) has described an application to its quantitative measurement in biological fluids. In this approach a 7 ft × $\frac{1}{4}$ in. i.d. column packed with 3% CDMS on 85-100 mesh Diatomite CQ is used at operating and injection port temperatures of 200 and 240°C, respectively. Plasma or urine, made alkaline with sodium hydroxide and extracted with hexane, is dried and evaporated, then butobarbitone in ethanol is added as internal standard and 3-5 μl is injected into the gas chromatograph. The ratio of peak height of methaqualone to butobarbitone was found by Berry to be linear over the range 0.1 to 1 μg and the two compounds had retention times of about 12 and 6 minutes, respectively.

In human volunteers given a single therapeutic dose of methaqualone, levels up to 0.226 mg% were measured in the plasma 1 hour after ingestion and the drug could still be detected 7 hours later. Only a small amount of free methaqualone was found in urine.

VI. PETROLEUM PRODUCTS

The need to identify or compare petroleum products arises principally from cases of arson, in which the fuel may have been used as a fire accelerant. Larceny of fuels is also not uncommon. Specific identification of the products is made difficult because they are complex mixtures of large numbers of different chemical compounds and, further, the composition of the product from any one refinery is variable. The composition of material from the scene of a fire which has been exposed to the atmosphere and to heat, is likely to have changed dramatically. Yet the variation in composition of the products may be of value in the more usual type of problem, that of comparing two samples which are thought to have a common origin. If the samples in question have an unusual component, or if a normal component is present in unusual proportions, it may be possible to give a definitive answer. The range of petroleum products available consists mainly of mixtures of hydrocarbons and the individual products within the range can be described by their paraffinic, olefinic, naphthenic and aromatic contents. Paraffins, followed by aromatic hydrocarbons are the most abundant constituents.

One of the major problems in the investigation of arson is the recovery of sufficient material for analysis from the wood, paper, fabric, and so on, onto which it has become absorbed. Methods of isolation include vacuum distillation, solvent extraction, and air flushing, but, if sufficient material is available, the distillation method of Macoun (115) is still favored. Here the material thought to contain the accelerant is saturated with alcohol in a wide-mouthed flask, the flask is set aside for some time, and sufficient water is added to produce an approximately 25% solution of alcohol. The mixture is slowly distilled and the distillate is collected in a burette containing chromic acid; hydrocarbons separate from the aqueous phase and can be removed for analysis. Petroleum and white spirits are said to be completely recovered by the procedure, whereas the recovery of paraffin is of the order of 50%, this being the more volatile fraction of the paraffin.

An alternative approach is to analyze directly a headspace sample of the vapor from the suspected accelerant. In preparation for this, the material containing the accelerant can be enclosed in a glass jar or nylon bag, which can be heated prior to sampling. Nylon is preferred to the more commonly available polythene bag for large items since the latter material allows significant diffusion of smaller molecules.

Most gas chromatographic methods for the identification of petroleum products make use of columns on which the paraffinic constituents are separated on the basis of boiling points (116–121). In early work, for example, Lucas (120) used silicone fluid on firebrick at two temperatures (100 and 160°C) and was able to show a significant difference between a number of different petroleum products. Of 28 gasolines examined, all but two exhibited significant differences in composition, and even these were later found to have come from the same refinery within an interval of one week. Aviation gasolines were particularly easy to differentiate from other products because, first, the chromatograms exhibited a smaller number of component peaks and, secondly, 60% of the sample eluted in the octane region, compared with 15% for an ordinary gasoline.

Ettling and Adams (121) have attempted to determine the amounts and types of hydrocarbons that are likely to be coextracted with an accelerant from charred materials such as wood, textiles, or paper. The materials were dipped in the accelerant, ignited in a muffle furnace at 600°C, and immediately transferred to jars. Subsequently, headspace and extracted samples were analyzed with a SE-30 column coupled with flame ionization detection and subject to temperature programming. Wood, polyester fabric, and wrapping paper generally yielded small amounts of hydrocarbons, whereas cotton, wool, and vinyl-coated fabrics yielded appreciable amounts. Newspaper, too, produced considerable amounts, which were thought to have orignated from the printing ink. Notwithstanding these observations, the authors concluded that, with care, accelerant peaks in the chromatograms could be distinguished from those of the residues from the materials examined.

Recently, Leung and Yip (121) have emphasized the importance of the aromatic hydrocarbon fraction in assisting the differentiation of petroleum products from one another. Extending the scope of earlier, related work (123), they used a packed column consisting of a mixed liquid phase of Bentone 34 and di-*n*-decylphthalate and investigated:

1. The identification of both fresh and partially evaporated petroleum products.

2. The detection of gasoline in the presence of other petroleum products.

3. Changes produced in petroleum products by absorption in soil.

It was concluded that since most evaporated petroleum products had relatively higher aromatic contents than the original products, due to preferential evaporation of the lower boiling aliphatic hydrocarbons, the conditions described were particularly valuable for identification. The chromatograms obtained allowed identification of gasolines alone in concentration as low as 10 ppm in air and 0.2% when mixed with other products. Changes produced in light petroleum products, other than gasoline and naphtha, by absorption in soil for periods up to one month were not considered to be significant.

VII. PAINTS AND POLYMERS

The investigation of road accidents, cases of safe-breaking and burglary-type crimes, in general, is greatly facilitated by the analysis of paints, particularly, and plastics, to a lesser extent. As with the analysis of petroleum products, emphasis is usually on comparison of samples rather than on specific identification.

Classically, the forensic examination procedure for paints involves color comparison, both of top and undercoat layers; an attempt to obtain a mechanical (jigsaw-type) fit between samples, solvent tests; and identification of the inorganic pigments by X-ray diffraction and spectrographic analysis. In recent years, ATR infrared spectroscopy and pyrolysis–gas chromatography have been used to obtain further parameters for comparison. The latter technique has particular merit when only small samples are available but, to its disadvantage, it is a destructive procedure (unlike infrared spectroscopy). In addition, reproducibility of results is difficult to attain, and interlaboratory comparison of pyrograms is rarely possible. Pyrolysis–gas chromatography has been the subject of a number of recent reviews (124–126).

One of the earliest forensic applications of the technique was described by Jain, Fontan, and Kirk (127). Of 34 samples of paint analyzed by them, only two showed the same pattern of pyrolysis products, these being identical in vehicle composition. The sample was pyrolyzed in a quartz tube, heated by a platinum wire coil, and gas chromatography of the products was on a 6.5-ft glass column packed with 9.1% diisodecylphthalate on 70-80 mesh, DMCS-treated Chromosorb G. The operating temperature of the oven was 70°C. A small piece of paint 3 to 4 mm long and about the diameter of a thread was placed inside the quartz tube and pyrolyzed at "white heat" for 20 sec. Typically, about 20 component peaks were obtained in each pyrogram, although none was unequivocally identified. Similar experiments were described by Groten (128), who presented the results of pyrolysis–gas chromatography of more than 150 different materials, including plastics, elastomers, resins, and natural products. A platinum heating coil was also used, but pyrolysis occurred at various temperatures and chromatography was on two columns: a 12-ft, 20% Carbowax on Diatoport P operated at 150°C and an 8-ft, 5% water-insoluble Ucon on Haloport F column working at 100°C. There were few cases of exact identity within the collection of pyrograms obtained, although there were definite family resemblances. Many nylons, for example, gave similar patterns.

The most recent contribution to this field of work is a reference collection of pyrograms of some 65 paints and plastics (129). With emphasis on reproducibility and discrimination between samples, it was established that identical pyrograms of the same materials can be obtained on different instruments, and an index retrieval system based on the three major pyrolysis components of the sample was presented. All major peaks in the pyrograms were identified by gas chromatography–mass spectroscopy. A Curie point pyrolyzer attached directly to the column was used, and pyrolysis was at 610°C, held for 10 sec. Separation was on twin 5 ft × 4 mm i.d. Poropak Q columns (50-80 mesh) with temperature programming, between 100 and 200°C, at 8°/min. On completion of the program, the temperature was held at 200°C for up to 25 min. A chromatogram obtained under these conditions from a typical alkyd paint is reproduced in Figure 8.

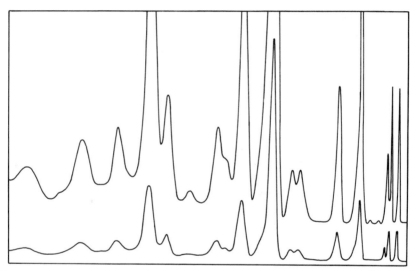

Fig. 8. Chromatogram obtained following pyrolysis of a typical alkyd paint following the method of May, Pearson, and Scothern (129).

References

1. A. S. Curry, *J. Pharm. Pharmacol.*, **7**, 969 (1955).
2. A. C. Maehly, *Methods of Forensic Science*, Vol. 1., F. Lundquist, ed., Interscience, New York, 1962, p. 539.
3. D. J. Blackmore, *Analyst*, **95**, 439 (1970).
4. E. J. Van Kampem and H. M. Klouwen, *Rec. Trav. Chim. Pays-Bas*, **73**, 119 (1954).
5. C. A. Ainsworth, E. L. Schloegel, T. J. Domanski, and L. R. Goldbaum, *J. Forens. Sci.*, **12**, 529 (1967).
6. H. A. Collison, F. L. Rodkey, and J. D. O'Neal, *Clin. Chem.*, **14**, 162 (1968).
7. R. M. McCredie and A. D. Jose, *J. Appl. Physiol.*, **22**, 836 (1967).
8. S. Nobel and A. Ricker, *Clin. Chem.*, **13**, 276 (1967).

9. A. M. Dominguez, H. E. Christensen, L. R. Goldbaum, and V. A. Stembridge, *J. Toxicol. Appl. Pharmacol.*, **1**, 135 (1959).
10. F. L. Rodkey, *Ann. N.Y. Acad. Sci.*, **174**, 261 (1970).
11. L. R. Goldbaum, E. L. Schloegel, and A. M. Dominguez, *Progress in Chemical Toxicology*, Vol. I, A. Stolman, ed., Academic Press, New York, 1963, p. 30.
12. K. Porter and D. H. Volman, *Anal. Chem.*, **34**, 748 (1962).
13. R. Bonnichsen, A. C. Maehly, and M. Moeller, *J. Forens. Sci.*, **11**, 186 (1966).
14. R. Bonnichsen and A. C. Maehly, *J. Forens. Sci.*, **11**, 414 (1966).
15. G. G. Esposito and M. H. Swann, *J. Paint Technol.*, **39**, 509, 338 (1967).
16. C. T. Dollery, D. S. Davies, G. H. Draffan, F. M. Williams, and M. E. Conolly, *Lancet*, 1164 (1970).
17. J. Fischl and M. Labi, *Israel J. Med. Sci.*, **2** (1), 84 (1966).
18. G. D. Divincenzo, F. J. Yanno, and B. D. Astil, *Am. Ind. Hyg. Ass. J.*, **32** (6), 387 (1971).
19. R. J. Nelms, R. L. Davis, and J. Bond, *Am. J. Clin. Pathol.*, **53**, 641 (1973).
20. F. B. Hall and C. H. Hine, *J. Forens. Sci.*, **11**, 404 (1966).
21. R. Maes, N. Hodnett, H. Landesman, G. Kananen, B. Finkle, and I. Sunshine, *J. Forens. Sci.*, **14**, 235 (1969).
22. D. W. Robinson, *J. Pharm. Sci.*, **57** (19), 185 (1968).
23. G. Machata, *Blutalkohol*, **4**, 345 (1967).
24. P. Herbert, *J. Med. Lab. Technol.* (London), **25** (3), 233 (1968).
25. R. H. Gadsden, K. B. H. Risinger, and E. E. Bagwell, *Can. Anaes. Soc. J.*, **12** (1), 90 (1965).
26. A. Brachet-Liermain, L. Ferrus, and J. Caroff, *J. Chromatogr. Sci.*, **9**, 49 (1971).
27. R. N. Harger and R. B. Forney, *Progress in Chemical Toxicology*, Vol. I, A. Stolman, ed., Academic Press, New York, 1963, p. 54.
28. R. N. Harger and R. B. Forney, *Progress in Chemical Toxicology*, Vol. III, A. Stolman, ed., Academic Press, New York, 1967, p. 1.
29. R. N. Harger, *Toxicology-Mechanism and Analytical Methods*, Vol. II, C. P. Stewart and A. Stolman, eds., Academic Press, New York, 1961, p. 86.
30. H. Ward-Smith, *Methods of Forensic Science*, Vol. IV, A. S. Curry, ed., Interscience, London, New York, Sydney, 1965, p. 1.
31. H. J. Walls and A. R. Brownlie, *Drink, Drugs and Driving*, Sweet and Maxwell, W. Green and Son, London and Edinburgh, 1970.
32. M. Wolthers, *Acta Med. Leg. (Liege)*, **9**, 325 (1956).
33. R. Bonnichsen and Maire Linturi, *Acta Chemica Scand.*, **16**, 1289 (1962).
34. G. Machata, *Mikrochim. Acta*, 691 (1962).
35. A. Brahm-Vogelsanger and H. J. Wagner, *Dtsch. Z. ger. Med.*, **55**, 137 (1964).
36. A. S. Curry, G. W. Walker, and G. S. Simpson, *Analyst*, **91**, 742 (1966).
37. W. M. McCord and R. H. Gadsden, *J. Gas Chromatogr.*, **2**, 38 (1964).
38. E. Osterhaus and K. Johannsmeier, *Blutalkohol*, **2**, 1 (1963).
39. J. E. Wallace and E. V. Dahl, *Am. J. Clin. Pathol.*, **46**, 152 (1966).
40. A. S. Curry, G. Hurst, N. R. Kent and H. Powell, *Nature*, **195**, 603 (1962).
41. G. Machata, *Mikrochim. Acta*, 262 (1964).
42. P. Skrabvánek and J. Novak, *Blutalkohol*, **3** (6), 271 (1966).
43. D. J. Blackmore, *J. Forens. Sci. Soc.*, **8**, 73 (1968).
44. Gas Chromatographic Intoximeter, Cal Detect Inc., Richmond, Calif.
45. C. M. Hoffman, R. L. Brunella, M. J. Pro, and G. E. Martin, *J. Ass. Offic. Anal. Chem.*, **51**, 580 (1968).
46. D. D. Singer, *Analyst*, **91**, 127 (1966).

47. B. C. Rankine, *J. Sci. Food Agric.*, **18**, 583 (1967).
48. G. I. de Becze, H. F. Smith, and T. E. Vaughn, *J. Ass. Offic. Anal. Chem.*, **50**, 311 (1967).
49. B. J. Gudzinowocz, *Gas Chromatographic Analysis of Drugs and Pesticides*, Marcel Dekker, New York, 1967.
50. A. S. Curry, *Poison Detection in Human Organs*, Charles C. Thomas, Springfield, Ill., 1969.
51. H. Leach, *Isolation and Identification of Drugs*, E. G. G. Clarke, ed., Pharmaceutical Press, London, 1969, p. 59.
52. J. Janák, *Nature*, **185**, 684 (1960).
53. E. Brochmann-Hanssen and A. Baerheim Svendsen, *J. Pharm. Sci.*, **50**, 804 (1961).
54. E. Brochmann-Hanssen and A. Baerheim Svendsen, *J. Pharm. Sci.*, **51**, 318 (1962).
55. K. D. Parker and P. L. Kirk, *Anal. Chem.* **33**, 1378 (1961).
56. W. J. A. Vanden Heyvel, E. O. A. Haahti, and E. C. Horning, *Clin. Chem.*, **8**, 351 (1962).
57. G. Machata and H. J. Battista, *Mikrochim. Acta*, **4**, 866 (1968).
58. N. C. Jain and P. L. Kirk, *Michrochem. J.*, **12**, 249 (1967).
59. H. Leach and P. A. Toseland, *Clin. Chim. Acta*, **20**, 195 (1968).
60. G. W. Stevenson, *Anal. Chem.*, **38**, 1948 (1966).
61. K. D. Parker, J. A. Wright, A. F. Halpern, and C. H. Kline, *J. Forens. Sci.*, **8**, 125 (1968).
62. E. Brochmann-Hanssen and T. O. Oke, *J. Pharm. Sci.*, **58**, 370 (1969).
63. H. V. Street, *J. Chromatogr.*, **41**, 358 (1969).
64. A. Baerheim Svendsen and E. Brochmann-Hanssen, *J. Pharm. Sci.*, **51**, 494 (1962).
65. J. Grove and P. A. Toseland, *Clin. Chim. Acta*, **29**, 253 (1970).
66. E. Frantzen, O. E. Hansen, and M. Kristensen, *Acta Neurol. Scand.*, **43**, 440 (1967).
67. H. Tenckhoff, D. J. Sherrard, R. O. Hickman, and R. L. Ladda, *Am. J. Dis. Child.*, **116**, 422 (1968).
68. F. A. Laubscher, *J. Am. Med. Ass.*, **198**, 1120 (1966).
69. C. J. A. Schulte and T. A. Good, *J. Pediatr.*, **68**, 635 (1966).
70. N. M. A. Viukari and K. Aho, *Brit. Med. J.*, **2**, 51 (1970).
71. H. Kutt, R. Brennan, H. Dehejia, and K. Verebely, *Am. Rev. Resp. Dis.*, **101**, 377 (1970).
72. K. Sabih and J. Sabih, *Anal. Chem.*, **41**, 1452 (1969).
73. G. Grimmer, J. Jacob, and H. Schaefer, *Arz. Forsch.*, **19**, 1287 (1969).
74. D. H. Sandberg, G. L. Resnick, and C. Z. Bacallo, *Anal. Chem.*, **40**, 736 (1968).
75. J. MacGee, *Anal. Chem.*, **42**, 421 (1970).
76. T. Chang and A. J. Glazko, *J. Lab. Clin. Med.*, **75**, 145 (1970).
77. C. Winek, *Clin. Toxicol.*, **3**, 541 (1970).
78. S. Winsten and D. Brody, *Clin. Chem.*, **13**, 589 (1967).
79. B. S. Finkle, *J. Forens. Sci.*, **12**, 509 (1967).
80. I. Sunshine, R. Maes, and R. Faracci, *Clin. Chem.*, **14**, 595 (1968).
81. P. Grieveson and J. S. Gordon, *J. Chromatogr.*, **44**, 279 (1969).
82. B. Widdop, *J. Chromatogr.*, **47** (1970).
83. L. F. Prescott, *J. Pharm. Pharmacol.*, **23**, 111 (1971).
84. J. Grove, *J. Chromatogr.*, **59**, 289 (1971).
85. C. R. Kingston and P. L. Kirk, *Anal. Chem.*, **33**, 1794 (1961).
86. L. T. Heaysman, E. A. Walker, and D. T. Lewis, *Analyst*, **92**, 450 (1967).
87. B. Caddy, F. Fish, and W. D. C. Wilson, *J. Pharm. Pharmacol.*, **19**, 852 (1967).
88. H. V. Street, *J. Chromatogr.*, **48**, 291 (1970).
89. H. M. Stone and H. M. Stevens, *J. Forens. Sci. Soc.*, **9**, 31 (1969).
90. P. Lerner, *Bull. Narc.*, **21**, 39 (1969).
91. D. A. Patterson and H. M. Stevens, *J. Pharm. Pharmacol.*, **22**, 392 (1970).
92. A. H. Beckett, G. T. Tucker, and A. C. Moffat, *J. Pharm. Pharmacol.*, **19**, 273 (1967).

93. A. H. Beckett and A. C. Moffat, *J. Pharm. Pharmacol.*, **20**, *Suppl.*, 485 (1968).
94. P. A. Toseland and P. H. Scott, *Clin. Chim. Acta*, **25**, 75 (1969).
95. R. B. Bruce and W. R. Maynard, *Anal. Chem.*, **41**, 977 (1969).
96. J. S. Noonan, P. W. Murdick, and R. S. Ray, *J. Pharmacol. Exp. Ther.*, **168**, 205 (1969).
97. E. Gordis, *Biochem. Pharmacol.*, **15**, 2124 (1966).
98. C. E. Wells, *J. Ass. Offic. Anal. Chem.*, **53**, 113 (1970).
99. C. R. Fontan, C. W. Smith, and P. L. Kirk, *Anal. Chem.*, **35**, 591 (1963).
100. A. MacDonald and R. T. Pflaum, *J. Pharm. Sci.*, **52**, 816 (1963).
101. A. C. Celeste and M. V. Polito, *J. Ass. Offic. Anal. Chem.*, **49** (3), 541 (1966).
102. R. B. Bruce, J. E. Pitts, and F. M. Pinchbeck, *Anal. Chem.*, **40** (8), 1246 (1968).
103. P. Kabasakalian, M. Taggart, and E. Townley, *J. Pharm. Sci.*, **57** (4), 621 (1968).
104. P. Kabasakalian, M. Taggart, and E. Townley, *Can. J. Pharm. Sci.*, **3** (1), 18 (1968).
105. N. C. Jain and P. L. Kirk, *Microchem. J.*, **12**, 242 (1967).
106. F. Marcucci, R. Fanelli, and E. Mussini, *J. Chromatogr.*, **37**, 318 (1968).
107. J. A. F. de Silva and C. V. Puglisi, *Anal. Chem.*, **42**, 1725 (1970).
108. L. B. Foster and C. S. Frings, *Clin. Chem.*, **16**, 177 (1970).
109. H. Matthew, A. T. Proudfoot, S. S. Brown, and A. C. A. Smith, *Brit. Med. J.*, **2**, 101 (1968).
110. A. A. H. Lawson and S. S. Brown, *Scot. Med. J.*, **12**, 63 (1967).
111. M. Akagi, Y. Oketani, and M. Takada, *Chem. Pharm. Bull. (Tokyo)*, **11**, 62 (1963).
112. R. Nanikawa and S. Kotoku, *Yonago Acta Med.*, **10**, 49 (1966).
113. J. Bogan, *Int. Ass. Forens. Toxicol. Bull.*, **4** (3), 4 (1967).
114. D. J. Berry, *J. Chromatogr.*, **42**, 39 (1969).
115. J. M. Macoun, *Analyst*, **77**, 381 (1952).
116. F. T. Eggertsen, S. Groennings, and J. J. Holst, *Anal. Chem.*, **32**, 904 (1960).
117. D. M. Lucas, *J. Forens. Sci.*, **5** (2), 236 (1960).
118. D. K. Albert, *Anal. Chem.*, **35**, 1918 (1963).
119. L. E. Green, *Anal. Chem.*, **36**, 1512 (1964).
120. V. F. Gaylor, C. N. Jones, J. H. Landerl, and E. C. Hughes, *Anal. Chem.*, **36**, 1606 (1964).
121. B. V. Ettling and M. F. Adams, *J. Forens. Sci.*, **13** (1), 76 (1968).
122. K. Leung and H. L. Yip, *Int. Microfilm J. Leg. Med.*, **5** (1), 1 (1970).
123. S. Spencer, *Anal. Chem.*, **35**, 592 (1963).
124. R. L. Levy, *Chromatogr. Rev.*, **8**, 48 (1966).
125. S. G. Perry, *Advan. Chromatogr.*, **7**, 221 (1969).
126. G. M. Brauer, in *Thermal Characterisation Techniques*, Dekker, New York, 1970.
127. N. C. Jain, C. R. Fontan, and P. L. Kirk, *J. Forens. Sci.*, **5** (2), 102 (1965).
128. B. Groten, *Anal. Chem.*, **36** (7), 1206 (1964).
129. R. W. May, E. F. Pearson, and M. D. Scothern, in press.

Applications of Digital Computers
in Gas Chromatography

DAVID A. LEATHARD, *Sheffield Polytechnic,*
Sheffield, England

I. INTRODUCTION

One of the earliest descriptions of the application of digital computers to gas chromatography (GC), which appeared in 1963 (1), involved recording chromatograms on magnetic tape for subsequent processing by a computer. In less than a decade, there has been a very rapid growth of interest in computerized GC, as evidenced by the proceedings of major international conferences (2–9) and by the numerous local symposia organized on the topic. Perhaps the most pressing reason for this interest is the growing realization that the cost of an analyst's time can be more significant than the cost of his equipment. In this sense, the arrival of the computer in the laboratory is a logical extension of the arrival of instrumental methods in the preceding 30 years. Laboratory automation, in general, is being introduced at a rapid pace, and computerized gas chromatography is just one aspect of this. Reasons for the current interest in computers for GC include the following:

1. The rapid growth of GC itself;
2. The general fall in the price of digital computers, and the growing availability of commercial remote time-sharing systems;
3. The simultaneous rise in labor rates;
4. The desire for better reproducibility through the elimination of human error and judgment in routine measurements and calculations;
5. The need for faster results in process control laboratories to avoid off-grade product;
6. The development of new GC techniques, such as temperature programming, pressure programming, multicolumn valve switching, and automatic injection, which demand better instrumental control;
7. The use of retention indices and methods, such as simulated distillation, which require novel calculations;
8. The application of GC theory, which involves detailed knowledge of peak shapes;
9. The possibility of applying curve-fitting methods to the envelopes of overlapping peaks, which may allow faster analyses in future.

Several detailed reviews of digital computer applications in chemical laboratories have been published (10–12). There are also numerous articles surveying general principles and trends, as well as the advantages and disadvantages of various systems of coupling instruments to the computer (13–26). For example, the needs of automated clinical laboratories (20), the desirability of building the computer into the instrument (21), and the available methods and motivations for automating analytical chemistry (18) have been discussed. The design features of several computerized multi-instrument laboratories have also been published (27–29). A small number of computer-operated and computer-optimized chemical experiments have been carried out in electrochemistry (30–32), in kinetics (33, 34), and in more general fields (35–37).

It is not yet possible to foresee the full extent of the coming computer "revolution" or to evaluate all the scientific, economic, and social implications of laboratory automation. In the data-processing field of the big computer, much of the early excitement and unrealistic use of the new "toys" has ended, and the very rapid expansionary era has given way to a more modest rate of growth. For the immediate future, expansion in the field of process control is likely to be restricted to certain well-defined areas (126). In the laboratory, however, we appear to be seeing the beginning of a very rapid expansion in the use of small computers in many different applications. In particular, computers are applied to gas chromatography in areas ranging from industrial quality control (38–41), to thermodynamic and surface area measurements (42), and to undergraduate experiments (43).

Although at the beginning of 1972 the computer did not appear to have made a significant impact in the majority of GC laboratories, the way was being prepared—witness the mounting pile of published literature, sales material from computer and analytical instrument companies, and numerous symposia and meetings of discussion groups. Some analysts are worried by the impending arrival of the computer; others eagerly search for valid ways of convincing their organizations of the economic desirability of installing one. Unlike other forms of electronic instrumentation, computers still seem able to raise human emotions. Some scientists and managers adopt fairly deeply committed positions, either for or against computers. Those who have actually installed computers often seem to feel personally associated with their particular systems. This being the case, it is probably difficult properly to evaluate some published work, and it can be argued that it is important for anyone writing about this field to declare his position. In this instance, the writer is more "for" than "against" computers, and he has been closely involved with a (successful) on-line laboratory automation and remote reporting system serving more than 20 gas chromatographs.

One of the biggest stumbling blocks in the way to a clear understanding of the use of computers in connection with analytical instrumentation is the uninhibited love and use of jargon by the computer trade. Every group of people with similar interests requires a special vocabulary for effective communication; unfortunately, however, computer jargon comes in a variety of dialects and is often ambiguous. Recently, there has been a real effort by many computer workers to overcome this communication barrier. Nevertheless, consider the following terms: *hardware, software, bit, byte, CPU, dedicated, real-time, real, integer, multiplexer, ADC, 16K, high-level language, macro assembler, fixed core, on-line, 14-bit resolution, indexed sequential files*. These are everyday computer words, and they must be understood by anyone wanting to make full use of a computer as an effective and efficient laboratory tool. The best way to learn "computerese" is to be guided by a friendly native, possibly on an exchange basis, for civilizing expressions such as plates per foot, katharometer, and retention index. A useful glossary of about 50 computer automation terms has been published (44).

Despite the current interest in computerized GC, the topic is almost completely outside the practical experience of many gas chromatographers. Therefore, this chapter begins with a brief pedagogical description of what a digital computer is, what it can do, and how it does it. The various options available to laboratories contemplating the introduction of some measure of automation for GC are then reviewed, and practical aspects of computer installation are discussed.

II. DIGITAL COMPUTERS

The digital computer is important because of its ability to carry out *data processing*. This consists of taking data as *input*, using the data in a previously planned *processing* operation, and producing the results of the processing step as some form of *output*. (Figure 1.) The input and output steps are carried out by a wide variety of devices, commonly referred to jointly as "input–output" or "I/O," which are often physically separated from the part of the computer that is responsible for the actual processing. Processing itself involves three kinds of operation: *storage, control*, and *arithmetic–logic*. At the heart of any digital computing system is a *stored program*, which resides in the main storage unit of the computer and controls the individual processing steps in a planned sequence. The essential features of a digital computer can be illustrated by a simple example that retains a man for control, arithmetic, and logic but uses pigeonholes for storage.

A. The Stored Program

Suppose that a GC laboratory has only one potentiometric recorder, and that the chart-drive motor breaks down. At any instant it will be possible to see what voltage is being registered, but no permanent record of the GC trace will be available. One unlikely but possible solution to the problem is as follows.

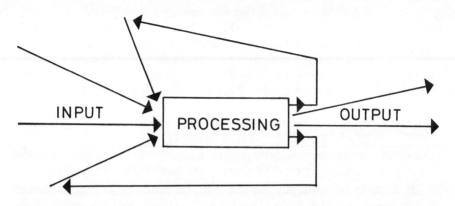

Fig. 1. Schematic representation of the data processing function of a digital computer.

A laboratory technician is supplied with a blackboard, chalk, an eraser, a stopwatch, and a set of numbered pigeonholes, each containing a card. His operating instructions are: "Take out a card from a pigeonhole, do exactly as it says, then return it to the same pigeonhole. Start with card number 4, then take cards 5, 6, 7, etc., in strict numerical order, unless specifically told to do otherwise. Do all your calculations on the blackboard and erase everything except the answer. Similarly, whenever told to write something on a card, first erase everything written on that card already. If you make a single mistake you will be sent to hospital."

The initial contents of the first 15 cards are listed below.

CARD 1 (blank)

CARD 2 Write the recorder reading to the nearest 0.1 division on the blackboard. Take out next the card corresponding to the number on card 3.

CARD 3 (blank)

CARD 4 Zero the recorder and set the attentuation to × 1. Inject sample A into the gas chromatograph. At the same time start the stopwatch.

CARD 5 Write 7 on card 3. Take out card 2 next.

CARD 6 (blank)

CARD 7 Write the number from the blackboard on to card 1.

CARD 8 Write 9 on card 3. Do nothing until the stopwatch reaches the "0" mark again, then take out card 2.

CARD 9 Subtract the contents of card 1 from the number on the blackboard.

CARD 10   If the new answer on the blackboard is greater than zero, take out card 11 next. Otherwise take out card 8 next.

CARD 11   You have taken this card out because the voltage is rising. If the number on the blackboard is greater than the number on card 13, take out card 12. Otherwise take out card 8.

CARD 12   A peak has just begun. Write the time on card 6. The voltage is already on card 1. Take out card 14 next.

CARD 13   5.0

CARD 14   Write on a sheet of paper:
PEAK START   THRESHOLD

CARD 15   Write the contents of card 6 under PEAK START and the contents of card 13 under THRESHOLD.

If these instructions are carried out, the resulting worksheet should show the time at which the first peak gives a voltage response greater than the threshold value of five chart divisions above the baseline.

These 15 cards in pigeonholes can help to highlight most of the important facts about a digital computer. In the first place, it is obvious that the set of cards is useless without the associated operating instructions. In particular, it is very important for the technician to start with card 4; and then follow strict numerical order, unless instructed by a particular card to *branch* to a different card. Two distinct kinds of branching instruction are present in the pigeonhole program. Cards 2, 5, 8, and 12 contain *unconditional* branch instructions, whereas the branch instructions on cards 10 and 11 are *conditional*. In order to carry out conditional branch instructions, it is necessary for the computer to be able to make *logical decisions* of the form: "If A then B, otherwise C." These are usually in the form of mathematical inequalities, of voltage levels being above or below a certain value, or of a switch being on or off.

The technician should understand each of the instructions. For example, if someone unfamiliar with laboratory instruments attempted to follow the instructions, he would be unlikely to progress beyond the first instruction to "zero the recorder." To use some computer jargon, we can say that "zero the recorder" is one of the instructions from the *instruction set* of a laboratory technician. Computers vary widely in the diversity of instruction sets. A program is simply a *logically connected* series of instructions from the instruction set of the particular computer concerned. The instructions need not be *physically* connected, because the branching instructions enable one or more storage locations (pigeonholes) to be bypassed. In the foregoing example, cards 1, 3, 6, and 13 do not contain instructions, and it is essential for the logical arrangement of the instructions on the other cards to prevent the technician from ever attempting to carry out an instruction from one of these cards. For example, if a misprint had resulted in card 5 reading "Write 7 on card 2. Take out card 3 next," the technician would find himself trying to carry out an instruction that simply says (blank). Like the technician with his pigeonholes, the computer can only distinguish instructions from data, by recognizing that the instruction "blank" is not in the valid instruction set and that it must, therefore, cause the program to be aborted. There is a very close analogy between the pigeonhole and the main storage unit of the computer. Each compartment of the storage unit is numbered to enable it to be identified, and each can contain an instruction, or merely some data. The contents of any storage location can be erased and replaced as directed by an instruction in another storage location. Thus the effect of the unconditional branch on card 2 is altered by cards 5 and 8, which overwrite the contents of card 3. In sophisticated programming it is common to replace complete instructions in a similar way.

Card 9 asks for an arithmetic operation to be carried out. This is done with the aid of a blackboard and of the contents of card 1. In a computer, the blackboard becomes one or more *registers*, where numbers are held before, during, and immediately after an arithmetic instruction is carried out. The actual arithmetic is performed by arithmetic electronic circuits similar to those used to make the logical decisions.

Card 2 calls for a recorder reading, and cards 14 and 15 call for a document to be written. In order to execute such instructions, a computer must receive or send coded electrical signals from or to suitable I/O devices (peripherals), such as analog-to-digital converters or electric typewriters.

The logic of the pigeonhole program is represented in the flowchart in Figure 2. From this it is apparent that the instruction on card 2 can be requested after either card 5 or card 8. This is characteristic of a *subroutine*, which is a group of instructions designed to perform a task that might be wanted under a number of independent circumstances (in this case, reading the position of the recorder pen).

In summary, the digital computer is a device that can store and operate on a number of variables held as discrete discontinuous values. Arithmetical and logical operations can be performed, and facilities are available for converting the values of the variables to and from different external forms through a variety of input and output interfaces.

B. Hardware

The term *hardware* refers to the mechanical, electrical, magnetic, and electronic devices and components comprising the physical computer equipment. Figure 3 is a block diagram of the hardware of a conventional data-processing computer. The three stages of data processing (input, processing, and output) are directed by the *control unit* in accordance with *instructions* contained in the *main storage unit*.

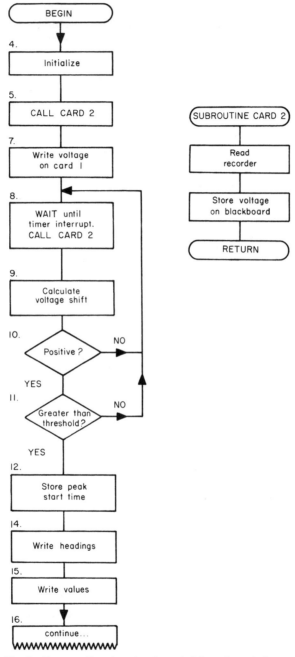

Fig. 2. Program flowchart for the technician–pigeon-hole example.

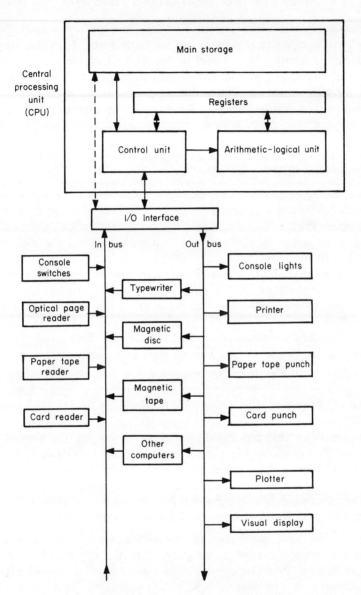

Fig. 3. A conventional data processing digital computer, suitable for off-line GC work.

Main storage is often known as "core storage" or simply "core" because most computers at present use magnetic cores. These are rings of ferromagnetic material about 1 mm in diameter which can be magnetized in a few microseconds. The magnetic field is created by electric currents passing through wires on which the cores are strung by a lacemaker's craft. Unless deliberately changed, the direction of magnetization in each core is retained indefinitely after the magnetizing current is removed. Each core can therefore represent one *binary digit*, or *bit*, representing 0 or 1. Solid-state monolithic integrated circuits are now being used instead of cores in the new "fourth-generation" computers.

Main storage is divided into locations, sometimes known as *words* or *bytes*, each containing a certain number of bits. Each location is identified by a number called an *address*, which is used in instructions to enable the control unit to access the contents of any desired location. (A few computers allow individual bits to be addressed.) Storage is any device into which data can be entered and from which data can later be retrieved; hence the commonly used synonym of "memory." The control unit has direct electronic access to main storage, but there are other forms of computer storage which are only accessible by means of some mechanical input–output device. The term can be applied to magnetic devices such as discs and tapes, as well as to punched paper tape and cards.

If an instruction calls for the processing of data, the necessary electronic steps are carried out by the *arithmetic logical unit*, which has access to the data in various *registers*. The data are normally held in main storage, intermixed with instructions, and are transferred to the registers by the control unit. If an instruction calls for input or output, the control unit initiates the proper electronic signals to the various I/O devices.

Few real machines are equipped with the range of I/O devices shown in Figure 3, the schematic diagram of a digital computer having only conventional input–output equipment. For on-line coupling to GC, additional process I/O features are required as indicated in Figure 4.

The control unit, main storage, registers and arithmetic logical unit together make up the *central processing unit* (CPU), which is usually housed in a single box. Input–output hardware can be housed with the CPU, located in a neighboring box, or even situated many miles away.

The control unit organizes the flow of instructions to be executed in a strict sequential progression from one address to the next in main storage, until a branch instruction is found. As mentioned earlier, an instruction has no features that distinguish it from a piece of data. Each is simply a string of binary digits stored at a main storage address. The branching facility makes it possible, however, to intersperse data and instructions in

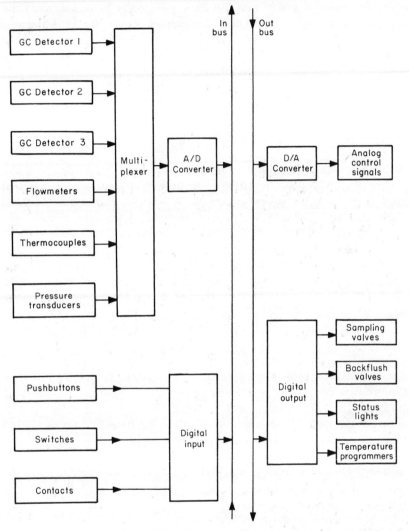

Fig. 4. Additional input/output features of process computer systems suitable for on-line GC work.

main storage in such a way that the data are always bypassed when the control unit goes to get the next instruction. It is the *programmer's* responsibility to ensure that this is the case. Programming errors are often responsible for a computer coming to a grinding halt as the result of trying to execute a nonsensical instruction that was in fact meant to be data. It is here that hardware and *software* meet.

C. Software

A computer program is a set of logical instructions (input, processing, output, or branching) operating on data in a way that is intended to give a desired result. *Software* is the term used to refer collectively to a set of programs written for a given computer. A computer is useless without software, and this distinguishes it from electronic calculators, where the logic is all "wired in."

In order to generate software, a means must be found to translate the logic of the program (as shown on flowcharts such as Figure 2) into meaningful binary instructions resident in main storage. This translation usually involves an extraordinary number of detailed steps. An error in the original logic or in any one of these steps will produce a major or minor error in the final program; and with human programmers, such errors are inevitable. Good testing procedures serve to locate major errors, but minor errors often lie dormant for months after a program has supposedly been tried and tested, only to be discovered when a particular combination of conditions first occurs. When this happens, there must be considerable doubt about the validity of results that had been obtained with the program before the error was recognized. This is a fundamental problem, and anybody using programs of even slight complexity must remain actively aware of it. Every conceivable test must be made in an attempt to "beat the system" before it can be accepted as being equally reliable as the manual method it is replacing. The bigger the program, and the fewer the number of independent users of the *identical* program, the more probable it is that there are hidden bugs.

There are still a few computers that must be programmed directly in machine code (i.e., by writing down the very numbers that the control unit will recognize for the particular instruction). Such machine language programming is also possible on most modern machines, but it is so time-consuming that it is usually reserved for changing the contents of a few core locations in an attempt to correct an error.

There is a progressive trend to make the translation from logic to machine code less labor-intensive for the programmer (89). At the very least, a symbolic language is usually provided with modern computers;

thus, for example, an instruction to add can be written as A or even ADD instead of as the actual binary instruction of, say, 10111101. In such a case, the computer manufacturer provides a general-purpose program that will "translate" the symbols into the appropriate machine code. Symbolic languages in which one symbolic instruction is translated as one machine instruction are known as assembler languages, and the program that does the translation is known as an assembler. Such languages are described as "low-level," to distinguish them from "higher-level" languages in which a single symbolic statement can be written to specify a large number of individual machine language instructions. The next stage "up" from simple assembler is a macro-assembler language; a "macro" is a small subroutine, written once only in full, which can be specified by a single symbolic statement. Both assembler and macro-assembler languages have a very close relationship with the machine language of the particular model of computer for which they have been devised. Good assembler programs can only be written by those with intimate knowledge of the corresponding instruction set. Seldom can a program written in such a language be successfully translated to run on any other computer.

High-level languages such as FORTRAN and ALGOL are not machine dependent, or at least not to a significant extent. A program written in FORTRAN IV for one particular computer, could be run on other computers that were larger or of similar size with the minimum of modification. Some manufacturers have added their own improving touches to internationally agreed FORTRAN specifications; but in such cases the variations from standard FORTRAN are usually clearly explained in the manuals. The small and medium-size computers do not support all the features of FORTRAN. A single FORTRAN statement will generate a number of machine instructions that vary somewhat from machine to machine. Each manufacturer provides a program to translate standard FORTRAN into the particular machine language required. In this case, the translation process is known as *compilation*.

The result of compiling a FORTRAN program is a block of machine code which is ready (in all but a few minor details) to be executed in full by the computer in the same way as output from an assembler. Recently, languages have been developed at a level even higher than that of a compiler. These are known as *interpreter* languages. The essential difference is that an interpreter does not translate the complete program into machine code. Instead, the necessary linkages are established to enable the computer to use standard subroutines to execute each "line" of the program quite independently of the rest of the program. Thus the original form of the high-level coding is preserved after interpretation and can be modified even while the program is being executed line by line.

With an interpreter language, the effects of working at a higher level are even more apparent than on changing from assembler to compiler level. Storage requirements are greatly increased, and the speed of execution is greatly reduced. The great advantages to be set against these disadvantages are: (*a*) the language is very easy to learn; (*b*) programs can be written, tested, corrected, and modified with great ease; and (*c*) it is easy to establish an *interactive* communication link between the computer and the user. These facilities make interpreter languages very attractive for use in scientific research work.

In principle, it is always possible to make more efficient use of core storage by writing in assembler rather than in a compiler-level language such as FORTRAN. (Figure 5.) In practice, this requires considerable experience by the programmer and a huge increase in coding effort. Nevertheless, in a small computer, it may become essential to use assembler in order to be able to get the program to fit. This situation can be aggravated because FORTRAN compilers on small computers tend to produce less efficient machine coding than do the optimizing compilers for bigger machines. It seems likely that the combined effects of the rapidly falling cost of hardware core storage and the rapidly rising labor cost of assembler programming will soon remove these space restrictions. Meanwhile, some computer firms are extending the capabilities of macro assemblers to simplify the programming effort. At the same time, special routines are being provided in FORTRAN to make it easier to carry out fundamental machine operations such as bit manipulation, Boolean logic, and modification of the contents of specific addresses. Improvements are also being made which enable input–output devices to be used from FORTRAN without dragging in the few thousand words overhead which is commonly associated with the use of conventional READ and WRITE statements.

Even if there is sufficient core space for a program generated by a compiler or interpreter, the time taken by the computer to execute the program will be longer than the time required for execution of a more efficient assembler-generated program. This time factor may be relatively unimportant if the program involves a considerable amount of conventional input–output, where the limiting factor is perhaps the typing rate rather than the time taken to process the program. However, in real-time applications, where a single voltage signal must be monitored at frequent intervals, or where a group of signals must be continuously monitored on a repetitive basis, timing considerations can attain prime importance. It then becomes a major priority to ensure that the number of machine instructions is kept to the minimum. This necessitates assembler programming, often of a sophisticated kind that can best be provided by the specialist software teams of the computer manufacturer concerned.

FORTRAN
Programmer writes: $I = J + K$
FORTRAN compiler generates machine language:

Word 67	1100	0100	0000	0000
Word 68	0000	0000	0100	0001
Word 69	1000	0100	0000	0000
Word 70	0000	0000	0100	0010
Word 71	1101	0100	0000	0000
Word 72	0000	0000	0100	0000

Assembler

Programmer writes		Translation by assembler			
LD	J	Word 67	1100	0000	1111	1101
A	K	Word 68	1000	0000	1111	1101
STO	I	Word 69	1101	0000	1111	1010

Fig. 5. Simplified illustration of the characteristics of FORTRAN and assembler programming languages. The processing operation required is to add together the two numbers in core locations 65 and 66 and put the answer in location 64, assuming that I, J, and K refer to core locations 64, 65, and 66, respectively.

There are very important differences in the attention devoted to software development by various manufacturers. As a general rule, the more advanced and "state of the art" is the hardware, the worse will be the software help provided. Conversely, a great deal of advanced software techniques and experience are usually available for well-established computers, even if the electronics are old-fashioned.

A judicious balance must be sought so that the required application can be implemented as economically as possible. It may well prove to be false economy to buy a cheap, up-to-date piece of hardware if it becomes necessary to devote several man-years to getting the programming sorted out to do a particular job (which a more expensive, but longer-established computer could do within a week or two of delivery).

D. Time-Sharing

A computer sometimes spends most of its time doing nothing. Thus if 500 instructions each lasting 2 μsec have to be carried out 200 times every second, the usage rate is only 20%. This rate may rise sharply at times of increased activity (e.g., when reports are being produced or when complicated calculations have to be performed). Nevertheless, averaged over the working day, a dedicated GC computer may well be less than 50% loaded. This makes "time sharing" possible, whereby another application takes advantage of the spare computer capacity. It would in fact be very undesirable for a computer to be routinely 100% loaded with GC work, because any increase in activity might lead to loss of data.

As already indicated, computer jargon is often ambiguous, and the term *time sharing* is used in two distinct ways:

1.   On-line systems sometimes use a medium-sized computer with a large disc memory and a sophisticated "time sharing" or "multiprogramming" software operating system to schedule the tasks on a priority basis. Typically, acquisition of data from multiple GCs and other analytical instruments have the highest priority. The calculation and reporting of the analytical results have an intermediate priority, and background work has the lowest priority. Examples of background work are updating GC computer methods, compiling FORTRAN programs, or executing programs completely unconnected with the laboratory work (e.g., engineering or accounting calculations, linear programs).   The background work is initiated in a similar way to that used on conventional data-processing computers. For example, a punched card deck may be put in the hopper of a card reader to be read as and when the computing time is made available by the operating system.

2.   Off-line systems produce paper tape digitized output that is subsequently fed through a tape-reader terminal in the laboratory, connected to a large commercial time-sharing computer network. The time-sharing idea here refers to the ability of several terminals scattered throughout the country to use the computer system virtually simultaneously.

As an example of the jargon difficulties, a paper at a computer automation symposium was entitled "Acquisition and Processing of Gas Chromatographic Data Using a Time-Shared Computer" (40) and referred to an on-line system. A second paper was called "Time-Sharing—A Powerful Approach for Gas Chromatographic Data Reduction" (45); it described the second, or off-line, type of approach.

III.   DIGITIZATION

A digital computer can process only discrete *digital* data. In any application of computers to GC, it is therefore essential to convert the *analog* voltage signals from the detector into a suitable digital form. Two aspects of analog-to-digital (A/D) conversion need to be considered:

1. How are the data to be digitized?
2. How are the digitized data to be fed to the computer?

In this section, various kinds of digitizers are considered, together with the form of the digitized output. Methods available for inputting the digital data into a computer are discussed in Section IV.

The conventional digitizer consists of a ruler or planimeter and a human brain, which together can carry out the necessary A/D conversion from the trace on a recorder chart into tabulated data. The usual output device is a pencil, which is used to store the data in the form of arabic numerals on a sheet of paper. These data can then be manually punched into holes in cards for further processing by a local data-processing computer. Data that are digitized manually in this way usually consist of integrated peak areas and retention times, because it is too tedious to tabulate the complete form of the chromatogram. In fact, the peak areas are frequently *not* integrated, but merely *calculated* from data such as the height of the peak and the width at half-height. When computerization is contemplated, a choice must be made between (*a*) digitizing only the areas and retention times and (*b*) obtaining a *complete* digital record of the chromatogram.

A. Complete Digitization of Chromatograms

The main factors to be considered for complete digitization are the *sampling rate, attenuation–amplification, legibility* and *available digitizers*.

1. Sampling Rate

A timer (possibly in a computer) is usually used to actuate the digitizer to give a constant number of readouts per second. The optimum sampling rate depends on the following factors: the digital smoothing and area calculation methods to be used by the computer, the accuracy and precision demanded of the calculated retention times and peaks areas, the shape of the peaks, and the magnitude and bandwidth of the noise. It has sometimes been claimed that the fastest possible sampling rate should be used. It is now generally recognized, however, that speed gives no advantage and, indeed, may be disadvantageous because of the more pronounced effect of high-frequency noise spikes and the difficulty of distinguishing wide peaks from baseline. Peak areas are usually calculated simply by summing successive digitized voltage values, using the trapezoidal rule or Simpson's rule. In the following discussion, a Gaussian peak of standard deviation σ is assumed to have a width of 8σ (i.e., twice the conventional basewidth). The width at half-height $w_{1/2}$ is approximately 2.4σ.

The theoretical error for Simpson's rule applied to a Gaussian peak has been computed as a function of the number of digitized points (46). According to this work, the error is about 0.9% for 10 samples/peak, and falls to 0.1% for 13.4 samples/peak.

TABLE I
Theoretical Area Recoveries of Gaussian Peaks at Various
Sampling Rates using Trapezoidal Rule (47)

Distance between scans (σ)	Samples per peak	Area recovery, %
<1.0	>7–8	99.99–100.01
1.5	5–6	99.97–100.03
2.0	3–4	98.56–101.44
>3.0	<2–3	<77.7->122.3

Table I lists theoretical area recoveries for the trapezoidal rule at various sampling rates for model Gaussian peaks (47). Samples taken at intervals exceeding 1.0σ give calculated areas that are significantly greater or less than the true area. Sampling at intervals of less than 1.0σ (i.e., in excess of 7 or 8 samples/peak) theoretically gives areas within 0.01% according to Table I. It follows that a sampling rate of $2.4/w_{1/2}$ should be adequate. On this basis, a constant sampling rate of 1 point/sec should suffice for all peaks with $w_{1/2}$ of 2.4 sec or more. For very fast peaks with $w_{1/2}$ of 0.5 sec, a rate of 5 points/sec would be needed.

In practice, however, a faster scan rate is required because of the effect of skewed peak shapes, noise, and digital smoothing techniques employed by the computer. In one study of digitization of actual GC peaks (47), it was found that a sampling rate of 5 points/sec allowed good trapezoidal area recoveries of between 100.5 and 100.35% for $w_{1/2}$ of 2.0 sec, giving about 33 points/peak. Poorer recoveries of between 101.16 and 102.27% were found for $w_{1/2}$ of 1.2 sec (i.e., with 20 points/peak). These areas were calculated from smoothed data. Simpson's rule has been applied to real aromatic peaks showing significant tailing on Apiezon L at temperatures between 80 and 138°C (46). Again, between 20 and 40 points were needed for 0.01% accuracy; but even 13.4 points/peak gave 0.1% accuracy, which is in surprisingly good agreement with theory. In this work, no digital smoothing was applied, and sampling rates were no faster than 1 point/0.6 sec. In other work using Simpson's rule applied to smoothed data (48), a maximum scan rate of 1 point/0.3 sec was used. Errors of less than 0.5% were found provided at least 9 or 10 points were taken across $w_{1/2}$ (i.e., about 30 points/peak). A detailed discussion of the effect of peak shape and noise on the optimum sampling rate has been published (49).

In theory, only three points are required to define a pure Gaussian peak if a curve-fitting method is used (Section V.C). This is about half the ideal number of points required for the trapezoidal rule according to Table I. In practice, 8 to 16 points/peak have been reported to give good curve-fitting results for peaks without excessive tailing (50, 51).

Retention time precision obviously depends on the sampling rate; but from this point of view, any rate greater than 1 point/sec should be adequate for most routine work. Some special applications requiring precision of ± 100 msec or better require a sampling rate of at least 10 points/ sec (42, 52, 53).

To summarize, it appears that a sampling rate giving 40 points/peak (12 points over the width at half-height) is adequate, unless retention time precision requires faster sampling. A rate of 5 points/sec is usually more than fast enough (39, 41, 46, 48, 54) and allows accurate digitization of peaks with $w_{1/2}$ of 2.4 sec or more.

Most commercial on-line processing routines reduce the scan rate during the course of isothermal runs as the peak width increases. If this is not done, it can be difficult for the peak detection logic to distinguish small, wide peaks from the baseline (40). Some packages adjust the sampling rate automatically according to the width of the detected peaks. Others leave the user to specify at what time, either absolute or relative to a reference peak, he wishes the scan rate to be decreased. An alternative to physically decreasing the scan rate is, of course, to use only selected data points when searching for peaks.

2. Range Problems

Signals covering a voltage range of about 10^6 are commonly encountered for a single position of the range switch of modern GC detector electronics, although some older equipment has a range of only 10^3. Thus one-tenth of a chart division on a 1-mV recorder usually represents 1 μV if the attenuator is set at $\times$ 1, whereas full scale represents 1 V with attenuation of $\times$ 1024. Often a separate unattenuated output is available for data-handling equipment. Such an output may give 10 V when 1 V is supplied to the recorder output attenuator, but the useful signal range is still usually limited to about 10^6 by the inherent noise level of the equipment. In contrast, the range of most electronic analog-to-digital conversion equipment is only about 10^4 (e.g., from 1 to 9999 for decimal presentation, or from 1 to 16,383 for binary presentation in 14 bits). Furthermore, such equipment usually gives full-scale output for input signals in the range 1 to 10 V, so that preamplification of GC signals below 0.1 or 1 mV is essential if the signals are to register at all. Preamplification by a constant factor, however, causes large detector signals to saturate the analog-to-digital converter.

Various options are available to accommodate this amplification–attenuation problem. A simple solution, which is probably acceptable for many routine purposes, is to apply a constant preamplification factor of about 1000 to the GC signal and be content with the limited range of the digitizer (48, 55). Manual range changing can be used (56), but this is obviously undesirable. A simple automatic timing system can be employed to switch the range or attenuator at predetermined points in the chromatogram (57). Such a device demands very stable conditions as well as an application requiring repeated runs with very similar samples. Moreover, there is no indication in the digital chromatogram itself that any range change has occurred. If the digital record is to be processed off-line, it is therefore more satisfactory to use a digitizing system that automatically attenuates or removes attenuation as the signal approaches the top or bottom of the digitizer range. These attenuation changes can be automatically recorded on the digitized chromatogram by an extra character (58).

In on-line systems, a number of parallel amplifiers, each with a different gain, can be used so that the appropriate digital signal is selected by the software depending on the trend of the previous points (38, 41, 59). Alternatively, programmable gain amplifiers can be used, which also allow software selection of a suitable gain (39, 60). However, high-speed auto-ranging amplifiers are now readily obtainable (42, 47, 61–63); these devices provide a high-level voltage, together with a digital code (usually in two or three bits) representing the gain. Such amplifiers can be employed both in on-line and off-line work.

3. Legibility

"Digitized data" can take a number of forms, such as arabic numerals on a sheet of paper, holes in paper tape, or voltage levels at electrical contacts. The question of the legibility of such data by a computer is particularly critical for off-line work. The two most important points to be considered are the nature of the available input devices for the computer and the codes that the computer library programs can understand for data fed in through these devices. For example, it is possible (but very unlikely) that the computer could understand a figure of 61 mV by reading a six and a one typed by the digitizer as "61" on a sheet of paper, which is fed to an optical character reader. The same data could be punched by the digitizer as a set of holes in paper tape representing "6" followed by a second set of holes representing "1." The characters are now in a "serial" rather than a parallel form. Alternatively, the decimal number of 61 could be represented by six binary-punched characters representing "1," "1," "1," "1," "0," and "1," by two octal characters "7" and "5," or by two hexadecimal characters "3" and "D." The actual format must be decided by reference to the standard input routines of the computer.

Assuming that binary coded decimal (BCD) punched tape is used, between six and ten characters are required for each data point, including attenuation and an "end of point" character (64). Provision must be made for either automatic or manual punching of "start of run" and "end of run" characters to enable the computer to recognize the complete chromatogram (65, 66). This is particularly important if a single length of paper tape contains several chromatograms, possibly produced from a series of samples introduced by an automatic injection system. Provision of these "start" and "end" characters can be rather difficult with some so-called automatic systems, which were designed as general-purpose data loggers. A chromatogram of about 3000 points will generate perhaps 20,000 BCD characters. It would take about 2000 sec to read such a tape through a normal time-sharing terminal, but only about 20 sec through a fast paper tape reader attached to the computer. However, many large computer installations have no paper tape input facilities. It then becomes necessary either to use expensive automatic card punching equipment instead of paper tape punches in the laboratory, or to transfer the paper tape data to punched cards (46, 65) before using the computer.

4. Digitizers

An encoder attached to the shaft of the balancing motor of a conventional potentiometric recorder provides a cheap digitizer that has been used for amino acid analysis (67–69) as well as for gas chromatography (58). The dynamic range of such a device is limited to that of the recorder (i.e., about 100), and the maximum sampling rate is 2 or 3 points/sec. The accuracy is obviously limited because the electrical detector signal is digitzed only after passing through two separate conversion steps.

Automatic analog-to-digital conversion is usually achieved with an electronic voltage-to-frequency converter that produces a series of pulses at a frequency accurately proportional to the input voltage. These devices, which are at the heart of most digital integrators, are also found in data loggers, digitizers, and integrating digital voltmeters.

In one of the earliest reports of computerized GC (1), the pulse train was recorded on magnetic tape, after being rectified and filtered. The tape (with its essentially analog signal) was then played back at a speed up to 20 times as fast as the recording speed, and the number of pulses in each constant time interval was counted electronically and transferred directly to a computer. This system has been used to find detection limits for peaks of different widths using three flame ionization detectors (70). However, the range of signals is limited to about 10^4, since no attenuation information is present on the tape. Results are also affected by tape stretch in the same way as an analog strip chart chromatogram is affected by extension of the paper. Moreover, many computers cannot handle this kind of magnetic tape without transferring the "counts" into true digital form.

It is much more usual to count the number of pulses in a fixed time interval directly from the voltage-to-frequency A/D converter before transferring them to punched paper tape, incremental magnetic tape, or directly to a computer. This can be done using a commercial digitizer (65, 66) or an integrating digital voltmeter. A type of digital voltmeter without a voltage-to-frequency converter can be used equally well (48, 52, 56, 71). One low-cost circuit converts the usual visual *parallel* BCD readout of such voltmeters into *serial* 11-bit ASCII (American Standard Code for Information Interchange) suitable for punching digit by digit into paper tape (72). A normal digital integrator can be modified to act as a digitizer at a rate of about 1 point/sec by disabling the slope detector and baseline correction logic and providing a trigger pulse at regular intervals (73). An internal timer pulse is available for recent digital integrators, and this digitizing mode can be selected instead of the usual integrating mode by means of a switch.

B.   Programs Using Digitized Retention Data

The previous discussion has centered round the production of a detailed digital record of the chromatogram. Increasing attention is being given to computer calculations starting with peak areas and retention times as the raw data. Many of these applications have used digitized data that were entered by hand.

Apart from straightforward calculations of response factors (74, 75), these include programs for storing and retrieving retention data to aid peak identification (76–79), carrying out rather complicated calculations using data grouped from multiple chromatograms (80), drawing histograms of the chromatograms (81), calculating simulated distillation data (82), optimizing choice of mixed stationary phases (83), minimizing analysis time (84), calculating retention indices (85), and calculating parameters in programmed temperature work (86).

Of greater general interest, however, is the use of a computer to identify peaks from the retention times in well-characterized routine runs and to report the component concentrations after carrying out a suitable calculation involving response factors and possibly internal standard concentration. In small laboratories, it is often more economical to calculate peak areas and retention times with a digital integrator, rather than to use a computer to do this on a fully digitized chromatogram. On the other hand, computerization of the subsequent tedious calculation process is usually easy to justify on economic grounds. This circumstance is leading to a rapidly growing use of "hybrid" systems.

C. Digital Integrators for Hybrid Systems

A digital integrator combines analog and digital techniques to compute the digitized peak areas. A built-in clock and associated electronics are usually incorporated to produce digitized retention times as well as areas. A detailed discussion of the design and operation of digital integrators has been published (87).

The analog signal from the GC detector is amplified before being fed to a voltage-to-frequency converter. The desired integrated area is obtained simply by counting the total number of pulses produced during a peak, in contrast to the discrete series of samples produced by a digital voltmeter. Mechanical counters used in early integrators have a maximum count rate of about 150 sec^{-1}. Statistical fluctuations could therefore become important for peaks lasting only a few seconds or representing only a small percentage of the sample. Electronic counters are now used even in integrators costing only about $2500 giving count rates of 10^6 sec^{-1} for 1-V signals. Each count represents 0.5 μV-sec, and this resolution is more than adequate for most practical GC runs. More sophisticated integrators are available having even faster count rates.

The first derivative of the input signal (used to detect the start, maximum, and end of a peak) is obtained by an *analog* slope detector. Decisions about the detection of peak events must be made *instantaneously* and cannot be revised in the light of information obtained slightly later in the chromatogram. This is the basic weakness of an integrator. In contrast, a computer operates on a string of values of the digitized signal held in memory. However, more and more sophisticated logic circuitry is being added to modern integrators, including digital memory features which are now sometimes used to control baseline correction. In fact, some recently introduced "integrators" have peak detection logic wired in which is almost identical with the software logic of on-line computers. Memory facilities are also provided, enabling the user to enter various parameters into core storage. The digitized values of peak areas and retention times, together with other information (e.g., that a peak is a shoulder) can be fed directly into an on-line computer. Much more usually, however, it is punched into paper tape for processing through a time-sharing terminal or at a computer near the laboratory.

One advantage of hybrid systems over complete digitization systems is that a visible record of peak areas and retention times is available, even if the computer is not. This has obvious attractions for routine quality control laboratories. A further advantage is provided by the compact form of the digitized data, which consist merely of peak areas and retention times instead of thousands of individual voltage values. This small bulk of output from a digital integrator means that use of commercial time-sharing terminals with data rates of only 10 characters/sec is more practicable. Rapid access to a large time-sharing computer for calculating results can therefore be gained through one or more typewriter–punched-tape terminals in the laboratory itself. The main disadvantages are the inherent limitations of normal digital integrators and the possible expense involved, for if time-sharing terminals are used *very* extensively, the cost can be much greater than that of having a computer in the laboratory.

There are indications that many small GC laboratories will use the hybrid off-line approach. Commercial systems (45, 88) can be had, costing about $7000 per GC for a "good" digital integrator, a code conversion unit (e.g., from parallel BCD to serial ASCII 8-level), a typewriter–paper-tape unit (Teletype ASR-33) for output, and suitable general software for use on the time-sharing network. Normally, an additional time-sharing terminal must be purchased or rented.

IV.   METHODS OF COUPLING COMPUTERS
WITH GAS CHROMATOGRAPHY

Digital computers can be used to automate three distinct areas of gas chromatographic analysis: (*a*) calculation and reporting of results, (*b*) digitization of the analog detector signal, and (*c*) control of the physical components of the gas chromatograph.

A computer is the only alternative to manual calculation for (*a*), and all GC computer systems involve automation of this process. Where (*a*) is the *only* area to be computerized, the system is usually said to be "off-line," and the raw or integrated digital data from (*b*) must be fed to the computer from some sort of intermediate storage medium. The majority of gas chromatography laboratories can benefit from the use of an off-line computer.

"On-line" systems involve computerization of (*a*) and (*b*), but (*c*) is still often carried out manually. One on-line computer usually serves a number of gas chromatographs, and laboratories having at least six routine GC analyses in progress at one time are most likely to benefit. However, laboratories undertaking fundamental GC research may find it worthwhile to dedicate a computer to a single gas chromatograph to achieve detailed data acquisition and control of (*a*), (*b*), and (*c*).

A growing number of so-called hybrid systems have been described in which (*b*) is automated with a digital integrator, which is coupled either on-line or off-line to a computer for stage (*a*). (See Section III.C).

Various approaches to GC analysis are compared schematically in Figure 6. One useful tabular comparison (90) includes selection criteria

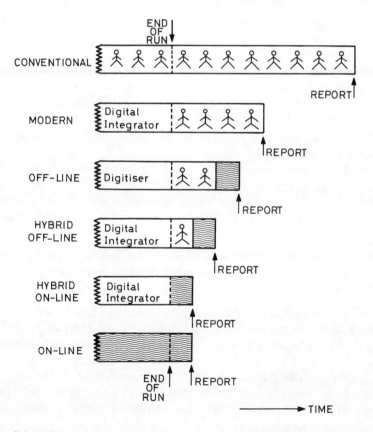

Fig. 6. Schematic summary of various GC analysis methods. The extent of computer involvement is indicated by the shaded areas.

covering economic considerations and describes the laboratory and the analyses to be undertaken. The cost of computers has fallen rapidly in recent years, sometimes in actual cash as well as in real terms. Dramatic price reductions are to be expected during the next few years, especially in memory. Nevertheless, it will be many years before most laboratories will be able to justify investing in an on-line computer.

A.   Off-Line Systems

In an off-line GC–computer system, data must be stored in digital form on an intermediate storage medium for a significant length of time before being processed by the computer (64, 73).

The *advantages* of an off-line system, as compared with an on-line system may be summarized thus:

1.   The cost is greatly reduced if only a few GCs are involved;
2.   The laboratory is not dependent on the good health of a particular computer installation;
3.   "Hard" copy of the chromatographic data is available for later calculation in the event of a computer or programming failure;
4.   Access to more powerful computers is possible. This is particularly valuable for nonroutine work;
5.   New or amended routines can be quickly and easily programmed in high-level languages;
6.   It is a good way to gain initial experience of computerized GC;
7.   If the complete chromatogram is digitized, it can be inspected several times by the computer, which can itself select the best parameters for peak detection. This multipass method is not possible with small on-line systems when the raw data are compressed during the course of the run.

Off-line systems suffer from the following *disadvantages* compared with on-line systems:

1.   There may be a marked increase in elapsed time, possibly of 24 hr or more, between the end of a GC run and production of results. This delay can be reduced to a matter of minutes if a commercial time-sharing terminal is placed in the laboratory itself;
2.   Conventional data-processing computer systems of the kind found in most large organizations do not usually have paper tape input facilities;
3.   If a hybrid system is used (i.e., employing a digital integrator), the peak detection and baseline fitting logic will be inferior;

4. If the complete chromatogram is digitized point by point, objection 3 does not apply, but the large amount of punched tape produced makes use of the 10 character/sec time-sharing terminals impracticable. There may therefore be a greatly increased turnaround time, as indicated in item 1.

5. There is no possibility of computer control of GC functions, such as injection or temperature programming. However, fairly cheap digital controllers can be used for this purpose;

6. Transmission of data down telephone lines from time-sharing terminals is not 100% reliable, a loss of perhaps 1 character in 2000 can be expected.

Many published uses of off-line systems employing digitized total chromatograms have been for nonroutine work. For example, one such arrangement (65) has served in very detailed analyses of peak shapes. Another (66, 91) has been used for highly precise measurements of heats of solution and adsorption (66, 92). The same system has also been employed to study the effect of molecular sieve micropores on peak shape (55, 93) and in applications of ensemble averaging to improve detection limits by a factor of 100 for a flame ionization detector (94). For the last application, the paper tape data were transferred from paper tape onto digital magnetic tape through a computer. The digitized chromatograms could then be sorted and retrieved more readily. Another nonroutine area in which off-line digitization has become popular is the study of computer methods for the resolution of overlapped peaks (Section V.C).

In a detailed study of contributions of instrumental effects to peak broadening, the digital chromatograms were stored as octal code in core memory (95). This is a "pseudo"-off-line system because no data reduction is done until the end of the experiment; but at the same time, no manual manipulation of the digitized chromatogram is required. A similar pseudo-off-line approach is used for routine work in some apparently on-line systems (29, 48, 96).

The off-line full digitization method has been used for routine work with fatty acid methyl esters with paper tape (97) or incremental magnetic tape (98) as intermediate storage. Paper tape has also been used in this way for investigation of tobacco smoke (56). The full digitization method has been recommended for routine work in small laboratories, and total cost for a digital voltmeter, interface, amplifier, and paper tape punch has been quoted as $2500 (48). This is cheaper than a digital integrator system, but it provides better accuracy than either an integrator or a small on-line computer system. The penalty is having to deal with many yards of paper tape, and possible slow turnround times, before seeing peak areas calculated. Incremental magnetic tape recorders are becoming more widespread, including cassette versions. On these rather expensive devices, which offer greater density of information and faster input than paper tape, 30 hr of chromatograms can be stored on a single 1200-ft reel of tape. The reel can then be processed to give reports within 20 min (98).

An interesting example of the added security for routine GC work provided by off-line systems, compared with on-line systems, has been reported for liquid chromatography (99). The signals from the retransmitting recorder slidewires were digitized by a data logger and also sent on-line to a small data-processing computer. The digitized data saved many man-hours of calculation on several occasions when the computer was out of action for a full working day. However, such a system would be much less attractive in an environment where *results*, as well as experiments, could not wait until the computer was repaired. In this context, the "hybrid" systems described in Section III.C can be very useful.

Many small laboratories already have digital integrators that usually can be modified to produce serialized output suitable for driving a paper tape punch attached to a typewriter (Teletype). The punching speed is 10 characters/sec, which is fast enough to handle one peak every 2 sec or so. A typewritten output of the peak area and retention time is produced at the same time (100–104). When the run is over, computer calculation of results can be obtained by feeding the tape directly into the tape reader of an in-house computer (104). Alternatively, a terminal to a private (102) or commercial (100, 101) remote time-sharing computer can be used. For in-house use, the integrator output can be stored on a magnetic tape instead of paper tape (105), allowing faster playback.

All the power of the data-processing computer can be used for peak identification and application of response factors. The usual procedure is to initialize the system by feeding in names and expected retention times of each component, followed by the composition of a standard blend. The tape output from the chromatogram of the standard blend can then be entered, to enable the computer to work out response factors. The subsequent processing of any number of unknown samples merely requires some identifying information to be provided, followed by the corresponding paper tape. Peak identification, retention index calculation, reporting, merging, storing and summarizing results can easily be achieved.

Standard GC programs are provided for many commercial time-sharing systems, either by the computer firm or by the integrator firm (45, 88, 106–108). These are normally furnished free of charge with the commercially available integrator–teletype systems. It is a relatively simple matter to write special programs in a high-level language such as BASIC, which can be learned in half a day. It is difficult to generalize about the cost of remote time-sharing systems, but the factors to be considered include fixed charges, rental for the terminal, hourly connection charge, cost of the telephone call, and charges for storing data in the computer files, as well as the actual cost of CPU time. Costs and economic justifications for various hybrid systems have been discussed by several workers (73, 100, 102, 103). One published application (101) describes the analysis of multicomponent organic mixtures in water by pyrolysis–GC. The commercial time-sharing computer used a least-squares method programmed in BASIC to compute the composition of the original mixture by comparing the observed pyrograms with those produced in calibration runs with pure materials. This usage indicates the attractions of having a powerful, easily programmed computer at the end of a telephone line.

Programmable desk calculators can be used for simple calculations. A solenoid deck is available which will depress the keys of such a calculator as the punched tape is read through a standard unit. This allows the calculation of area normalization and relative retentions (109). Calculators can also be interfaced directly to digital integrators (110).

B. On-Line Systems

In an on-line system, the intermediate off-line step of storing digital data for a significant length of time before processing by the computer is removed. Instead, the detector output is fed directly to the computer through an electronic interface. As indicated earlier, most on-line computers serve multiple gas chromatographs. If this is the total capability of the computer, the system is said to be *dedicated*. If the computer is somewhat larger and can handle other types of laboratory instruments, as well as conventional computer batch background work, it is often called a *time-shared* system. Several reviews and comparisons of the advantages and disadvantages of dedicated and time-shared systems have been published (12, 18, 26, 90, 111, 112, 188, 189). The attractions of a time-shared system for simultaneous development of extended laboratory automation and management information systems have been stressed (113, 114). The general instrumentation problems involved in on-line connection of GC and computers have been discussed (115–117), as well as the economic considerations that led to the choice of certain particular on-line systems (118, 119).

1. Passive and Active Applications

The various general types of on-line computer applications in laboratories have been surveyed and classified as *passive* or *active* (12). Passive applications, in which the computer serves essentially as a data logger, can be divided into four areas: simple data handling and processing, complex data processing, file searching and pattern recognition, and visual display of the raw or correlated data. Active applications involve some control of the experiment—either by a simple fixed program or in response to decisions made by the computer in the light of conditions prevailing during the course of the experiment. Thus, in addition to simple automation, active applications include computer interaction with the experiment, and possibly with the experimenter, as well as iterative optimization of the experiment.

Most of the on-line computer GC systems involve mainly "passive" simple data processing, with a little "active" routine automation of injection and valve-switching devices. The other categories are only rarely encountered in GC at the moment. For example, passive complex data processing, such as deconvolution of overlapping peaks, is usually carried out off-line (Section V.C), although a few simplified on-line deconvolution techniques have been published (120–122). Similarly, file searching and pattern recognition, such as identification from retention data for nonroutine samples, is usually done on off-line machines.

The ability of an on-line computer to display data visually during the course of a run is rarely used in GC, although a facility to monitor the run can be extremely valuable when first setting up a routine GC computer method. However, oscilloscopic displays have been used in coupled GC–mass spectrometer computer systems (123). Sophisticated active applications involving interaction between the chromatograph, the experimenter, and the computer have been described (52, 124, 125), but such work as yet has little routine application. Instead, conventional temperature and pressure control systems are usually employed.

2. Examples of

Existing On-Line Systems

On-line computers play an increasingly important role in the control of process plant, and this field was the subject of critical annual reviews in *Industrial and Engineering Chemistry* until publication of that journal ceased in December 1970 (126). Each review contained a classified bibliography giving the titles of about 700 papers, including some of relevance to laboratory automation in general and to GC in particular. A review of process control by gas chromatography (127) discussed laboratory analyses and process chromatographs coupled to computers. Computer control of a cumene synthesis reactor based on GC on-line analysis has been described (128). Several other process applications of computer-controlled chromatographs have been discussed (129–141), including a large-scale commercial system incorporating 20 process gas chromatographs (142).

Many of the first on-line GC laboratory systems were based on the third-generation solid-state process control computers. The IBM 1800 is typical of these time-shared systems (14, 59, 143–145) and appears to be the one most widely used for routine GC analyses in laboratories having 20 or more gas chromatographs. Users include Monsanto (38), Standard Oil of New Jersey (146), Procter & Gamble (41), Sun Oil (40), du Pont (147), Esso Chemical (118), and BASF (148, 149). Time-shared GC systems based on other computers, including PDP 10 (29, 96, 150), CDC 1700 (122, 151), and EAI 640 (120, 152–154), have been described (155–158). Rather less versatile multichannel dedicated systems based on a wide variety of machines also exist (159–161). These include PDP 8 (39), IBM 1130 (61, 162), English Electric M2112 (163), and Elliott 905 (57, 164). There are reports of time-shared laboratory systems for instruments other than GC. These include a Honeywell DDP-516 on-line to spectrophotometers, a spectropolarimeter, a Raman spectrometer, and an ultracentrifuge (28).

When a gas chromatograph is coupled to a mass spectrometer, some form of on-line data reduction system is almost a necessity (12, 165–170). A time-sharing system for high-resolution mass spectroscopy (MS) and low-resolution GC–MS is used at M.I.T. (123, 171, 172). About 400 spectra are accumulated in the course of a 30-min GC run. This system has served in such diverse applications as the rapid identification of a drug in the urine of a comatose patient (173) and the identification of trace acidic components in shale (174). Smaller dedicated computers have also been used on-line for GC–MS work (166, 175, 176). The sampling rate is obviously faster than for GC alone, and serious noise problems have been reported in an attempt to transmit the analog signals directly to a remote computer (177). Magnetic tape recording was used in some of the early computerized GC–MS off-line systems (178, 179). Recently, the development of such an off-line system was reported (180); this multiuser system has processed 60,000 spectra in two years.

Many of the on-line systems for GC laboratories have been tailor-made for the individual organization. This may involve only the firm itself in a home-made effort (57, 164). Alternatively, use can be made of the service offered by some software houses, which specialize in providing complete systems built around a computer of the customer's choice (61). More usually, however, standard hardware and software items have been bought off the shelf from a computer manufacturer, and modifications have been made to accommodate the requirements of the particular laboratory. However, since requirements change over the years in all but the most static or the most far-seeing laboratories, it is usually necessary for the laboratory to acquire a considerable amount of expertise in programming and systems analysis. This is probably best done by scientists and technicians rather than by computer specialists. One of the great attractions of time-shared systems is the ease and speed with which programs can be modified. In small systems, even to move a comma in a report format can be a significant undertaking. Standard GC program packages are available for some on-line computers, either without charge or at a monthly rental. The actual programming of these packages is usually "supported" by the computer firm concerned, who also supply considerable help and advice. Nevertheless, the responsibility for the success of the whole installation is left to the user, unless he is prepared to pay a considerable premium for an individual turnkey operation.

Recently, however, the GC instrument companies have entered the laboratory small computer market with some force. They offer modular, standardized on-line systems at low prices, including installation, performance guarantees, and operator training. Such systems can be used in the same way as an integrator (i.e., simply by reading the instruction manual), and no programming skill is necessary. Nevertheless, a few of these systems do have a considerable "spare" computing power, which can be put to good use if there is some computer skill in the laboratory. Because of the nature of this turnkey mass-marketing operation, there must be strict similarity between units sold to customers with very different detailed applications. To many of the smaller laboratories, however, the disadvantages of a rigid report format, and limitations on ways of processing the results, are likely to be outweighed by the advantages of an automatically typed report of the composition of a sample and the freedom from worry about the computer overhead.

On-line GC turnkey systems using complete digitization have been described by Perkin-Elmer (60, 181, 182), Pye-Unicam (62, 183), and Varian (47, 184, 185); Hewlett-Packard are marketing a *hybrid* on-line system based on their digital integrators and 2114 computer (186). There will probably be few published reports of experience by users, except perhaps at local symposia and in instrument manufacturers' newsletters. However, one laboratory has described how programs were added to the Varian turnkey software by interfacing a cassette tape recorder (187). This procedure saved GC results on magnetic tape, which was used to produce automatically typed summary reports, including comments entered by the analyst through a keyboard. There are obvious attractions in this mini-time-sharing approach. Even in some large systems, the results are lost as soon as they have been typed out once, so that clerical manipulation of the figures is necessary from that stage on.

3. Hierarchical Systems

Even in large laboratories it seems likely that future installations will use dedicated computers instead of time-shared systems. The growing importance of the minicomputer is already apparent in process control (126). The dedicated minicomputer can stand alone, or it can be coupled by a slow-speed data link to a remote supervisory computer, which is thereby relieved of the repetitive scanning, timing, and calculation routines. The supervisory computer can be a conventional data-processing machine or a process time-shared computer, with most of the process input–output features removed. This central computer controls and receives data from one of more satellite minicomputers and manages the communication links between them, thus allowing interaction of information obtained from a number of satellites. The characteristics of these hierarchical computer complexes have been discussed in relation to analytical laboratories (157, 190, 191), and some actual GC applications have been described (142, 148, 149).

4. Signal Transmission

The cheapest (and therefore the most common) way of interfacing a GC detector to a computer is through a pair of wires carrying the voltage signal to a digitizing system housed near the computer. By using a multiplexer, the output of a large number of chromatographs can be monitored by the computer using only a single amplifier and A/D converter. However, this approach is susceptible to earth loops, mains pickup, and other noise that can badly distort microvolt signals produced by trace components. Since noise problems are aggravated by the use of long lengths of wire, a centralized GC laboratory is desirable if this cheap interfacing method is adopted. Careful installation planning, assisted by expert engineering advice and the use of analog and digital filtering techniques can then reduce noise to an acceptable level (38, 154, 184). A typical precaution to minimize the effect of earth loops is to ensure that the computer and all the GC equipment share a single earth point not used by other equipment. Metal carrier gas connections were found to be the principal source of earth loops in one system, and typical noise levels were reduced from 1200 to 20 μV by replacing the metal lines with nylon (154). Cables carrying the detector signals should be screened with an earth wire and enclosed in conduits away from mains cables, for minimum mains pickup. In the absence of conduit, however, an increase in noise level of only 50 μV has been reported between a GC and a computer 1000 ft away (154). Noise levels are best studied with the aid of an oscilloscope—a chart recorder is a very good filter for noise with a frequency greater than 5 Hz!

The noise usually consists of high-frequency spikes superimposed on a 50- or 60-Hz mains ripple. The high-frequency components can be very easily removed with a simple filter. Since much of the high-frequency noise is produced in the detector itself, the filter can be situated at the GC end of the wire.

In order to avoid the noise problems inherent in transmitting micro-volt signals, a few systems use individual amplifiers at each GC to raise the signal to the 1 to 10-V level. This introduces range problems, as discussed in Section III.A, which can be overcome in a number of different ways. For example, if an autoranging amplifier is used at each chromatograph (62), the range is transmitted as a separate piece of digital information. Alternatively, a simple timing device can be used for routine runs (57). Range changing at the detector can be avoided by running several voltage signals at different amplifications from the detector to the computer. The optimum voltage level can then be selected by software (41, 161). Obvious-ly, each of these high-level voltage solutions to the noise problems involves much greater cost per GC channel than the low-voltage method.

A different solution to the noise problem is to actually digitize the detector signal at the GC and then transmit the digital information as high-voltage pulses. This need not be an expensive method, and it is used in one of the commercial turnkey systems intended for up to eight chroma-tographs (60). A programmable gain amplifier, controlled at each GC by software, is situated between the detector and a voltage-to-frequency con-verter. The voltage pulses are transmitted to a counter register in the purpose-built computer CPU, where they are accumulated. The magnitude of the detector output for any particular GC is calculated by the software by subtracting two successive values from the corresponding register. The accumulated integrated area above electrical zero is available at any time directly from the register.

Two different methods are available for enabling a computer to moni-tor the detector signals of several chromatographs. The most common is the multiplexer approach, where relay or solid-state switches are used to switch in the selected detector voltage to the CPU. This involves a radial arrangement of cables, with a separate channel for each GC. (A separate channel is also needed for each GC in the digitized approach, but the digital data are fed continuously into individual registers in the CPU and do not need to be multiplexed.) An alternative approach for voltage trans-missions is to regard the GC as an extension of the I/O of the computer itself, and hence to attach each GC directly to the "in bus" of the comput-er (cf. Figure 4). Each GC is given an address in the same way as other peripheral computer equipment and remains switched "off" until that address is transmitted by the computer down the "out bus," to be decoded by a special address unit incorporated into the GC. A GC therefore obtains access to the "data highway" only when it has recognized the address code. This means that switching rates are slower than is possible with multiplexers, but still sufficiently fast for scanning GC peaks. An attraction of the data highway system is that the cable can be installed once and for all, and additional chromatographs can be connected to it.

That there is no "best" way at present of solving the signal transmission problem is clearly indicated by the different approaches adopted by the turnkey systems. Thus Varian (47, 184, 185) use the simple low-voltage method with multiplexed radial lines and suitable screening methods to reduce noise, Pye (62, 183) use a high-voltage data highway, and Perkin-Elmer use the digitized method (60, 182).

5. Typical Operating Procedures

Many on-line systems now use essentially one-way communication between the gas chromatograph and the computer. Typically, the GC operator informs the computer of the nature of the next sample for a particular GC by pressing a number of keys on a typewriter keyboard or on a purpose-built operator's console. (The design of the console and the simplicity of the operations should be major considerations in routine laboratories. Elaborate software precautions need to be taken to detect obvious miskeying, so that a run that has been wrongly entered is prevented from continuing.) The information entered may be simply a GC channel number, an identifying sample code, and the time at which the sample was taken. The computer then selects the appropriate method from a library of all the standard runs and signals acceptance of the information by turning on a light at the particular GC concerned. Alternatively, in a nonroutine environment, more detailed information can be supplied by the operator (e.g., run time, scan rate, calculation method, and reference peak times).

A button at the GC itself is pressed to signify the start of the run. This usually causes a hardware-forced branch instruction to be executed by the computer (known as an interrupt). A well-designed dedicated system will respond instantaneously to this interrupt so that the button can be pressed at the very moment of injection of the sample. If a time-shared computer that performs functions other than monitoring GCs is working on a higher-priority task at the moment that the button is pressed, a delay of several seconds may occur before the interrupt is recognized. In this case, injection must be delayed until a signal is given by the computer, such as turning on a "run" light. This is usually the final communication *from* the computer until the end of the run.

There is growing interest, however, in "closing the loop"; that is in using the on-line computer in an active capacity throughout the run (90). Perhaps the most obvious application is computer initiation of sample introduction by an automatic injection unit, possibly using one of the commercial units which are now available (152). Other GC functions that lend themselves to computer control include valve switching and temperature programming. In these areas, however, computers have to compete with purpose-built control units with self-contained automatic logic. At present, these purpose-built units seem to be preferred, possibly because of cabling convenience. To what extent programmed logic will eventually see routine service for control of GC is still unknown. It may even be that microcomputers attached to each GC and linked by a single cable to the local minicomputer will come into use.

Meanwhile, there is no doubt that a gas chromatograph *can* be completely controlled by a computer. This control includes flow rate and temperature, as well as sample introduction and valve switching (42, 55, 124, 125, 184). A useful on-line flowmeter has been described in this connection (192). By making the programming suitably interactive, it is even possible to have the computer outperform any purpose-built control unit by carrying out a series of experiments unprompted by external suggestions from the operator. For example, experiments to determine the shape of a van Deemter plot of column efficiency against flow-rate have been computerized (125). Similarly, a series of runs could be organized by a computer to determine the fastest analysis possible for a given separation on a particular column, subject to constraints specified before the first run was carried out.

V. SOFTWARE FOR GAS CHROMATOGRAPHY

Some indication of the variety of off-line computations to which GC data have been subjected was given in Section III.B. In addition, off-line computers have been used in simulation studies of peak shapes (193) and for pattern recognition [e.g., searching for a match between library data and observed pyrolysis GC chromatograms for microorganisms (194)]. The potential variety of off-line computations is endless, but it seldom involves direct computer analysis of raw chromatographic digital data, because the data are normally punched by hand. However, three areas of software development are of more direct relevance:

1. Detection of GC peaks in complete digitized chromatograms.
2. Identification and production of quantitative analysis reports for the detected peaks.
3. Resolution of overlapped peaks.

A. Peak Detection Algorithms

It takes several man months of programming and several weeks of exhaustive testing to develop a satisfactory set of algorithms that can be applied with confidence to noisy digitized data. Therefore, the vast majority of computer users employ a commercial package system in which the bugs are kept at a minimum. It is nevertheless clearly important to have complete confidence in the validity of the algorithms that are used, and such trust can be achieved only by learning enough assembler language to understand the program listings. Errors will almost certainly be found, but these are rapidly being reduced over the years as the number of users increases.

The logic and memory facilities of an on-line computer are available in "real time" (i.e., while the GC run is still in progress). This enables the computer to calculate peak areas and retention times during the run, so that only crucial peak data need be stored for use in the later identification and calculation phases. Most on-line systems use this approach, which minimizes requirements for core storage and disc or tape storage. It also saves time, and enables calculation and reporting of results to start somewhat earlier than is possible if the complete chromatogram has to be scanned, as is the case with off-line systems. As mentioned earlier, a few on-line systems virtually operate off-line, storing complete digitized chromatograms on disc or tape until the end of the run. In some cases this is regarded as essential (29), and may indeed be mandatory if curve-fitting methods are adopted; but a bigger computer installation is required than would otherwise be the case for GC work alone.

The raw digitized data are usually subjected to digital filtering (47, 195) before being smoothed. Smoothing is almost invariably performed by what has become the "classic" Savitsky–Golay method (196). This least-squares technique is used on-line (65, 94), off-line (48, 92), and in modern digital integrators with built-in memory facilities (197). The smoothed data are further processed to give first and second differences, which themselves are commonly smoothed. The extent of smoothing can usually be specified by the user when setting up a particular run. Thus a noisy katharometer detector usually has computer methods defined with fairly large smoothing factors applied. The smoothed first and second differences are used to make decisions about whether a peak is being eluted and, if so, on what part of the peak the detector is situated at that moment. Some early programs used only the first difference, but it then becomes difficult to distinguish a peak start from a sloping baseline. Even today, some workers believe that second differences are too noisy to give useful information; but suitable smoothing techniques appear to overcome this objection.

Special difficulties are introduced by the existence of a shoulder on the front or on the back of a peak. In such cases it is necessary to withhold judgment about the nature of the peak until some time after the second difference has changed sign—or, rather, has passed through some threshold value that is usually specified by the user in the same way that smoothing factors are allotted. There are a large number of types of position on a peak, in terms of whether it is on the front or back of the peak, whether the slope is concave up or concave down, or whether a change in position is merely suspected (e.g., from baseline to concave up). This explains the complexity of the logic algorithms required. Details of peak detection methods have been given in a few instances (38, 59, 197).

B. Routines for Identification and Reporting

Identification of peaks from the retention times can be carried out at leisure after the GC run is over. Furthermore, if an obvious error in a computer report is observed by an operator, he can alter the necessary parameters and instruct the computer to recalculate the results from the peak areas and retention times retained in memory. This sort of facility requires considerable storage capacity if runs can be recalculated in this way several hours after the actual run; but at the same time, the necessity for carrying out another actual run is removed.

The usual method of identifying peaks is by time windows specified by the user against the name of the component or components known to elute in that region. Because of the long-term variation in retention times, it is very useful to be able to make these time-window allocations with respect to a reference compound that can be clearly identified by, for example, its large area. As the reference peak time is found to change by the computer, so will the time-window limits be updated. A single reference peak usually suffices for an isothermal run, but multiple reference peaks are essential for temperature programmed work and for capillary columns, at least until flow and temperature control is improved beyond that found in most routine instruments.

The calculation and reporting of results is a trivial task for the computer, if it has correctly assigned names to the individual detected peaks. Methods are usually available to allow calibration factors to be calculated automatically by injecting a standard blend of known composition. Most methods appear to assume, however, that calibration factors are independent of concentration; for accurate work, therefore, separate methods must be defined for each range of concentrations encountered in a given type of sample. A system has been described, however, in which nonlinear response can be allowed for (57).

C. Resolution of Overlapping Peaks

The achievement of good separation has been one of the main aims of gas chromatographers. Much effort has been spent on developing GC columns and conditions permitting good resolution of components that are normally difficult to separate. Interest has centered on being able to obtain a qualitative and quantitative analysis from the retention time and area (or height) of each peak directly from the recorder trace, with the minimum of interference from peaks due to adjacent components. Achieving separation is emphasized primarily because of the convenience, speed, and reproducibility with which the resulting well-resolved peaks can be quantitatively treated by hand.

In any discussion about computers and GC, the subject of overlapped peaks will surely be included, often producing rather heated debates. Many workers maintain that the only satisfactory method is to develop better chromatographic procedures, which avoid overlapped peaks. However, there is a limit to the number of components that can be completely resolved by a column, even in principle; thus overlapped peaks are often the only alternative to some form of sample pretreatment to remove certain components. Moreover, even today some separations are "difficult." Not many chromatographers can put hand on heart and claim they do not carry out an analysis unless the minimum resolution is 1.5. Therefore, there appears to be a real need for methods of resolving overlapped peaks other than by the counsel of perfection.

One possible method (which is little used because of the time required) involves calibration with a large number of mixtures containing known concentrations of the overlapping components, thus building up a file of overlapped peak shapes. Another time-consuming way of analyzing overlapped peaks is to add a reference peak which itself is eluted in the overlap region (198–200). The most common manual approach, however, is to use simple peak separation methods such as dropping a perpendicular from the valley floor to the baseline for peak envelopes showing valleys. Peak skimming is often used for shoulders on the front or back of a peak. These simple manual methods (Figure 7) are also those most commonly employed in on-line computer systems.

The perpendicular drop method always overestimates the smaller area, and Figure 8 shows the percentage error involved at various resolutions and ratios of heights for Gaussian peaks with standard deviation σ (122). The resolution is defined in the standard way as the ratio of the difference between the true retention times of the peaks to the average basewidth. Thus a resolution of 1.0 corresponds to a separation of 4σ, whereas a resolution of 1.5 requires a separation of 6σ. Also visible in Figure 8 are

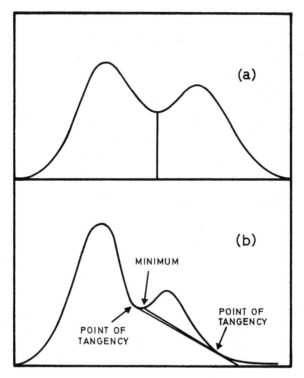

Fig. 7. Two commonly employed methods of resolving overlapping peaks; (*a*) dropping a perpendicular; (*b*) peak skimming.

the shoulder limit (where the minimum between the two observed peaks is replaced by an inflection point) and the detection limit (where this inflection point disappears to leave only a single broad peak) (122). It is interesting to note that for a peak height ratio of about 0.45, two peaks can be closer together (separation of about 1.5σ) than for any other height ratio and yet still be observed.

The peak skimming method grossly underestimates small peaks and shoulders, often by 50% or more, at resolutions less than 1.0 (120). Since the error is strongly dependent on the peak height ratio, the simple expedient of allowing for the error in the calibration factor is not satisfactory if a wide range of concentrations is encountered in the samples to be analyzed.

There are important differences between the manual and computer peak skimming methods. The manual approach is usually to sketch in the curved tail of the larger peak, or to draw a straight line which is tangential

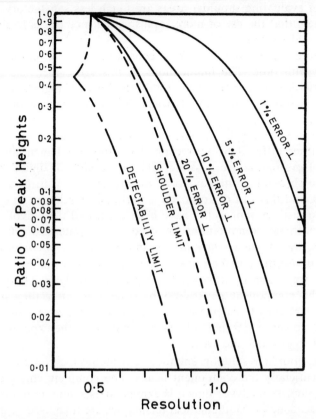

Fig. 8. Error curves (adapted from ref. 122) for the perpendicular drop method for Gaussian peaks, as a function of peak height ratio and resolution. The shoulder and detection limits are also shown.

By courtesy of *Analytical Chemistry*.

to the peak envelope at the start and finish of the shoulder or small peak. Neither of these approaches is very suitable for real-time computer solutions, and it is customary for an on-line computer to simply construct a straight line from the minimum or inflection point to be tangential to the tail (Figure 7). The computer therefore tends to produce estimates of the shoulder area even lower than those obtained by the manual methods.

Some computer packages combine the peak skimming and perpendicular drop approaches and allocate some fraction of the area under the tangent to the smaller peak. The fraction is determined empirically.

In general, the perpendicular drop method can be used with reasonable success, provided suitable calibration factors are employed. Simple methods of evaluating shoulder areas are much less satisfactory, however, and it is here that the use of more sophisticated methods is likely to show most practical success.

A chromatogram is a waveform consisting of a continuous variation of voltage with time. The implications of this self-evident statement are easily overlooked as a result of the established way of regarding a chromatogram as a series of individual peaks. Years of practice of evaluating chromatograms by finding the position of the peak maxima (retention times) and area or height of peaks directly from the recorder trace can make it rather difficult to appreciate that other fundamentally different methods of evaluation are possible. The availability of digital computers enables the chromatogram waveform, or any section of it, to be analyzed as a whole, and this sort of approach is particularly relevant to overlapped peaks. Two distinct approaches have been used, and they involve some of the few nontrivial computations involved in computerized GC. The two approaches are known as *curve fitting,* and *Fourier resolution enhancement.*

The curve-fitting method uses a computer to fit a number of parameters to the waveform by least-squares regression analysis. The parameters describe the component peaks and baseline, and after they have been fitted, they are used to calculate retention times, peak areas, and so on. In contrast to conventional methods, no integration of the experimental data is required to evaluate the area.

In the Fourier resolution enhancement method, a computer serves to sharpen artificially the component peaks. The complete chromatogram is subject to Fourier analysis; thus it is expressed as a summation of terms of different frequencies. Mathematical techniques are then used to modify the Fourier terms by a peak shape factor, before the chromatogram waveform is regenerated. After this treatment, peaks have the same retention times and area as in the "real" chromatogram, but resolution is improved as a result of the peak sharpening.

Advocates of these two methods point out that they make fuller use of the information available in a chromatogram, also allowing chromatographic runs to be carried out more quickly by deliberately sacrificing resolution.

1. Curve Fitting for Symmetrical Peaks

Curve fitting (201) involves fitting parameters in a chosen function so that the squares of the differences between the calculated values of the function and the observed values of a variable are minimized. The method is used in a variety of analytical techniques, including nuclear magnetic resonance (NMR) (202), infrared (203, 204), and Mössbauer spectroscopy

(205). For these techniques, and for GC, it is usual to represent the overall function as a sum of individual functions, one for each peak, and each having the same general form. The assumption of linear addition of component peaks is expected to be less valid for katharometers than for flame ionization detectors (206).

As an example, consider the equation for a pure Gaussian GC peak, where the variation of response y with time t is given by

$$y = h \exp - \frac{1}{2}\left(\frac{t - t_R}{\sigma}\right)^2$$

where h is the height of the resolved peak, t_R is the corresponding retention time, and σ is the standard deviation of the peak. (The basewidth of a Gaussian peak measured as the distance between the points of intersection of the baseline with the tangents to the inflection points is 4σ; the distance between the inflection points is 2σ.) Thus three parameters (h, t_R, σ) are needed to fit a symmetrical Gaussian peak. An envelope considered to be produced by N of these peaks overlapping would simply be the sum of N Gaussian functions and would therefore involve $3N$ parameters. The curve-fitting method must simply optimize the values of each of these $3N$ parameters to give the best fit (122, 207). Having fitted the parameters, the true area of each peak can readily be calculated as $\sqrt{2\pi}\, h\sigma$.

Simplified curve-fitting methods (120, 208) assuming Gaussian or skew Gaussian peaks are used in some commercial on-line systems. Several other simplified methods of allocating overlapped peak areas have been described (121, 209, 210, 216) based on various assumptions (e.g., that the retention index and peak shape are known) (211).

True nonlinear, iterative curve-fitting methods are well established (see Ref. 51), and suitable off-line computer programs are readily available. These programs do not in themselves require a large amount of core, and storage requirements are dictated mainly by the total number of data points n and the total number of parameters m to be fitted. Since commonly employed calculation procedures involve the storage of an $n \times m$ matrix, use of a large number of data points could quickly lead to storage problems in a small computer. It is possible to avoid storage of the $n \times m$ matrix (50) by suitable matrix manipulation methods, and the matrix storage requirement is then reduced to $m \times m$. Even then, however, use of an excessive number of data points could result in very high processing times of several tens of minutes or more. The use of between 8 and 16 points/peak for groups of up to 4 peaks has been found to be satisfactory (50, 51) with rather more for badly tailing peaks.

Any iterative curve-fitting procedure demands a considerable amount of judgment from the user. In all cases it is necessary to provide an initial estimate for each parameter. For the usual Newton–Raphson least-squares method, it is essential that these estimates be as close as possible to the best-fit values. If this is not the case, the basic assumption that the function to be fitted is linear in the region between the first estimate and the true value will be invalid, and the curve-fitting process may fail. Failure can take the form of an inability to find the inverse of a matrix; it could also appear as divergence or convergence to a false solution. The first two forms of failure can easily be recognized, but the third possibility must always be kept in mind. It is less likely to occur if the number of parameters to be fitted is kept as small as possible, and the importance of doing this has often been stressed (50, 212).

A good but meaningless fit can nearly always be produced by allowing any number of peaks to be fitted to a particular envelope (50). It is therefore very important to know how many peaks really are contained in an observed waveform. This puts very strong demands on the peak detection logic, especially for peaks showing considerable overlap. Figure 9 shows how a good fit of three peaks to an envelope resulted after the peak detection program had found a spurious peak. It was known in this case that in fact only two peaks were present.

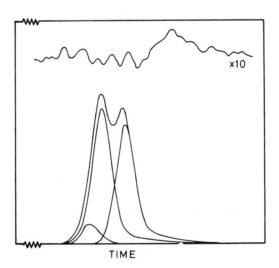

TIME

Fig. 9.   Illustration of need to know how many peaks are actually in the peak envelope (50). The peaks were fitted by a computer on the assumption that three peaks were present; as indicated by the error curve above, the fit is good. In fact, only two peaks were present.
By courtesy of *Analytical Chemistry*.

As indicated previously, a Gaussian peak requires three parameters corresponding to height, width, and position, with $3N$ parameters needed to fit N peaks. This is true for *any* function, provided all peaks in the chromatogram are geometrically similar and the baseline is constant. Baseline position can itself be included as a parameter, making $3N + 1$ parameters to be fitted by the curve-fitting program. Alternatively, a smooth baseline with continuous first and second derivatives can be fitted to pass through all experimentally observable sections of baseline before the curve-fitting proper begins. The mathematical technique of spline interpolation (206) has been recommended for this purpose. The assumption of a constant baseline can then be justified if the value of the spline function at each point is subtracted from the observed detector signal.

Chromatographic knowledge can be used to limit the number of parameters by requiring that the peak width be directly proportional to retention time over any group of overlapped peaks. Thus the N individual width parameters are reduced to the single parameter of the number of theoretical plates, giving $2N + 2$ parameters in all. This restriction to a constant plate number has been reported to give greatly improved convergence (50).

By a modification of the least-squares approach, the number of parameters can be reduced to N; thus only the amount of each component has to be fitted, assuming that the baseline has been fitted and that the form of the peak envelope for each component is known (206).

Various calculation techniques can aid convergence, such as fixing some parameters until the variation in other parameters between successive iterations becomes small, and use of damped least-squares methods (201, 212).

Use of the height and time of the apparent maximum as initial estimates for h and t_R has been reported to give satisfactory convergence for Gaussian peak models (122), but these estimates do not necessarily guarantee convergence for skew models (144). For shoulders, initial estimates of h and t_R have been taken as the average values of y and t at the inflection points, whereas for all peaks in an envelope, a good initial estimate for σ is half the maximum distance (time) between inflection points (122). For very fused peaks where no inflection points are apparent, it becomes difficult to choose good initial estimates, and results must be treated with great caution. "Improve your chromatography" would still seem to be the best advice here.

If necessary, a "forcing factor" can be applied to diagonal terms in the simultaneous equation matrix to ensure that the results will converge (144). It is important to note that convergence cannot be tested simply by demanding a good fit with the observed data. It is possible for a set of "absurd" parameters to give a very reasonable fit to the peak envelope, as illustrated in Figure 10. The dotted trace was obtained with a height of 6662 units for the third peak at a time of 279.5 sec, compared with an observed maximum in the peak envelope of only 5542 at the later time of 285.0 sec.

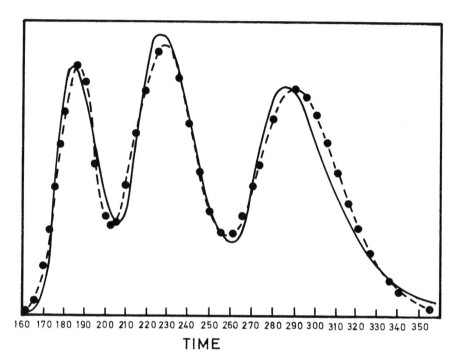

TIME

Fig. 10.   Illustration of convergence to a false solution; see text (144).
By courtesy of *Analytical Chemistry*.

Again, Figure 11 shows a good but nonsensical·fit where a double peak has been interpreted as a low, very broad peak, followed by a tall, sharp peak. Clearly, restriction of parameters by imposing a constant plate number would have prevented this false fit.

When properly used, with a critical eye and proper convergence tests, the least-squares method is probably the best way of generating an analytical or empirical expression that closely approximates the "true" peak outline of each peak in the envelope. The method, however, can give no better results than the model allows. In particular, the actual outline of a GC peak is sure to differ significantly from that of an assumed pure

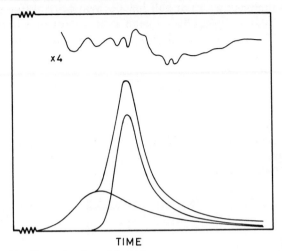

x4

TIME

Fig. 11.   Example of good fit with erroneous parameters; see text (50).
By courtesy of *Analytical Chemistry*.

Gaussian model. However, if highly overlapped peaks are resolved on the
assumption of a Gaussian model, it is likely that the *ratio* of estimated
areas for the two peaks will be better than that obtained by any manual
method. The actual areas can then be obtained by normalizing the estimated areas up to the observed total area. Thus if h_i and σ_i are the
estimated heights and standard deviations, respectively, for the ith peak in
an envelope of total area A, the individual peak areas are given by

$$A_i = \frac{h_i \sigma_i}{\sum_i h_i \sigma_i} A$$

where systematic errors in the individual estimates of h_i and σ_i are
removed. This approach has been used on-line (122), but it is not clear how
long the calculation took after the run was over.

In practice, even for apparently symmetrical peaks, significant inac-
curacy is introduced by use of a pure Gaussian model (207). In general,
then, it is essential to use more realistic model functions. Cauchy
(Lorentzian) functions (201, 203, 204, 212) are symmetrical, but they have
more pronounced tails at the front and back than do Gaussian peaks:

$$y = \frac{h}{1 + \left[\dfrac{2(t - t_R)}{w_{1/2}} \right]^2}$$

where $w_{1/2}$ is the peak width at half height. Symmetrical functions that are combinations of Cauchy and Gaussian peaks, with or without the same values of $w_{1/2}$ have also been considered (e.g., Ref. 212), but most interest has been devoted to asymmetrical functions.

2. Curve Fitting for Asymmetric Peaks

For asymmetric peaks, a fourth parameter can be added to h, t_R, and σ, as a skew factor. Skewness can be introduced empirically into a Gaussian peak in many ways (212, 213). For example, the term $(t - t_R)/\sigma$ can be replaced by

$$\frac{1}{b} \ln\left[1 + \frac{b(t - t_R)}{\sigma} \right]$$

where values of b greater than zero give rise to sharpening of the front of the peak and tailing at the back but do not alter the position or height of the peak maximum (212).

An alternative way of introducing skewnness is to consider that the symmetrical Gaussian chromatographic shape is *convoluted* with another function because of imperfections in the apparatus. In particular, there is physical justification for assuming that an exponential decay is a suitable convoluting function for effects in the injection and detection systems. In this case (51, 143, 144) the fourth parameter is the time constant for the exponential decay τ. This has the effect of delaying and reducing the height of the peak maximum, in contrast to the empirical skew term b mentioned previously. The shape (skewness) of the peak is characterized by the value of τ/σ. The area of the peak remains the same as that for a pure Gaussian peak (i.e., $A = \sqrt{2\pi}\, h\sigma$) and A has been used as a parameter in place of h (51). This avoids complications introduced because h is a function of τ and σ. Use of these exponentially convoluted Gaussian peaks has been reported to eliminate errors down to separations of 2σ, compared with 3σ for a pure Gaussian model (51).

Asymmetric peaks can also be generated by combining the front half of a Gaussian peak with the back half of a second Gaussian peak with different σ (120, 122, 144). This again needs four parameters per peak: h and t_R are the same for both halves, but the standard deviations of front and back are different. This idea can be extended by combining a pure Gaussian front half with a skewed Gaussian back half (five independent parameters), or even by combining two skew Gaussian half-peaks (six parameters) (144).

A much simplified method based on two Gaussian half-peaks has been described (120). Since extensive calculation is not required, the procedure can be carried out during the course of the run in the same way as conventional perpendicular drop and peak skimming methods. The assumption is made that the standard deviations of front and back halves of the first peak σ_{f1}, σ_{b1} and of the second peak σ_{f2}, σ_{b2} are known from calibration with pure components. It is further assumed that the true retention times t_{R1} and t_{R2} are the same as the times of the observed peak maxima t_1 and t_2. This assumption becomes progressively less justifiable as the degree of overlap increases. The observed heights at the peak maximum are given by

$$y_1 = h_1 + h_2\exp-\frac{1}{2}\left(\frac{t_1-t_2}{\sigma_{f2}}\right)^2$$

$$y_2 = h_1\exp-\frac{1}{2}\left(\frac{t_1-t_2}{\sigma_{b1}}\right)^2 + h_2$$

and it is therefore a simple matter to calculate h_1 and h_2, the true peak heights. The peak areas are then given by

$$\sqrt{\pi/2}\,(\sigma_{f1}+\sigma_{b1}) \quad \text{and} \quad \sqrt{\pi/2}\,(\sigma_{f2}+\sigma_{b2})$$

It has been demonstrated that this method is consistently more accurate than the perpendicular drop or peak skimming method. Results have been obtained using this approach (120) showing no loss of accuracy on increasing the chromatographic temperature from 80 to 155°C, with a reduction in the run time from 11 to 3 min, for analysis of a mixture of C_1, C_3, and C_4 alcohols. However, even for the fastest run, the amount of overlap was rather small (resolution marginally less than 1).

Because of the wide variety of factors determining peak shape in GC, it is unlikely that any one kind of function can be made to give a good fit to all peaks. There is a useful alternative approach in which the peak shape function is determined empirically by the computer itself from a single calibration peak of a suitable pure compound (50, 214). This method has been used successfully to fit $2N+2$ parameters down to separations of about 1.3σ, and at this stage there are not even any inflection points to indicate the presence of a double peak. This compares with a useful limit of 2σ for the exponential deconvolution method, and 3σ for a pure Gaussian function (51).

This empirical approach can be taken even further by determining the function representing *each* component peak in the chromatogram (206). The observed chromatogram F then becomes simply the sum of the terms:

$$F = \sum_{i=1}^{N} a_i f_i$$

where a_i is the weight percentage of the ith component, which has the empirical form f_i. Algorithms for solving this equation for the individual weight percentages have been developed, both for the usual least-squares method and for the much faster orthogonalization procedure, where F is regarded as a sum of orthonormal functions that can be calculated from the values of f_i (206). With this method, only N parameters need to be fitted; but this obviously requires either frequent calibration to find f_i or exceptionally good control of chromatographic conditions.

3.   Resolution Enhancement

When Fourier transform *resolution enhancement* techniques are app-lied to GC peaks, the chromatograph is regarded as a black box that transforms an impulselike injection waveform into a series of widened peaks appearing at different delay times (215). The chromatogram is first transformed from the time domain into a corresponding Fourier function in the Fourier domain. Then the Fourier function is simply divided by a peak shape function, before being returned to the time domain by an inverse Fourier transform. By proper choice of the peak shape function, it is theoretically possible to convert the observed chromatogram into a set of ideal injectionlike impulses occurring at the elution times of the component peaks. The areas of the peaks are unchanged by these operations. It is claimed (215) that operating in the Fourier domain allows more relevant use of all the information in the GC trace than does the least-square fitting approach.

Peak shape presents the major problem here, as in the least-squares approach (50). The shape function used for sharpening must be very carefully chosen. For example, a symmetrical Gaussian sharpening func-tion produces no significant sharpening of the tails of model peaks consist-ing of two Gaussian halves of different widths. Of course, the Fourier transform of a real peak can be used to characterize the shape of that peak, and work in this direction is reported to be in progress (215). This approach has obvious similarities to the use of pure compounds to characterize peak shapes in the empirical least-squares-fit approach.

For ideal peaks, and in the absence of noise, it seems probable that the least-squares and Fourier transform methods will give indistinguishable, excellent results. Any comparison between them must therefore be made on the grounds of practical utility for real, noisy, digitized non-Gaussian peaks. When applied to a few real traces containing reasonably symmetrical peaks (215), the Fourier method produced significant peak sharpening, but far from complete resolution. Apparently, the noise problem can be dealt with satisfactorily by removing frequencies above a suitable cutoff point. This cutoff frequency can be determined for each run by investigating the power spectrum that is readily obtainable from the Fourier function itself. The main effect of removing high frequencies from the Fourier function is to set a limit to the amount of sharpening that is possible. Each peak can be no sharper than the cutoff frequency itself.

4. Future Prospects of Curve Fitting

It is possible that curve fitting may revolutionize the whole concept of gas chromatography in laboratories that analyze large numbers of routine samples in which the identities of all components are known. By using very short columns to obtain poorly resolved multiple-peak envelopes, a great increase in throughput of samples could be achieved at the cost of employing a computer to carry out the curve fitting.

There are many factors to be taken into consideration, and a lot of work remains to be done before such a project would be feasible. The sort of questions that must be answered are as follows:

1. Is a single peak shape function a good enough model for all peaks in a chromatogram, or should a completely empirical method be used?

2. Can it be assumed without loss of accuracy that the retention time and overall shape of a peak is independent of the component concentration?

3. It is known that the chromatographic process for one component is affected by the presence of molecules of a neighboring component in the stationary phase. Can the effect of variable concentrations of closely neighboring components be neglected during the curve fitting?

4. Will a larger computer be needed than would be required for conventional GC automation?

5. Iterative curve fitting is the only nontrivial task that a GC computer is likely to be called upon to perform. Will the time taken for the curve-fitting calculation and production of report after the GC run is over be significantly shorter than the time saved by use of faster GC?

None of these questions has been unambiguously answered. The commercial software now available usually assumes that the retention time is a fixed property of a component, analogous to the frequency of a spectroscopic peak. The "shape" of a peak is usually defined by running the pure compound, but with no allowance for concentration dependence. Future work may indicate the need to use more sophisticated models and extensive calibration tables, which will certainly make "yes" the answer to question 4 and very probably to question 5 as well.

It is clear that even for pure Gaussian peaks a good curve fit cannot be achieved in "real time" (i.e., during the course of the run). This means that extensive raw data must be held until the GC run is finished. Under these circumstances, we see already that the answer to question 4 tends to be "yes," unless an on-line time-shared computer has already been selected as the most economic for that particular laboratory.

For the immediate future, curve fitting must remain an experimental tool to be used with great caution on a few difficult-to-separate peak doublets. Only when there is a considerable amount of GC (as opposed to computer) experience in this area, can the daring step of deliberately destroying resolution be taken.

References

1. H. W. Johnson, *Anal. Chem.*, **35**, 521 (1963).
2. Sixth International Symposium on Gas Chromatography, Rome, 1966. Proceedings published in *Gas Chromatography, 1966*, A. B. Littlewood, ed., Institute of Petroleum, London, 1967, pp. 353, 376, 388.
3. Seventh International Symposium on Gas Chromatography, Copenhagen, 1968. Proceedings published in *Gas Chromatography, 1968*, C. L. A. Harbourn, ed., Institute of Petroleum, London, 1969, pp. 297, 319, 330, 338, 346, 367.
4. Eighth International Symposium on Gas Chromatography, Dublin, 1970. Proceedings published in *Gas Chromatography, 1970*, R. Stock, ed., Institute of Petroleum, London, 1971, pp. 204, 247, 257, 280, 292, 302, 310.
5. Ninth International Symposium on Gas Chromatography, Montreux, 1972.
6. American Chemical Society meeting, Chicago, 1967. Papers published in *J. Chromatogr. Sci.*, **5** (12), 1967
7. American Chemical Society meeting, New York, 1969. Papers published in *J. Chromatogr. Sci.*, **7** (12), 709, 714, 720, 725, 731, 740 (1969); **8**, 1, 13, 20, 31, 39, 46, 57 (1970).
8. American Chemical Society meeting, Washington, D.C., 1971. Papers published in *J. Chromatogr. Sci.*, **9** (12), 706, 710, 718, 722, 729, 735 (1971); **10** (1), 1, 8, 14, 22, 27.
9. Institute of Petroleum Gas Chromatography Discussion Group, meeting of computer subgroup, Mainz, 1972.
10. C. W. Childs, P. S. Hallman, and D. D. Perrin, *Talanta*, **16**, 629 and 1119 (1969).
11. N. R. Kuzel, H. E. Roudebush, and C. E. Stevenson, *J. Pharm. Sci.*, **58**, 381 (1969).

12. S. P. Perone, *Anal. Chem.*, **43**, 1288 (1971).
13. H. M. Gladney, *J. Comput. Phys.*, **2**, 255 (1968).
14. C. H. Sederholm, P. J. Friedl, and T. R. Lusebrink, presented at Pittsburgh Conference on Analytical Chemistry and Applied Spectroscopy, Cleveland, March 1968.
15. H. M. Gladney, *Proceedings of IBM Scientific Computing Symposium on Computers in Chemistry*, IBM, New York, 1969, p. 157.
16. G. Wiederhold, *Proceedings of IBM Scientific Computing Symposium on Computers in Chemistry*, IBM, New York, 1969, p. 249.
17. K. Jones and A. Fozard, *Chem. Brit.*, **5**, 552 (1969).
18. R. E. Mahan, *Automation in Analytical Chemistry, Methods and Motivation*, U.S. Atomic Energy Commission, BNWL-SA-3341, 1970.
19. J. R. Scherer and S. Kint, *Appl. Opt.*, **9**, 1615 (1970).
20. S. R. Gambino, *Anal. Chem.*, **43** (1), 20A (1971).
21. M. Margoshes, *Anal. Chem.*, **43** (4), 101A (1971).
22. R. Kaiser, *Allg. Prakt. Chem.*, **21**, 406 (1970).
23. I. Wehling, *Chem. Ind. (Düsseldorf)*, **22**, 308 (1970).
24. J. T. Clerk and F. Erni, *Mitt. Geb. Lebensmittelunters. Hyg.*, **62**, 88 (1971).
25. G. Seifert, *Z. Chem.*, **11**, 161 (1971).
26. J. F. Krugers, *Chem. Weekbl.*, **66** (51), 26 (1970).
27. J. A. Despotakis, R. L. Fink, M. S. Itzkowitz, and M. P. Klein, *Computer-Based Laboratory-Wide Data Acquisition, Analysis and Retrieving System*, U.S. Atomic Energy Commission, UCRL-20132, 1970.
28. M. Shapiro and A. Schultz, *Anal. Chem.*, **43**, 398 (1971).
29. E. Ziegler, D. Henneberg, and G. Schomburg, *Anal. Chem.*, **42** (9), 51A (1970).
30. G. Lauer, R. Abel, and F. C. Anson, *Anal. Chem.*, **39**, 765 (1970).
31. S. P. Perone, D. O. Jones, and W. F. Gutknecht, *Anal. Chem.*, **41**, 1154 (1969).
32. D. O. Jones and S. P. Perone, *Anal. Chem.*, **42**, 1151 (1970).
33. G. E. James and H. L. Pardue, *Anal. Chem.*, **41**, 1618 (1969).
34. A. A. Eggert, G. P. Hicks, and J. E. Davis, *Anal. Chem.*, **43**, 736 (1971).
35. S. N. Deming and H. L. Pardue, *Anal. Chem.*, **43**, 192 (1971).
36. G. P. Hicks, A. A. Eggert, and E. C. Toren, *Anal. Chem.*, **42**, 729 (1970).
37. L. Ramaley and G. S. Wilson, *Anal. Chem.*, **42**, 606 (1970).
38. P. P. Briggs, *Cont. Eng.*, **14**, (September) 75 (1967).
39. H. R. Felton, H. A. Hancock, and J. L. Knupp, *Instrum. Contr. Syst.*, **40**, (August) 83 (1967).
40. A. J. Raymond, D. M. G. Lawrey, and T. J. Mayer, *J. Chromatogr. Sci.*, **8**, 1 (1970).
41. F. Tivin, *J. Chromatogr. Sci.*, **8**, 13 (1970).
42. M. F. Burke and D. G. Ackerman, *Anal. Chem.*, **43**, 573 (1971).
43. S. P. Perone and J. F. Eagleston, *J. Chem. Educ.*, **48**, 438 (1971).
44. J. M. Gill and S. P. Perone, *J. Chromatogr. Sci.*, **7**, 709 (1969).
45. M. Tochner, J. A. Magnuson, and L. Z. Soderman, *J. Chromatogr. Sci.*, **7**, 740 (1969).
46. K. Kishimoto and S. Musha, *J. Chromatogr. Sci.*, **9**, 608 (1971).
47. F. Baumann, E. Herlicska, A. C. Brown, and J. Blesch, *J. Chromatogr. Sci.*, **7**, 680 (1969).
48. L. L. Hegedus and E. E. Peterson, *J. Chromatogr. Sci.*, **9**, 551 (1971).
49. S. N. Chesler and S. P. Cram, *Anal. Chem.*, **43**, 1922 (1971).
50. A. H. Anderson, T. C. Gibb, and A. B. Littlewood, *Anal. Chem.*, **42**, 434 (1970).
51. A. H. Anderson, T. C. Gibb, and A. B. Littlewood, in *Proceedings of the Sixth International Symposium on Advances in Chromatography*, A. Zlatkis, ed., University of Houston, 1970, p. 75; *J. Chromatogr. Sci.*, **8**, 640 (1970).
52. M. F. Burke and R. G. Thurman, *J. Chromatogr. Sci.*, **8**, 39 (1970).

53. R. S. Swingle and L. B. Rogers, *Anal. Chem.*, **43**, 810 (1971).
54. J. T. Walsh, R. E. Kramer, and C. Merritt, *J. Chromatogr. Sci.*, **7**, 348 (1969).
55. A. K. Moreland and L. B. Rogers, *Separation Sci.*, **6**, 1 (1971).
56. H. F. Dymond and K. D. Kilburn, in Ref. 2, pp. 353 and 388.
57. D. R. Deans, *J. Chromatogr. Sci.*, **9**, 729 (1971).
58. K. Jones, A. O. McDougall, and R. C. Marshall, in Ref. 2, p. 376.
59. R. D. McCullough, *J. Gas Chromatogr.*, **5**, 635 (1967).
60. H. A. Gill, R. E. Lee, and R. D. Condon, presented at Pittsburgh Conference on Analytical Chemistry and Applied Spectroscopy, Cleveland 1971, paper 194.
61. D. A. Craven, E. S. Everett, and M. Rubel, *J. Chromatogr. Sci.*, **9**, 541 (1971).
62. P. Mitchell, *Column* (Pye Unicam Ltd.), **3**, 2 (1970).
63. G. S. Wilson and L. Ramaley, *Anal. Chem.*, **42**, 611 (1970).
64. A. B. Littlewood, *Z. Anal. Chem.*, **236**, 39 (1968).
65. E. Grushka, M. N, Myers, P. D. Schettler, and J. C. Giddings, *Anal. Chem.*, **41**, 889 (1969).
66. J. E. Oberholtzer and L. B. Rodgers, *Anal. Chem.*, **41**, 1234 (1969).
67. W. L. Porter and E. A. Talley, *Anal. Chem.*, **36**, 1692 (1964).
68. A. Yonda, D. L. Filmer, H. Pate, N. Alonzo, and C. H. W. Hirs, *Anal. Biochem.*, **10**, 53 (1965).
69. R. D. B. Fraser, A. S. Inglis, and A. Miller, *Anal. Biochem.*, **7**, 247 (1964).
70. H. W. Johnson, *Anal. Chem.*, **37**, 1591 (1965).
71. M. Goedert and G. Guiochon, *J. Chromatogr. Sci.*, **7**, 323 (1969).
72. M. Janghorbani, J. A. Starkovich, and H. Freund, *Anal. Chem.*, **43**, 493 (1971).
73. J. E. Oberholtzer, *J. Chromatogr. Sci.*, **7**, 720 (1969).
74. J. Montastier, *Method. Phys. Anal.*, **7**, 19 (1971).
75. P. Talbot, E. D. Pellzzari, J. H. Brown, R. W. Farmer, and L. F. Fabre, in *Proceedings of the Sixth International Symposium on Advances in Chromatography*, A. Zlatkis, ed., University of Houston, 1970, p. 86.
76. P. C. Bentsen and R. M. Bethea, *J. Chromatogr. Sci.*, **7**, 399 (1969).
77. D. Lawson and S. Havilk. *J. Gas Chromatogr.*, **1**, 17 (1963).
78. O. E. Schupp and J. S. Lewis, *Res./Develop.*, **21** (5), 24 (1970).
79. C. E. West and T. R. Rowbotham, *J. Chromatogr.*, **30**, 62 (1967).
80. J. E. Buchanan and T. P. Maher, *J. Gas Chromatogr.*, **6**, 474 (1968).
81. J. E. Buchanan and T. P. Maher, *J. Chromatogr. Sci.*, **9**, 448 (1971).
82. T. H. Gouw, R. L. Hinkins, and R. E. Jentoft, *J. Chromatogr.*, **28**, 219 (1967).
83. R. S. Porter, R. L. Hinkins, L. Tornheim, and J. F. Johnson, *Anal. Chem.*, **36**, 260 (1964).
84. T. B. Rooney and W. Aznavourian, *Anal. Chem.*, **36**, 2112 (1964).
85. G. Castello and P. Parodi, *Chromatographia*, **4**, 147 (1971).
86. R. Rowan, *Anal. Chem.*, **39**, 1158 (1967).
87. F. Baumann and F. J. Tao, *J. Gas Chromatogr.*, **5**, 621 (1967).
88. H. W. Shira, presented at Pittsburgh Conference on Analytical Chemistry and Applied Spectroscopy, 1970. Technical Paper 43, Hewlett Packard.
89. R. E. Anderson, *J. Chromatogr. Sci.*, **7**, 725 (1969).
90. J. M. Gill, *J. Chromatogr. Sci.*, **7**, 731 (1969).
91. J. E. Oberholtzer, *Anal. Chem.*, **39**, 959 (1967).
92. R. A. Culp, C. H. Lochmuller, A. K. Moreland, R. S. Swingle, and L. B. Rogers, *J. Chromatogr. Sci.*, **9**, 6 (1971).
93. J. E. Oberholtzer and L. B. Rogers, *Anal. Chem.*, **41**, 1590 (1969).
94. L. G. Lorenz, R. A. Culp, and L. B. Rogers, *Anal. Chem.*, **42**, 979 (1970).

95. T. H. Glenn and S. P. Cram, *J. Chromatogr. Sci.*, **8**, 46 (1970).
96. G. Schomburg, F. Weeke, B. Weimann, and E. Zeigler, *J. Chromatogr. Sci.*, **9**, 735 (1971).
97. A. W. Boyne and W. R. H. Duncan, *J. Lipid. Res.*, **11**, 293 (1970).
98. G. J. Nelson, *J. Am. Oil Chem. Soc.*, **48**, 210 (1971).
99. P. Vestergaard, L. Hemmingsen, and P. W. Hansen, *J. Chromatogr.*, **40**, 16 (1969).
100. H. W. Jackson, *J. Chromatogr. Sci.*, **9**, 706 (1971).
101. I. Lysyj, P. R. Newton, and W. J. Taylor, *Anal. Chem.*, **43**, 1277 (1971).
102. H. E. Pattee, E. H. Wiser, and J. A. Singleton, *J. Chromatogr. Sci.*, **8**, 668 (1970).
103. K. Rathgeb, *Chromatographia*, **4**, 270 (1971).
104. P. B. Stockwell, W. Bunting, R. Sawyer, and P. H. C. Ingram, in Ref. 4, p. 204.
105. H. Guenzler, *Chem. Ing. Tech.*, **42**, 877 (1970).
106. L. Mikkelson, *Anal. Advan.* (Hewlett Packard), **2** (2), 2 (1969).
107. R. C. Master, *J. Am. Oil. Chem. Soc.*, **48**, 191 (1971).
108. M. Tochner and L. Z. Soderman, presented at American Chemical Society Meeting, Atlantic City, N.J., 1968 (General Electric).
109. B. L. Walker, *Anal. Biochem.*, **37**, 44 (1970).
110. J. C. Winfrey and C. E. Bethel, *J. Chromatogr. Sci.*, **9**, 353 (1971).
111. G. F. Farrar, *Oil Gas. J.*, **68**, 83 (1970).
112. S. P. Perone, *J. Chromatogr. Sci.*, **7**, 714 (1969).
113. A. W. Munsen and E. L. Schneider, *J. Am. Oil Chem. Soc.*, **48**, 220 (1971).
114. E. L. Schneider and A. W. Munsen, *J. Am. Oil Chem. Soc.*, **48**, 217 (1971).
115. F. Baumann, D. L. Wallace, and L. G. Brendon, "Interfacing Chromatographs to the Computer," presented at Pittsburgh Conference on Analytical Chemistry and Applied Spectroscopy, Cleveland, 1969, paper 199 (Varian Aerograph).
116. H. Brusset, D. Depeyre, and J. P. Petit, *Chim. Anal. (Paris)*, **53**, 207 (1971).
117. G. B. Marson, in *Ciba Foundation Symposium on Gas Chromatography in Biology and Medicine*, R. Porter, ed., Churchill, London, 1969, p. 187.
118. D. R. Hodges and J. M. Barraclough, presented at meeting of the Institute of Petroleum Gas Chromatography Discussion Group, Nottingham, 1969. Reported by A. G. Douglas, *J. Chromatogr. Sci.*, **8**, 169 (1970).
119. J. G. Lyons, in Ref. 4, p. 302.
120. H. A. Hancock, L. A. Dahm, and J. F. Muldoon, *J. Chromatogr. Sci.*, **8**, 57 (1970).
121. C. D. Scott, D. D. Chilcote, and W. W. Pitt, *Clin. Chem.*, **16**, 637 (1970).
122. A. W. Westerberg, *Anal. Chem.*, **41**, 1770 (1969).
123. R. A. Hites and K. Biemann, *Anal. Chem.*, **42**, 855 (1970).
124. J. W. Frazer, L. R. Carlson, A. M. Kray, M. R. Bertoglio, and S. P. Perone, *Anal. Chem.*, **43**, 1479 (1971).
125. R. G. Thurmann, K. A. Mueller, and M. F. Burke, *J. Chromatogr. Sci.*, **9**, 77 (1971).
126. T. J. Williams, *Ind. Eng. Chem.*, **61** (1), 76 (1969); **62** (2), 28 (1970); **62** (12), 94 (1970).
127. I. G. McWilliam, in *Advances in Chromatography*, J. C. Giddings and R. A. Keller, eds., vol. 7, Dekker, New York, 1968, p. 163.
128. F. R. Kloft, *regelungstech. Prax. Process-Rechentechnik.*, **13** (1). 13 (1971).
129. M. C. Burk and W. H. Williams, *Analysis Instrumentation 1964*, L. Fowler, R. J. Harmon, and D. K. Roe, eds., Instrument Society of America, Pittsburgh, 1964, p. 327.
130. C. Grunberger, *M Regul. Automat.*, **32**, 113 (1967).
131. J. Hadley and J. L. Book, *Process Contr. Autom.* **13** (9), 24 (1966).
132. J. A. Jones and L. N. Levy, *Proceedings of the Tenth Annual Symposium of the Chemical and Petroleum Industries Division of the Instrument Society of America, 1969*, Instrument Society of America, Pittsburgh, II.1, 1970.

133. R. M. Keeler and M. C. Burk, Proc. Am. Petrol. Inst. Sect. III, **43**, 248 (1963).
134. F. C. Mears, *Hydrocarbon Process.*, **46** (12), 105 (1967).
135. F. C. Mears, *Oil Gas J.*, **65** (51), 90 (1967).
136. F. C. Mears, *Anal. Instrum.*, **7**, 59 (1969).
137. F. C. Mears, *Oil Gas J.*, **64** (Sept. 26), 80 (1966).
138. F. T. Ogle, *Chem. Eng. Progr.*, **61** (10), 87 (1965).
139. W. J. Rejan, *Instrum. Technol.*, **16** (11), 57 (1969).
140. M C. Simons and D. W. Spence, *Analysis Instrumentation 1963*, L. Fowler, R. D. Earnes, and T. J. Kehoe, eds., Instrument Society of America, Pittsburgh, 1963, p. 151.
141. S. R. Tait and V. S. Morello, *Instrum. Chem. Petrol. Ind.*, **5**, 57 (1968).
142. D. W. Stevens and R. Villalobos, *Anal. Instrum.*, **7**, 50 (1969).
143. H. M. Gladney, B. F. Dowden, and J. D. Swalen, *Anal. Chem.*, **41**, 883 (1969).
144. S. M. Roberts, D. H. Wilkinson, and L. R. Walker, *Anal. Chem.*, **42**, 886 (1970).
145. R. N. Sauer, in Ref. 3, p. 330.
146. E. B. Harris, J. F. Hickerson, and M. W. Morgan, *Use of a Real-time Monitoring Computer for Gas Chromatography and Mass Spectrometry Analysis*, presented at American Society for Testing and Materials, E-19 meeting, San Francisco, October 1967
147. R. A. Loscher, *Chem. Eng. Progr.*, **62** (10), 64 (1971).
148. F. Grohe, W. Hesse, R. Kaiser, and K. H. Schneckenburger, in Ref. 4, p. 247.
149. R. Kaiser, presented at meeting of the Institute of Petroleum Gas Chromatography Discussion Group, London, 1970. Reported by A. G. Douglas, *J. Chromatogr. Sci.*, **8**, 556 (1970).
150. G. Schomburg, F. Weeke, B. Weimann, and E. Zeigler, in Ref. 4, p. 280.
151. A. W. Westerberg, *Anal. Chem.*, **41**, 1595 (1969).
152. J. G. W. Price, J. C. Scott, and L. O. Wheeler, *J. Chromatogr. Sci.*, **9**, 722 (1971).
153. L. C. Torres and L. A. Dahm, *Am. Lab.*, **1**, 30 (1969).
154. W. O. Wilson and J. G. W. Price, *J. Chromatogr. Sci.*, **8**, 31 (1970).
155. J. W. Frazer, *Anal. Chem.*, **40** (8), 26A (1968).
156. D. J. Fraade, R. S. Davis, and W. Kipiniak, *Instrum. Chem. Petrol. Ind.*, **4**, 53 (1967).
157. J. W. Frazer, *Chem. Instrum.*, **2**, 271 (1970).
158. W. J. Ryan, *Ann. ISA Conf. Proc.*, **24** (pt 1), 517 (1969).
159. J. Baudisch, *Chromatographia*, **1**, 443 (1968).
160. B. Dewey, *Eastern Annual Symposium on Analytical Chemistry, 1968,* vol. 4, C. H. Orr, ed., Plenum, New York, 1970, p. 11.
161. H. D. Metzger, *Tech. Mitt.*, **63**, 259 (1970).
162. D. T. Winski and M. Rubel, *Drug Cosmet. Ind.*, **106** (5), 48, 156 (1970).
163. L. M. Collyer, L. H. C. Hawkins, and G. H. Thomson, in Ref. 4, p. 257.
164. D. R. Deans, in Ref. 4, p. 292.
165. E. J. Bonelli, *Am. Lab.*, **3** (2), 27 (1971).
166. R. Binks, R. L. Cleaver, J. S3 Littler, and J. MacMillan, *Chem. Brit.*, **7**, 8 (1971).
167. J. R. Chapman, *Chem. Brit.*, **5**, 563 (1969).
168. J. C. Craig, *Psychopharmacol. Bull.*, **6** (1), 34 (1970).
169. T. Nakamura, T. Sasaki, and S. Nakoshi, *Bunseki Kagaku*, **19**, 996 (1970); *Chem. Abstr.*, **73**, 80942d (1970).
170. P. Schulze and K. H. Kaiser, *Chromatographia*, **4**, 381 (1971).
171. R. A. Hites and K, Biemann, *Anal. Chem.*, **40**, 1217 (1968).
172. R. A. Hites and K. Biemann, *Advances in Mass Spectrometry*, vol. 4, E. Kendrick, ed., Institute of Petroleum, London, 1968, p. 37.
173. I. R. Althaus, K. Biemann, J. Biller, P. F. Donaghue, D. A. Evans, H.-J. Förster, H. S. Hertz, C. E. Hignite, R. C. Murphy, G. Preti, and V. Reinhold, *Experientia*, **26**, 714 (1970).

174. R. C. Murphy, M. V. Djuricic, S. P. Markley, and K. Biemann, *Science*, **165**, 695 (1969).
175. C. C. Sweeley, B. D. Ray, W. I. Wood, J. F. Holland, and M. I. Kirchevsky, *Anal. Chem.*, **42**, 1505 (1970).
176. D. H. Smith, R. W. Olsen, F. C. Walls, and A. L. Burlingame, *Anal. Chem.*, **43**, 1796 (1971).
177. G. R. Waller, H. Li, K. Kinneberg, R. Saunders, D. Simpson, and L. Mills, *Proc. Okla. Acad. Sci.*, **50**, 19 (1970).
178. W. J. McMurray, B. N. Greene, and S. R. Lipsky, *Anal. Chem.*, **38**, 1194 (1966).
179. S. R. Lipsky, W. J. McMurray, and C. G. Horvath, in Ref. 2, p. 299.
180. R. Reimendal and J. Sjövall, *Anal. Chem.*, **44**, 22 (1972).
181. F. W. Karasek, *Res./Develop.*, **22** (2), 26 (1971).
182. Papers presented at Pittsburgh Conference on Analytical Chemistry and Applied Spectroscopy, Cleveland, 1971: G. T. Paul and W. J. Sloughter, paper 195; D. J. Noonan and R. D. Condon, paper 196; E. W. March and E. C. Pieper, paper 197. Perkin-Elmer GC-175.
183. R. J. Maggs and A. S. Mead, in Ref. 4, p. 310.
184. F. Baumann, A. C. Brown, and M. B. Mitchell, *J. Chromatogr. Sci.*, **8**, 20 (1970).
185. F. Baumann, D. Wallace, E. Herlicska, R. Katzive, and J. Blesch, in Ref. 3, 346.
186. Papers by G. V. Peterson; E. Zerenner; M. Eccles and W. Kapuskar; J. Poole; L. Mikkelsen; all in *Anal. Advan.* (Hewlett-Packard), **3**, 1–28 (1970).
187. J. T. Frazer and B. T. Guran, *J. Chromatogr. Sci.*, **9**, 718 (1971).
188. K. Derge, *Chem. Ztg., Chem. Appl.*, **95**, 147 (1971).
189. C. E. Watson, *Anal. Instrum.*, **7**, 45 (1969).
190. C. E. Klopfenstein, presented at Pittsburgh Conference on Analytical Chemistry and Applied Spectroscopy, Cleveland, 1971.
191. C. H. Sederholm, presented at Pittsburgh Conference on Analytical Chemistry and Applied Spectroscopy, Cleveland, 1971.
192. R. G. Thurman and M. F. Burke, *J. Chromatogr. Sci.*, **9**, 181 (1971).
193. B. E. Bonnelycke, *J. Chromatogr.*, **61**, 322 (1971).
194. F. M. Menger, G. A. Epstein, D. A. Goldberg, and E. Reiner, *Anal. Chem.*, **44**, 430 (1972).
195. R. Merkel, *Contr. Eng.*, **18**, 92 (1970).
196. A. Savitsky and M. J. E. Golay, *Anal. Chem.*, **36**, 1627 (1964).
197. J. D. Hettinger, J. R. Hubbard, J. M. Gill. and L. A. Miller, *J. Chromatogr. Sci.*, **9**, 710 (1971).
198. V. Cejka, M. H. Dipert, S. A. Tyler, and P. D. Klein, *Anal. Chem.*, **40**, 1614 (1968).
199. P. D. Klein and B. A. Kunze-Falkner, *Anal. Chem.*, **37**, 1245 (1965).
200. P. D. Klein, *Separation Sci.*, **1**, 511 (1966).
201. R. D. B. Fraser and E. Suzuki, *Anal. Chem.*, **38**, 1770 (1960).
202. W. O. Keller, T. R. Lusebrink, and C. H. Sederholm, *J. Chem. Phys.*, **44**, 784 (1964).
203. J. Pitha and R. N. Jones, *Can. J. Chem.*, **45**, 2347 (1967).
204. J. Pitha and R. N. Jones, *Can. J. Chem.*, **44**, 303 (1966).
205. B. J. Duke and T. C. Gibb, *J. Chem. Soc. (A)*, **1967**, 1478.
206. B. Goldberg, *J. Chromatogr. Sci.*, **9**, 287 (1971).
207. A. B. Littlewood, A. H. Anderson, and T. C. Gibb, in Ref. 3, p. 297.
208. J. C. Bartlett and D. M. Smith, *Can. J. Chem.*, **38**, 2057 (1960).
209. R. H. Muller, *Anal. Chem.*, **38**, (12), 121A (1966).
210. E. Proksch, H. Bruneder, and V. Granzner, *J. Chromatogr. Sci.*, **7**, 473 (1969).
211. C. Bosshard, O. Piringer, and T. Gaeumann, *Helv. Chim. Acta*, **54**, 1059 (1971).
212. R. D. B. Fraser and E. Suzuki, *Anal. Chem.*, **41**, 37 (1969).

213. E. J. Levy and A. J. Martin, paper presented at Pittsburgh Conference on Analytical Chemistry and Applied Spectroscopy, 1968.
214. A. H. Anderson, T. C. Gibb, and A. B. Littlewood, *Chromatographia*, **2**, 466 (1969).
215. D. W. Kirmse and A. W. Westerberg, *Anal. Chem.*, **43**, 1035 (1971).
216. S. M. Roberts, *Anal. Chem.*, **44**, 502 (1972).

Analytical Applications of Induced Phase Changes in Gas Chromatographic Stationary Phases

Peter F. McCrea, *Research Department, The Foxboro Company, Foxboro, Massachusetts*

I. INTRODUCTION

If phase changes in gas chromatographic stationary phases are thought of at all, users of analytical chromatographs undoubtedly regard them as detrimental to successful analyses. Thus, although it is hardly ever explicitly described as such, one phase transition religiously avoided in gas-liquid chromatography (GLC) involves volatization of the stationary phase. We have this requirement because a constant loading of sorbent on the solid support is a prerequisite to meaningful and reproducible solute retention data (1). At lower column temperatures, the latter being the most conveniently manipulated parameter affecting phase changes, lies another

87

condition generally avoided in GLC, namely, the freezing of the stationary phase. In the past, accordingly, this temperature has delineated the lower limit of the useful span of operation of an analytical GLC column (2). It is the purpose of this chapter to show that this limited view of phase changes is no longer valid and that many unique analytical benefits may be gained by purposely bringing about a controlled solid→solid or solid→liquid change in the state of a gas chromatographic stationary phase.

II.   THEORY OF NORMAL AND ANOMALOUS TEMPERATURE DEPENDENCE OF SOLUTE RETENTION

A.   General Theory of Solute Retention in Gas-Liquid Chromatography

Since GLC involves the distribution (partitioning) of a volatile solute between an involatile sorbent (stationary phase) and a moving gas (mobile phase), it is convenient to represent this process in terms of the partition coefficient K, of the solute–solvent system. It is well established (3) that the partition coefficient at infinite dilution K^∞ can be represented in terms of two experimentally determined parameters, namely:

$$K^\infty = \frac{V_R}{V_L} \tag{1}$$

where V_R is the fully corrected net retention volume of a solute eluted from a column containing V_L volume units of stationary phase.

Solute retention data are conveniently presented, independent of the the amount of sorbent, in terms of specific retention volume $V_g{}^T$ (i.e., the net corrected retention volume per gram of stationary phase at some temperature T). Convention and experience establish the utility of only two temperatures, either that of the column or 273.2°K. These are related by way of the following expression:

$$V_g{}^T = \frac{T V_g{}^0}{273.2} \tag{2}$$

Introducing the column temperature Eq. 1 becomes:

$$V_g{}^T = \frac{K^\infty V_L}{w} \tag{3}$$

where w is the weight of stationary phase in the column.

B. Temperature Dependence of Retention Volume

It has also been well established that plots of log V_g^0 as a function of reciprocal absolute temperature are frequently linear, particularly over narrow temperature spans (4). This behavior arises from the thermodynamical properties of the solute–solvent system which, because K is an equilibrium constant, are described by the equation

$$\log V_g^0 = \frac{\Delta H^s}{2.3RT} + \text{constant} \tag{4a}$$

where ΔH^s is the molar heat of vaporization of the solute from the solution.

If T is, hereafter, taken as the column temperature, we can also write

$$\log V_g^T = \frac{\Delta H^s}{2.3RT} + \text{constant} + \log \frac{T}{2/3.2} \tag{4b}$$

The last term in Eq. 4b introduces only trivial nonlinearity into plots of log V_g^T against $1/T$ if the temperature span of the experimental data is not large. This condition is expected to be maintained throughout the investigations described here and, since V_g^T is the more directly useful function for the present study, we can write with reasonable approximation

$$\log V_g^T = \frac{a}{T} + b \tag{5}$$

where a and b are experimentally determined constants for a given solute–sorbent system.

Comparison of Eqs. 3 and 5 shows that the primary contribution to temperature dependence of solute retention in GLC is from the partition coefficient, since the cubical expansion of the stationary phase is small compared with the variation of solvent solubility with temperature. For a conventional solvent, $\Delta K^\infty/\Delta T$ is invariably negative and $\Delta V_L/\Delta T$ is generally positive but very small; hence, solute retention volume decreases with increasing temperature. This familiar behavior is depicted in Figure 1A.

C. Anomalous Retention Behavior

Consider now the case of retention volume change over a range of temperature that includes a reversible transformation of the stationary phase such as a solid–liquid phase transition. The terms in Eq. 3 that

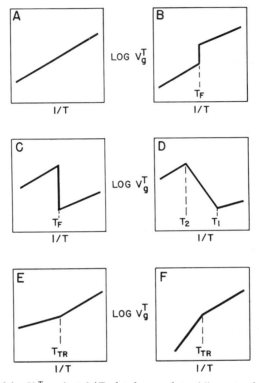

Fig. 1. Forms of $\log V_g{}^T$ against $1/T$ plot for a solute. (A) conventional GLC or GSC sorbent; (B),(C) sharp phase transition of sorbent; (D) extended temperature range transition of sorbent with resulting positive temperature dependence of retention; (E),(F) sorbent transition yielding a shift in the differential heat of adsorption.

determine the specific retention volume (i.e., K^∞ and V_L) would be expected to alter significantly because of both solubility and solvent volume changes. Thus the "constants" in Eq. 5 also would alter significantly as the physical state of the stationary phase changes in the region of the transition temperature.

As we learn later, many workers have firmly established that surface adsorption is the dominant mechanism of solute retention at temperatures well below the freezing point of the stationary phase. In this region, the a term in Eq. 5 will be proportional to the differential molar heat of adsorption of the solute at the surface of the sorbent. Ideally, if the system is now heated to a melting transition point, a pure substance should change rapidly from a solid to a liquid. If the absorbability of a solute is reduced in the process, the resulting temperature dependence would be similar to that in Figure 1*B*. However, vapor solubilities are generally higher in liquids than is the sorption of vapors onto the surface of a solid of the same volume, hence the solute retention behavior about a transition temperature T_F is much more likely to be of the form illustrated in Figure 1*C* if there is an infinitely sharp melting transition. Behavior of the latter type is rarely observed, however, since multicomponent mixtures and even "pure" stationary phases do not melt sharply; rather, they exhibit a very substantial disordering of crystal structure at temperatures below the melting point (5). These pre-transition phenomena can arise by way of several mechanisms, depending on the nature of the substance and its purity. Since Ubbelohde (5) has published an extensive critical review of the many contributory effects that result in extended range transitions, they are not detailed here.

Hence, for a system that liquefies progressively over a definable range of temperature, as far as solute solubility is concerned, we might expect to observe solute retention behavior similar to that depicted in Figure 1*D* where the range T_1 to T_2 is the transition region as determined by gas chromatography. At temperatures below T_1, solute retention will depend essentially on adsorption processes; but as the column temperature is raised to T_1 and above, liquid phase is generated in ever-increasing amount up to T_2. At the latter temperature, all the solid stationary phase has been converted to liquid; at even higher temperatures, solute retention follows the normal decrease with increasing temperature as predicted by Eq. 5. Providing the melt does not supercool, the reverse of the foregoing behavior should take place on cooling the column. Any stationary phase that manifests the aforementioned retention behavior is hereafter designated *a transition sorbent*.

It must be pointed out that throughout the transition region K is varying continuously and, since we assume K to be generally greater for solution than for adsorption, the absolute value of V_g^T increases on account of both the increment of liquid volume and the increment of the fundamental sorption parameter K. Occasionally, of course, these effects oppose each other; the net effects then depends on the relative magnitudes of the K and V_L terms. Thus the slope and linearity of the solute retention curve in the T_1 to T_2 transition region will depend on the rate of generation of the liquid, as well as its sorption and volumetric parameters as compared with the solid phase; in some cases, these properties will depend on the properties of the solute and the resulting solute–solvent interactions, as well.

In general, solid I→solid II phase transitions do not exhibit the pretransition effects previously mentioned, and the phase change typically occurs quite sharply; hence their use as calibration standards for differential thermal analysis (DTA) and the differential scanning calorimeter (DSC) (6). The form of solute retention to be expected at such transitions is depicted in Figures 1E and 1F, where well-defined slope changes occur, reflecting a net decrease or increase, respectively, in the differential molar heat of adsorption of the solute at the surface of the adsorbent. The extent to which this transition is discernible obviously depends on the nature of the crystalline rearrangement of the adsorbent and its effect on specific interactions with the adsorbates. The behavior typified in Figures 1A, 1E, and 1F might also be observed in a solid→liquid transition wherein the solute is essentially insoluble in the liquid stationary phase and hence is retained merely by adsorption at the gas-liquid interface.

III.   REPORTED OBSERVATIONS AT SORBENT TRANSITIONS

A.   Organic Melts

The earliest recorded solute retention studies at the sorbent solidification point involved the experimental determination of the freezing point of n-decane by using the substance as a stationary phase for various solutes and determining the temperature at which marked retention discontinuities were observed (7). In later, more extensive work (8), utilizing methane and n-decane as the mobile and stationary phases, respectively, the authors observed considerable pre-transition behavior as the temperature was increased to the sorbent melting point, even though the transition to complete GLC partitioning was well defined and reproducible. The former phenomenon was attributed to freezing point lowering because of surface tension forces exerted on the liquid trapped in the pores of the solid support. Utilizing specialized apparatus, pressure-induced freezing-point depressions were chromatographically obtained for the binary system methane–decane. The corresponding phase diagram at pressures approaching 2000 psi was constructed using these data.

Scott (9) has investigated the partition and adsorption of solutes in columns containing benzophenone, eicosane, and even-numbered fatty acids from C_{10} to C_{18}. Solute retention for the aromatic sorbent was observed to decrease to very low values as the temperature was lowered beyond the freezing point. With a further decrease in temperature, the elution order of a test mixture was rearranged, and adsorption occurred. Confirmation of the occurrence of the latter process was demonstrated by the good low-temperature correlation between solute retention and surface areas of the coated support at various sorbent loadings. This relationship was not observed for the eicosane or fatty acid sorbent columns, however, since temperatures well below the solidification-point-retention discontinuity resulted in solute retention volumes that were still somewhat proportional to the amount of stationary phase in the column. This observation is consistent with premelting disordering in long-chain *n*-paraffins and their derivatives as was determined by specific heat measurements and NMR, X-ray and infrared spectroscopy (10). The latter technique indicates that, for stearic acid, as much as 20% of the crystalline lattice is disordered at a temperature 30°C below the freezing point and, moreover, that the tendency toward extensive pretransition effects generally lessens with decreasing chain length. This observation is to some extent substantiated by Scott's finding that decreasing chain length of the fatty acids produces more pronounced gas-solid adsorption effects. Clearly, gas chromatographic solute retention is an extremely sensitive indicator of the physical state of the stationary phase and serves as additional proof that considerable temperature-dependent phenomena occur at temperatures far below the observed freezing point of many substances.

A number of authors (11–15) have reported solute retention shifts at the solidification point of the widely used polymeric stationary phase, Carbowax 20M. Dal Nogare (12) examined this stationary phase at temperatures above and below the sorbent melting range (50–68°C) by DTA and found that retention was proportional both to solvent weight loading on the melted sorbent, and to the surface area, at temperatures well below the freezing range. Unexpectedly, however, he observed GLC retention dependence proportional to weight loading at the low-temperature end of the retention shift (i.e., when the sorbent had supposedly solidified). This finding was later supported by the observation that sample-size-dependent retention volumes, a reasonable indicator of a surface adsorption retention mechanism, were manifest only at temperatures well below the retention shift (14). Thus, from these and other investigations (13, 15), it appears that GLC partitioning does occur in the traces of uncrystallized polymer that can exist at temperatures appreciably below the sorbent "melting point."

A further example of the universality of the retention shift at the sorbent melting point is given by a reported investigation of squalane at temperatures down to −180°C (14). The pertinent specific retention data for various solutes are plotted in Figure 2. Here we observe broad transitions over the −60 to −100°C region, with the most soluble substance, ethane, showing the earliest deviation from linear gas-solid chromatographic (GSC) behavior. Fluoroform, which has five times less retention in the melted sorbent than ethane, exhibits much less "sensitivity" to the premelting effect, even though all retained solutes achieve a condition of GLC linearity, indicative of complete melting, at exactly the

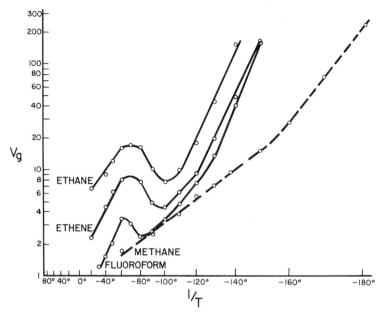

Fig. 2. Retention behavior of various solutes as a function of $1/T$ with a column of 20% w/w squalane on Chromosorb W (14).

By courtesy of the *Journal of Gas Chromatography*.

same temperature. Methane is practically unretained at this temperature and shows little retention shift. The k' dependence on the magnitude of the chromatographically observed extended transition is not unreasonable, since we would expect the more highly retained solutes to exhibit greater partitioning in the minute quantities of liquid present at the initiation of crystal disordering. With fluoroform, on the other hand, a considerable amount of liquid must be present before any significant amount of partitioning takes place.

An unexpected consequence of operating at temperatures far below the sorbent transition region is also seen in Figure 2; namely, the slope changes in the GSC region which are similar to those depicted in Figure 1E. These effects are even more prominent with Carbowax 20M columns, where the temperature at which the slope change occurs ranges from $-140°C$, for methane, to $-30°C$ for *n*-pentane (14). Apparently, the postulated fracturing of the crystalline sorbent at cryogenic temperatures (16) cannot satisfactorily explain these observations, since we would then expect all solutes to be influenced at the same temperature. Similarly, it is difficult to imagine any solid-solid transition occurring in the sorbent (cf.

Figures 1E and 1F) which could be so thermally dependent on the nature of the solute probe used to study the phenomenon. Thus it would appear that this anomaly, which occurs at a temperature roughly proportional to the boiling point of the solute, is not intimately related to the structural state of the sorbent, but rather to some as yet undetermined property of the solute. Obviously, further investigations are needed in this area to elucidate the cause of the phenomenon.

Claeys and Freund (15) have reported column efficiencies, solute elution orders, and specific retention volumes obtained at temperatures above and below the solid-liquid transition of various stationary phases. These sorbents included polymers of varying degrees of polarity and crystallinity (e.g., Carbowax 1540 and 20M, ethylene glycol succinate, and Halocarbon 6-00, as well as a polar crystalline compound, 1-hexadecanol). The largest positive retention shift at the sorbent solid-liquid transition was obtained with the highly crystalline alkanol column, whereas lesser shifts were exhibited by the polymer columns. The magnitude of a retention shift on sorbent melting is dependent on the solute retention in the melt, on whatever liquid or adsorbent surface contribution provides retention prior to the transition, and on the rate of generation of liquid with increasing temperature. It has been observed that polymer systems, in general, possess varying degrees of crystalline arrangement and that these defects result in extensively broadened melting ranges (17). Thus, even though pretransition effects certainly do occur for the alkanol stationary phase (10), the rate of generation of liquid phase with increasing temperature is apparently much higher than that which is observed in polymer sorbents.

Purnell, Wasik, and Juvet (18) have examined solute retention at temperatures above and below the melting range of stationary phases composed of mixtures of magnesium stearate and n-octadecanol. The primary aim of these workers was one of establishing the presence and the stoichiometry of organic complexes formed in the mixed sorbents. Figure 3 depicts toluene retention behavior in the pure alcohol column and in a mixed sorbent column. The most striking aspect of solute retention in the former column is the large span of temperature over which the transition from GSC adsorption ($T<35°C$) to partitioning in the completely lique-fied sorbent ($T>58°C$) takes place, even though the bulk sorbent was observed to melt rather abruptly in a 2°C span at about 57°C. The authors suggest that impurities may contribute to the chromatographically ob-served extended transition, and rightly so, since minor impurities forming a solid solution with the alcohol can broaden the transition region somewhat. However, this mechanism cannot, on its own, account for the 20°C transition span. Pre-transformation effects of the solid-solid reorien-tation genre are known to occur in long-chain polymethylene molecules

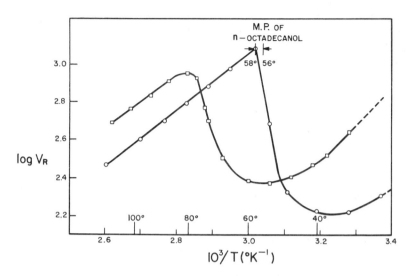

Fig. 3. Plot (18) of $\log V_R$ against $1/T$ for elution of toluene from columns containing: magnesium stearate–n-octadecanol, mole ratio 40:60 (square points) and n-octadecanol (round points).

By courtesy of *Acta Chimica, Hungarian Academy of Science.*

with methyl, hydroxyl, carboxyl, or other more polar group terminations (10). Furthermore, the temperature at which these transformations occur are markedly sensitive to impurities (19). Hence the chromatographically observed extended transition region might well arise from closely spaced solid-solid and solid-liquid events.

The retention data for the 40 mole % stearate column, also shown in Figure 3, indicate varying degrees of thermal activity in the sorbent from about 40°C, where departure from GSC linearity is indicated, up to the complete liquefication of the sorbent at about 80°C. Over approximately half this span (i.e., 20°C), the temperature dependence of retention is strongly positive. Here the broad transition region is due to the solid solution characteristics, or more likely, the mutual freezing point depression, which is commonly observed in organic mixtures. It is with systems such as these, combined with thermal analysis techniques, that we can begin to predict, to a very good approximation, the inflection points that should be observed chromatographically. This is because the onset of melting at the solidus line of a binary component system is generally much easier to observe and interpret than is the gradual disordering of a crystalline lattice due to pretransformation events in a relatively pure compound. In a subsequent investigation (20), a 12 mole % magnesium stearate in n-octadecanol column yielded initial and final deviations from GSC and GLC aromatic solute retention linearity at 40 and 77°C, respectively. DTA investigation of a portion of the conditioned coated support indicated the initiation of thermal activity at 39°C, followed by a continuously rising endothermal baseline, suggestive of solid solution or mutually depressed freezing point behavior. Superimposed on this rising baseline were minor endothermal events at 53, 56, and 69°C, and a final prominent endothermal peak at 76°C, followed by a return to the original baseline. Thus the extremes of the DTA thermogram delineate quite well the observed chromatographic data, even though the intermediate DTA events were not chromatographically observed (probably because their influence on overall solute retention was within the normal variance of chromatographic measurement).

The nature of the above-described, low-temperature transitions is not a mystery, however; the metallic salts of stearic and other fatty acids are usually liquid crystals and, as such, can exhibit an extremely detailed thermal behavior with as many as six transitions on going from a crystalline solid to an isotropic liquid (21). Furthermore, the addition of long-chain hydrocarbons to these metallic stearates has been shown to result in very complex phase diagrams (22, 23). Therefore, in view of the highly temperature-dependent identity of the sorbent on the column support, assuming that sorbent existing as a solid is relatively inert to the partitioning process, it is understandable why Purnell et al. (18) could only imply the possible existence of mixed solvent complex formation at 33 and 50 mole % stearate concentrations in n-octadecanol. Indeed, without a fully characterized phase diagram, observed solute retention behavior cannot be interpreted meaningfully.

Employing a stationary phase (the ureide of L-valine isopropyl ester) that had served for the separation of enantiomers of a variety of secondary amine derivatives, Corbin and Rogers (24) observed markedly increased separation factors when the column was operated at a temperature below the sorbent melting point. As the temperature was increased above the melting point, solute retentions increased by an order of magnitude; but the measured peak resolution was essentially unchanged, resulting in greatly reduced separation factors and unnecessarily long analysis times. On the liquid phase, the temperature dependence of α was relatively constant, whereas it altered substantially throughout the solid region. This is presumably due to pre-transition disordering or an extended melting range due to impurities.

Although the exact mechanism for enhanced separation factors on the solid stationary phase is not clear, it seems likely that the solid sorbent is structured such that it retains its potential for specific interactions, whereas the mobility of the liquefied partitioning fluid adds a great deal of nonspecific interaction without any increase in resolution.

In addition to the anomalous differential heat of adsorption shift discussed earlier, which is apparently related to solute properties (14), a recent investigation (25) has yielded a *post-transition* anomaly that appears to reflect the influence of the active solid support on the chromatographic retention of solutes. Although moderately heavily loaded columns (20% by weight) have yielded the expected upward retention shift at the sorbent melting point, Serpinet (25) has observed very different behavior from lightly loaded columns of octadecyl propionate (1% by weight on Chromosorb P), utilizing *n*-heptane and perfluorooctane as solutes. Figure 4 indicates that *n*-heptane undergoes the expected positive retention shift

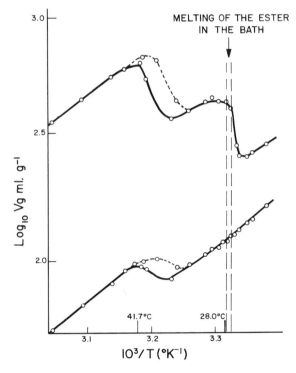

Fig. 4. Plot (25) of log specific retention volume versus $1/T$ for *n*-heptane (upper curve) and perfluorooctane (lower curve) with a column containing 1° w/w octadecyl propionate on Chromosorb P.

By courtesy of *Nature*.

at the 28°C melting point of the bulk stationary phase; then, surprisingly, there is a second retention increase in the 35 to 42°C region. A positive retention shift in the latter region is also observed for perfluorooctane, even though its behavior in the region of the bulk sorbent melting point only takes the form of a slight slope alteration. This behavior could be observed if the fluorocarbon retention on both solid and liquid sorbent were due primarily to adsorption, hence only an alteration in the differential heat of adsorption would occur at the first transition. The true anomaly, however, lies in the apparent *post*-transition retention shift at the higher temperature. The author has also observed this phenomenon in lightly loaded columns containing docosane, and with various fatty acid esters as the stationary phase (26).

Serpinet indicated that the phenomenon perhaps can be explained by the proposed distribution of stationary phase on high-surface-energy supports; that is, that the sorbent is distributed both as droplets in the support pores and as a thin film on the entire surface area, and that different melting behavior for the two forms of sorbent distribution could take place. The author further stated that the upper retention shift suggests that the somewhat polar film of sorbent on the surface melts at a higher temperature.

These conclusions seem to conflict with the result of Drew and Bens (27), who carried out scanning electron microscopy on chromatographic supports and observed little, if any, filling of observed pores on heavily loaded Chromosorb P, thus disagreeing with previous stationary phase distribution data obtained by indirect means [i.e., mercury porosimetry (28, 29)]. Furthermore, past observation (30) has brought to light the great difficulty encountered in achieving superheating of crystals even by a mere fraction of a degree above their transition temperatures; this alone, then, would not account for the apparent 14°C increase in melting point. Similarly, heteromolecular surface effects on the melting of adsorbed multilayers deposited on an active substrate have always resulted in a lowering of the transition temperature (31). Nevertheless, thermal analysis of docosane deposited at a 1% loading on unsilanized Chromosorb P has revealed (32) that endothermal peak activity does exist in the region of both chromatographically observed retention shifts.

Since the phenomenon is not observed when the surface of the support has been deactivated by silanization (25, 26), it is possible that strong chemisorption of the sorbent as a very thin layer has taken place and that this "bonding" is overcome at the higher temperature. The second retention shift could be due to the increased freedom of movement that characterizes the molecules on the surface after the breaking of the bond and permits the direct adsorption of solute molecules onto the surface silanol groups of the solid support. However, these anomalous, postfusion transitions have also been observed in a column that had been coated with 20% w/w glycerol and then 1% pentadecanol (33). It would seem that alkanol–surface interaction would be prevented by the prior deposition of the relatively heavy loading of glycerol, a known deactivator of solid supports (34). Thus the only alternative explanation is that an ordered film of the alkanol exists on the surface of the glycerol itself. This facet of the phenomenon seems to be inconsistent with the lightly loaded column observations described previously.

Substantiation for the film-support bonding hypothesis arises from thermal analysis and chromatographic examination, as a sorbent, of alkanols and polyethylene glycols that have been chemically bonded (35, 36) onto the surface hydroxyls of porous silica beads. Bulk properties of the starting material, such as solubility, vapor pressure, and transition temperatures, are significantly altered for the two former parameters and undetected with regard to the latter (37).

In order more fully to characterize and understand the mechanism underlying the post-transition event, more experiments must be carried out by way of high-sensitivity DTA, and employing wide variations in support surface area and activity, sorbent identity, and film thickness, and also on the solute probes used to study the phenomenon.

B.   Liquid Crystals

Many workers (38–46), employing liquid crystals as gas chromatographic stationary phases, have observed solute retention shifts at mesophase transitions. We expect that these events would be observed frequently because, unlike conventional isotropic sorbents, maximum solute selectivity is generally achieved when operating near a transition (47). The temperature at which the chromatographically detected transition occurs is, generally, in good agreement with bulk thermal measurements (39, 41–48). In some instances, however, retention behavior suggestive of the persistence of a somewhat ordered structure at temperatures above the transition have been reported (39, 47). The most significant retention shift generally occurs at the mesophase-isotropic liquid transition, although Dewar and Schroeder (39) have observed retention alterations at both the smectic-nematic and the nematic-isotropic transition of 4,4′di-n-hexyloxyazoxybenzene. Kelker (44) examined pretransformation effects in mesomorphic stationary phases and found that they bear a striking similarity, insofar as their chromatographic parameters are concerned, to solids that melt directly to give an isotropic liquid.

Although it is clear that analytical investigations involving liquid crystals, usually for the purpose of meta-para isomer separation, entail phase changes in the chromatographic sorbents, the recent review by Kelker (42) and other articles (39–41, 43, 45, 47, 49–50) capably outline the direction that should be taken to achieve optimal analytical results. It is the aim of this work to discuss several additional areas in which induced phase changes in mesomorphic sorbents have produced interesting results.

The optimum separation ratio for a given pair of isomers is generally obtained at a relatively low temperature in the mesophase region. This is due both to the decreased degree of ordering of the sorbent as the transition to the isotropic liquid is approached (38, 39) and to the normal negative temperature dependence of relative solute retentions found with most stationary phases (51). Thus the usefulness of mesomorphic sorbents is often limited by high transition temperatures or by narrow spans of liquid crystalline behavior. These shortcomings have been significantly reduced through the use of eutectic mixtures of nematic stationary phases to achieve both wider operating ranges and improved low-temperature selectivity. Kelker (52) and Schroeder (47) have examined mixtures of dialkoxyazoxybenzene nematic liquid crystals, and Figure 5 depicts the

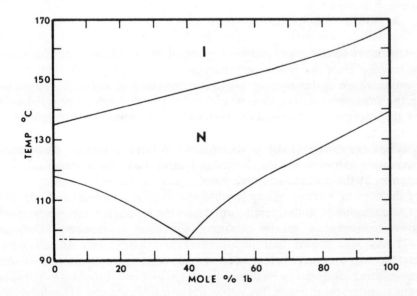

Fig. 5. Phase diagram (47) of the binary system 4,4′diethoxyazoxybenzene (1b) in 4, 4′dimethoxyazoxybenzene: I = isotropic liquid, N = nematic.

By courtesy of Plenum Press, New York.

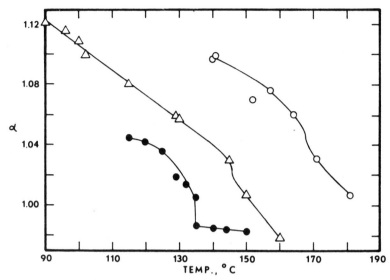

Fig. 6. Variation of relative retention (α) of p-xylene (m-xylene = 1.00) with column temperature using pure 4,4'dimethoxyazoxybenzene (solid circles), pure 4,4'diethoxyazoxybenzene (open circles) and their eutectic mixture (40 mole % DEAB) (triangles) as stationary phases (47).

By courtesy of Plenum Press, New York.

temperature-composition phase diagram of the dimethoxy and diethoxy homologs. Here we see a very substantial increase in the span of the nematic range at the 40 mole % eutectic composition and a 20°C lowering in the solid-nematic transition temperature. The chromatographic behavior of the two pure sorbents and their eutectic composition are plotted in Figure 6, where the temperature dependence of relative retention α for p-xylene (m-xylene = 1.00) is shown on the three columns. We observe decreasing values of α with increasing temperature and pronounced slope changes at the nematic-isotropic transition of the sorbents. The selectivity of the eutectic mixture at any fixed temperature is intermediate to the two pure compounds, but since it can retain the nematic structure to much lower temperatures, greater maximum selectivity is obtained. Schroeder (47) has also shown that high-melting and easily decomposed liquid crystals that are nevertheless selective, can be successfully combined with low-melting and less selective substances to yield excellent separation ratios under much more favorable operating conditions. The foregoing technique, of course, could be successfully applied to nonmesomorphic sorbents to overcome inconveniently high melting points of the pure compounds.

A recent investigation (53) into the effect of electric fields on a liquid crystalline stationary phase has yielded some interesting observations (i.e., field-induced changes in solute peak retention and symmetry). The authors attributed these effects to the alignment of the sorbent molecules in the direction of the radial field, as opposed to their normal wall-parallel configuration, thus exposing different sorbent functional groups. The magnitude of the retention increase on application of the field approached 150% for slightly retained components. Plotting their tabulated data of percentage increase in capacity ratio k' versus k' at a fixed potential (-500 V) leads to Figure 7, which reveals an apparent relationship between the solubility of a solute and the magnitude of the field-induced retention shift. Indeed, this phenomenon seems to be analogous to adsorption at the gas-sorbent interface, where the greatest effect is generally observed at low solute retention. Furthermore, the authors reported that the electric field effect is primarily observed in columns having a relatively thick sorbent

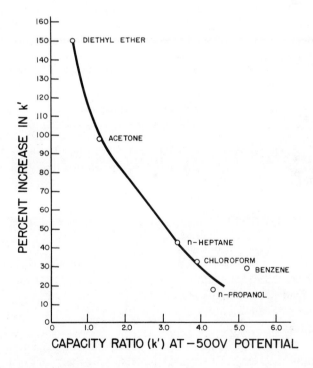

Fig. 7.   Plotted data from ref. 53 for solute retention in capillary columns containing a liquid crystalline sorbent under an external electric field.
   By courtesy of *Separation Science*.

film (i.e., coated by a 50–75% liquid crystal solution). This solution concentration is 10 to 15 times that normally employed (54) in coating capillary columns. This finding might be related to the somewhat irregular electrode spacing to be expected from the wire-in-capillary method used, but it could also be explained in terms of the effective linear dimensions of a swarm of mesomorphic molecules, which is of the order of magnitude of 1 μ (55). This dimension is also the value considered to be the optimum film thickness for capillary columns (54). Thus we would expect that film thicknesses at least an order of magnitude greater than the swarm dimension would be required for alignment in an external field. Unfortunately, films of this thickness do not generally result in high capillary column efficiencies.

Clearly, much work remains to be done in order to understand the mechanism by which the field-induced retention shifts occur. Future investigators should not overlook the possibility of utilizing magnetic fields, which can often be as effective as electric fields (56), to achieve molecular alignment. However, some difficulty may be experienced in obtaining the required magnetic field within a chromatographic column. The analytical potential of a chromatographic column whose "polarity" could be rapidly varied by means of a potentiometer should not be underestimated.

C. Solid-Solid Transitions

Guran and Rogers (57) have employed thallium nitrate as a GSC adsorbent in order to determine whether its orthorhombic-hexagonal transition could be detected by solute retention anomalies. Although large retention shifts at the 79°C $\gamma \rightarrow \beta$ transition temperature were observed for cyclenes and alkenes, it appears that this phenomenon is only indirectly related to the transition; namely, by way of desorption of surface and interstitial water. Perhaps the higher temperature $\beta \rightarrow \alpha$ transition would be a better choice for a chromatographically observed event, since this transition was found (by DTA) to be reproducible and rapidly reversible when the salt was coated on chromatographic support. Higher column temperatures may lead to sorbent–solute reactions, but these can usually be detected by using the techniques of Phillips (58).

A positive chromatographic observation of a solid-state transition was later reported by Guran and Rogers (59), who employed the thermo-chromic solids Ag_2HgI_4 and Cu_2HgI_4 as GSC adsorbents. The authors were able vividly to distinguish between the high-temperature α forms of the solids, which show adsorptive properties similar to metals, and the low-temperature β forms, where the compositional differences of the solids were readily apparent. In Figure 8, retention data are plotted as log capacity ratio versus reciprocal absolute temperature for C_8 hydrocarbons on the Ag_2HgI_4 column. A dramatic increase in the differential heat of adsorption is observed as the column temperature is lowered (increasing $1/T$) through the $\alpha \rightarrow \beta$ transition, and the magnitude of the shift is greatest for solutes containing double bonds, in keeping with the well-known interaction of olefins with silver ions. The other adsorbent showed significantly lesser ΔH shifts at its transition temperature.

Guillet (60) has described a technique for the determination of glass transition temperatures for polymeric stationary phases by observing slope reversals in retention volume–temperature plots for weakly interacting solutes. This topic is fully discussed in another chapter.

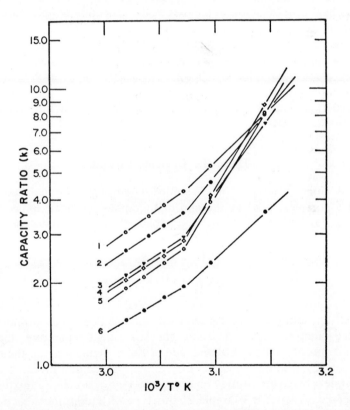

Fig. 8.   Capacity ratios (59) versus $1/T$ for C_8 hydrocarbons eluted from a column of Ag_2HgI_4. Adsorbates: 1, n-octane; 2, 1-octene; 3, 1,7-octadiene; 4, cyclooctene; 5, ethylbenzene; 6, cyclooctane.

By courtesy of the *Journal of Gas Chromatography*.

D.   Melting Transitions of Inorganic Salts

Many workers have examined solute retention in columns containing solid inorganic salts as GSC adsorbents (61–67), and others have utilized molten salt eutectic mixtures as a liquid phase (68–72); but apparently no observations have been made of any solute anomalies in the melting region of the sorbent. A brief investigation (73) was made by this author of the properties of a column containing the $LiNO_3$–KNO_3 eutectic mixture coated on Chromosorb P. The use of the eutectic composition allowed a reasonable transition temperature (121.5°C) for the sorbent, as well as yielding the maximum rate of melting with increasing temperature that can be achieved with any given mixture. The latter property has been found to be important for the chromatographic observation of inorganic salt melting transitions. The experimental data indicate a slight decrease in the differential heat of adsorption of alkanes at the transition temperature (of the form depicted in Figure 1E). If the transition had occurred over a substantial range of temperature, as would have been the case for an off-eutectic mixture, it is doubtful whether the transition could have been detected chromatographically. These transitions might even exhibit the large retention shifts shown by organic sorbents if less inert solutes were employed (i.e., solutes that complexed with the molten salt but not the crystalline adsorbent).

E.   Thermally Unstable Complexes

This author attempted to determine whether the thermal decomposition of a complex, used as the chromatographic stationary phase, would exhibit the behavior shown by uncomplexed sorbents undergoing phase changes (73). It was supposed that an equilibrium existed of the type denoted as class D by Purnell (74), that is,

$$A_n S_m \Leftrightarrow A_n + S_m \qquad (6)$$

in which an additive A reacts with the solvent S in some proportion to yield the complex $A_n S_m$. At some characteristic temperature, the equilibrium is forced to the right-hand side of Eq. 6. Depending on the relative extents of adsorption and solution by the decomposition products and by the complex, behavior entirely analogous to that of a melting mixture may be observed. A specific example of this type of equilibrium reaction was suggested by the work of Langar, Zahn, and Vial (75), in which they reported the existence of a solid-state 1 : 1 charge-transfer complex between hexamethylbenzene (HMB) and di-n-propyl tetrachlorophthalate (DNPT).

Langer observed that the decomposition temperature of this complex was 80.6°C. This system thus appeared to be of interest as the complex was a solid that decomposed to yield a solid (HMB, m.p. 165°C) and a liquid (DNPT, m.p. 26°C). Accordingly, the complex was prepared after the method of Langer et al. (75), deposited at a 25% w/w loading on Chromosorb P, and examined by DTA and as a chromatographic stationary phase. DTA thermograms indicated a relatively sharp endothermal event at 80.7°C, in good agreement with Langer's value. Subsequent runs on the same sample manifested the same transition temperatures and peak area, implying that the transition was completely reversible.

A rapid preliminary survey of the retention behavior of benzene and toluene indicated initial positive temperature dependence resulting in a small (ca. 10%) increase in retention over the range 70 to 75°C. The 75 to 80°C region, however, yielded a 5.5-fold gain in solute retention. Temperatures above the sorbent dissociation temperature yielded a normal negative slope. Unfortunately, accurate retention data could not be taken because it was observed that the sorbent was subliming from the column and solidifying in the cooler postcolumn tubing. Nevertheless, the sharp increase in retention coinciding with the complex dissociation confirmed Purnell's hypothesis that such transitions would be observed in gas chromatographic studies.

The work of Knight (76) and, later, McAdie (77) on urea–hydrocarbon inclusion complexes suggested the next system for investigation. McAdie found an increasing dissociation temperature for urea–n-paraffin complexes as the chain length of the hydrocarbon was increased; he also learned that decomposition of the solid complex yielded urea plus liquid n-paraffin, although there was evidence of urea sublimation at temperatures above 120°C. In view of the latter limitation, hexadecane was selected as the n-paraffin, since its reported (77) complex decomposition temperature (108.1°C) placed the column operation temperature below the sublimation region of urea. Preliminary DTA scanning runs indicated that the reversibility of complex formation was destroyed by the melting of urea at 135°C; hence all subsequent DTA and chromatographic observations were made well below that temperature.

In the DTA thermograms in Figure 9, curve A illustrates the thermal behavior observed with a sample consisting of 50-mesh crystals of the hexadecane–urea complex dispersed on the solid support and thermally unconditioned. The endothermal peak transition temperature (107.8°C) corresponds very well to the previously mentioned literature value. The decomposition is much more gradual than is that of the HMB:DNPT complex mentioned earlier, due, no doubt, to the different bonding mechanisms involved. A subsequent rerun on the same sample (curve B)

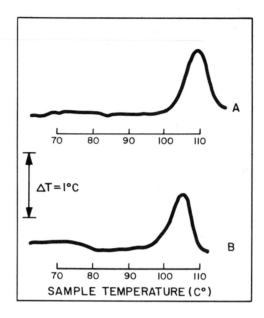

Fig. 9.   DTA thermograms (20) of *n*-hexadecane–urea complex crystals dispersed on solid support: *A*, thermally unconditioned sample; *B*, subsequent analysis.

illustrates a transition of similar symmetry and area but at a lower temperature, suggesting a possible interaction with the active support (Chromosorb P). Further thermal conditioning produced a continuing drift of the transition in the same direction. In spite of the varying thermal response of the support-coated complex, retention data for toluene were obtained from columns that had been conditioned for 4 hr above the complex dissociation temperature.

The solid curve in Figure 10 illustrates that extremely low values of solute retention are obtained when the solid complex acts as a GSC adsorbent in the region 95°C. In the 96 to 101°C interval, however, toluene retention increases by almost an order of magnitude, manifesting the decomposition of the complex over that temperature region with the resulting release of liquid hexadecane, which then acts as a stationary phase of vastly increased sorbing power. At temperatures above 102°C, solute retention assumes normal GLC temperature dependence. The remaining urea crystals are probably also coated with the paraffin at this temperature and hence act as a solid support. After a 4-hr hold at 102°C

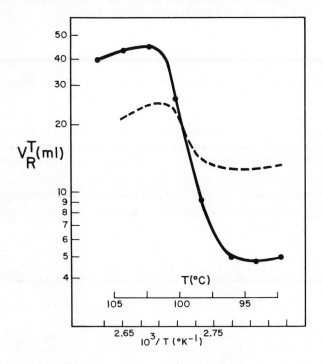

Fig. 10. Toluene retention on *n*-hexadecane–urea coated support (20): solid curve, initial run; dashed curves, repeat determination.

and a brief return to ambient temperature to allow complex re-formation, a repeat determination of toluene retention behavior was made (broken line in Figure 10). Here, solute retention increases by only a factor of 2 at the complex dissociation, because of an increase in the low-temperature retention and a decrease in the postdissociation retention. The former change may be due to larger amounts of free hexadecane no longer bound into the complex, whereas the latter downward shift in retention seems to indicate the loss of a significant amount of hydrocarbon sorbent from the column.

Although the lack of reproducibility of these two solvent–additive complexing sorbents severely compromises their successful use in analytical columns, it appears that a novel means of studying complexing reactions has been found. This, again, will require considerable investigation before it can be regarded as a viable proposition, however.

IV. APPLICATIONS OF TRANSITION SORBENTS
TO GAS CHROMATOGRAPHY

A. Temperature-Independent Retention

One of the most ubiquitous features of gas chromatographic solute retention is its temperature dependence, previously described by Eqs. 4 and 5 and depicted in Figure 1A. It occurred to us (78, 79) that the positive temperature dependence of solute retention exhibited by transition sorbents might be utilized to compensate the normal negative slope of conventional stationary phases, thus achieving temperature-dependent retention over some limited range of temperature. Figure 11 illustrates graphically the principle of temperature-independent columns.

We can describe solute retention behavior for the positive slope transition region by the equation

$$\log V_g^{\,T}(2) = \frac{c}{T} + d \qquad\qquad (7)$$

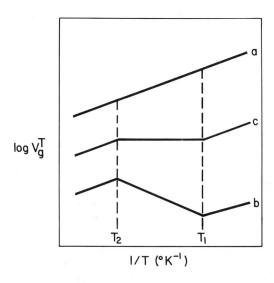

Fig. 11.   Illustration of concept of temperature-independent solute retention (80): curve a, conventional sorbent response for some solute; curve b, transition sorbent response for same solute; curve c, anticipated response for solute if sorbents **a** and **b** are combined in some way.
By courtesy of *Analytical Chemistry*.

which is of the same form as the equation describing the conventional sorbent, that is,

$$\log V_g^T(1) = \frac{a}{T} + b \tag{5}$$

Combining the two columns serially, the total resulting solute retention can be described as

$$V_R^T(1,2) = w_1 V_g^T(1) + w_2 V_g^T(2) \tag{8}$$

where w_1 and w_2 are the respective weights of conventional and transition sorbent.

The condition that the derivative of Eq. 8 with respect to temperature is equal to zero leads to the equation

$$\frac{w_1}{w_2} = -\frac{c}{a}\left(10^{(d-b)}\ 10^{(c-a)/T}\right) \tag{9}$$

which is an expression for the relative weight of conventional w_1 to transition sorbent w_2 which results in temperature-independent retention over the transition region. The apparent temperature dependence of Eq. 9 has been shown to be of little consequence (80), and the temperature term is assigned the value corresponding to the central temperature in the transition region.

Earlier studies (9, 18, 81) indicated that mixtures of fatty acids should yield the desired combination of extended melting transition and solute selectivity required for this application. Extensive thermal analysis of solid support coated with varying mixtures of stearic and oleic acid were carried out to construct the phase diagram (80). Chromatographic determination of the constants in Eqs. 5 and 7 were made on a conventional squalane column and on a 25 mole % oleic acid-stearic acid column; the resulting values are given in Table I. Note that the slopes (the c terms) for benzene and acetone on the transition sorbent column are quite similar, whereas they differ appreciably on the nonpolar squalane column, reflecting the different heats of solution of these solutes in squalane. These differences result in sorbent weight ratios (via Eq. 9) that seem to have an optimum value for each solute. Subsequent theoretical and experimental studies (80) demonstrated that this was not the case, since a substantial variation in the solvent weight ratio still results in temperature-independent retention, although over a somewhat reduced span.

TABLE I

Constants of Eqs. 5 and 7 for Elution of Benzene and Acetone from Approximately 20% ·/w Columns, Containing Squalane or 25 mole % Oleic Acid–Stearic Acid Mixture, in Temperature Range 53 to 66°C.

	Constants for squalane		Constants for acid mixture		
	$10^{-3}a$	$-b$	$-10^{-3}c$	d	w_1/w_2[a]
Benzene	1.500	2.178	2.949	11.015	1.31
Acetone	1.223	1.745	3.015	10.638	1.10

[a]The values of w_1/w_2 were calculated for the midpoint of the temperature range $(1/T_m = 3.005 \times 10^{-3})$.

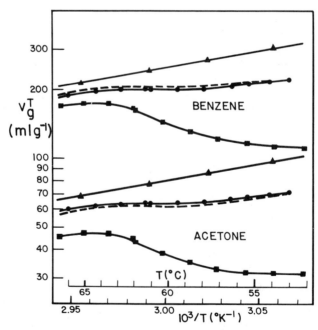

Fig. 12. $\text{Log} V_g{}^T$ as a function of $1/T$ for elution of benzene (upper four plots) and acetone (lower four plots) from columns containing: squalane (triangles); 25 mole % oleic acid–stearic acid (squares); series combination with $w_1/w_2 = 1.29$ (circles). Curves for series combination computed via Eq. 8 are indicated by dashes (80).

By courtesy of *Analytical Chemistry*.

Figure 12 depicts solute retention behavior for benzene and acetone in columns containing squalane and 25 mole % oleic acid-stearic acid, and in a series column with a solvent weight ratio (squalane-transition sorbent) of 1.29. Also shown is the theoretical retention curve computed by way of Eq. 8. The agreement between the latter and the experimentally measured curve is better than 2%, thus confirming the additivity of retention in series columns wherein the pressure drop is not excessive. The temperature independent span is rather substantial (i.e., some 6°C around 60°C allowing for a ±0.5% variation in retention volume). Acetone, even though its optimum solvent weight ratio is somewhat less than that of benzene, nevertheless exhibits near-constant retention over a 5°C span.

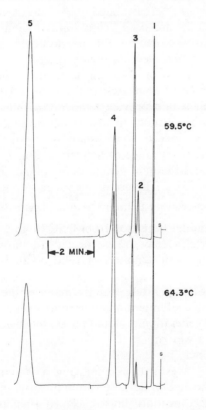

Fig. 13. Chromatograms (80) of representative mixture obtained in elutions at 59.5° and 64.3°C with series combination of squalane and 25 mole % oleic acid–stearic acid columns ($w_1/w_2 = 1.29$). Solutes: 1, air; 2, isopentane; 3, n-pentane; 4, acetone; 5, benzene. Corrected flow rates at outlet: 59.5°C run, 33.95 ml/min; 64.3°C run, 33.45 ml/min.

By courtesy of *Analytical Chemistry*.

As further evidence that temperature-independent retention can be achieved over a moderate span of temperature for sample components whose polarities range from that of acetone to the relatively inert alkanes, Figure 13 depicts chromatograms of such a solute mixture in a series-compensated column. There is little measurable shift in retention time on this column, although for a squalane column alone, the retention shift would be close to 20% over this range of temperature.

Alternate column configurations to the series method just employed include the striated column, which uses a single column of alternate slugs of conventional and compensator packing. This method, although possibly susceptible to sorbent zone migration that could alter its thermal response characteristics over a period, possesses the advantage that carrier gas compressibility corrections to calculated w_1/w_2 ratios are not required if more that four striations are used. Thus long columns with substantial pressure drops would not exhibit the nonadditivity of series column operation that has been observed (82) under these conditions.

After the initial success in obtaining temperature-independent retention in series and striated columns, research was oriented toward finding a transition sorbent that would provide temperature-independent retention without additional compensation. In undergoing fusion, most mixtures of organic substances and, indeed, some "pure" sorbents, exhibit constancy of retention over some short range of temperature, as in Figure 3, for the stearate-octadecanol and the pure octadecanol column and also in Figure 12 in the region below 55°C. The use of such regions on the log V_g versus $\frac{1}{T}$ plots for self-compensating columns is not practicable however, since these represent the gradual transition from GSC to GLC behavior and, as such, display the poor efficiency to be expected from a very small amount of liquid phase dispersed in a matrix of solid organic stationary phase. What we require, then, is a sorbent whose rate of generation of liquid phase with increasing temperature exactly matches the normal negative temperature dependence of retention. It is anticipated that, in order to attain this low rate of generation of liquid phase, the melting of the sorbent must take place over a wide range of temperature, hence substantial spans of constant retention may be realized.

Chromatographic investigation employing a 90 mole % oleic acid-stearic acid column showed no deviation from linearity in the region of the DTA-detected melting transition; thus it seemed obvious that a composition intermediate between 25 and 90 mole % should be self-compensated over some temperature range. Figure 14 depicts solute retention on columns containing 80, 60, and 50.6 mole % acid mixtures, with both the transition temperatures and the rate of generation of liquid phase increas-

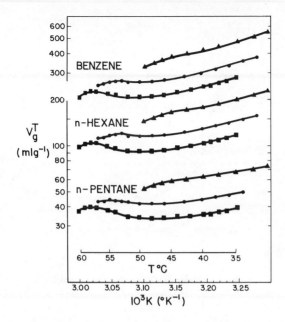

Fig. 14. Comparison (80) of $\log V_g^T$ dependence on $1/T$ for elution of benzene, n-hexane, and n-pentane from columns containing: oleic acid in stearic acid as follows: 80 mole % (triangles); 60 mole % (circles); 50.6 mole % (squares).

By courtesy of *Analytical Chemistry*.

ing with decreasing oleic acid content. Here we clearly observe relatively wide spans of retention constancy for both the 50.6 and 60 mole % columns, approaching 10°C (at ±1% retention variation) for the former column. The self-compensated column possesses several advantages over the series and striated configurations inasmuch as the former offers ease of fabrication and, generally, low sensitivity to stationary phase migration.

Column efficiency plots as a function of carrier gas velocity and column temperature for benzene eluted from columns containing pure squalane, from a 25 mole % acid mixture compensator column, and from a temperature independent striated column, appear in Figures 15 and 16. The former plot, made at an intermediate temperature in the transition region, indicates the slightly less efficient operation of the transition sorbent column due to its greater resistance to mass transfer. The striated column (like the series column, which has the identical curve) has essentially the same H_{min} as the squalane column, but is shifted to a higher velocity, which is advantageous to fast analysis times. Figure 16 reveals the identical nature of the columns when complete liquification of the transi-

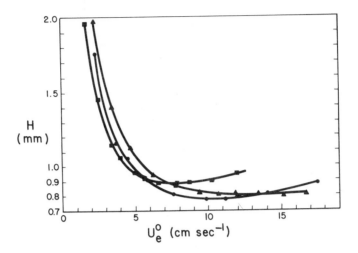

Fig. 15. Column efficiency (van Deemter) plots (80) for elution of benzene by hydrogen at 61°C from columns containing: squalane (circles); 25 mole % oleic acid–stearic acid (squares); striated squalane-mixed acids (triangles).

By courtesy of *Analytical Chemistry*.

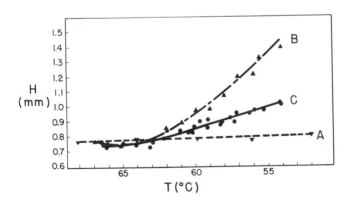

Fig 16. Minimum theoretical plate height as a function of temperature (80) for elution of benzene by hydrogen at 61°C from columns containing: *A*, squalane; *B*, 25 mole % oleic acid–stearic acid; *C*, series and striated combinations of *A* and *B* packings.

By courtesy of *Analytical Chemistry*.

tion sorbent is achieved in the region of 65°C. In the 63 to 57°C span of temperature independence, there is a small, but totally acceptable decrease in efficiency for the series and striated columns. The transition sorbent column efficiency decreases rapidly with temperature as the stationary phase solidifies; in fact, the plot of H versus T is remarkably linear from 56 to 32°C. Figure 16 suggests that the efficiency of a series or striated column is an additive function of the values for the component columns.

Self-compensated columns, such as the 60 and 50.6 mole % acid mixture columns shown in Figure 13, yield height equivalent to theoretical plate (HETP) values similar to the series and striated column data of Figure 16, and thus completely acceptable for analytical use. In addition to the oleic acid-stearic acid columns discussed here, substantial ranges of temperature-independent retention have been obtained in series columns with compensator systems comprising: stearic acid and cholesterol (the latter component serving to broaden the melting range of the acid), magnesium stearate and n-octadecanol, and methyl stearate and n-octadecanol, all in series with squalane columns. A dramatic comparison of the latter compensated system against squalane alone is presented in Figure 17. Here we see constancy of series column solute retention to within a few percentage points over a 9.2°C temperature span, whereas the pure squalane column yields solute retention shifts approaching 40%.

B. Transition Sorbents for Preparative Gas Chromatography

It is generally held that localized thermal excursions resulting from the passage of solute bands in chromatography columns can contribute to reduced column performance (83–86). These short-term thermal variations about the mean column temperature are generated by the exothermal latent heat of sorption of the solute in the stationary phase, followed by the endothermal desorption process. Scott (83) observed that excess localized plate temperatures arising from the aforementioned mechanism increased with flow rate, sample size, and decreasing values of the partition coefficient (i.e., increasing solute band velocity). Therefore, any attempt to increase the throughput of a preparative or industrial (large-scale) chromatograph will result in an increase in the magnitude of the thermal effects and a subsequent decrease in column performance, the latter manifested by increased band spreading and peak asymmetry.

Packed 1/8- and 1/4-in. o.d. columns do not normally exhibit this thermally induced decrease in column performance because of the extremely small samples employed in most conventional analytical work. However, in scaling up to a 1-in. diameter preparative column, with

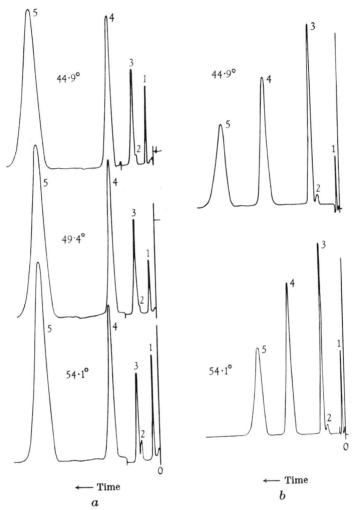

Fig. 17. Chromatograms (78) of mixture: 1, air; 2, isopentane; 3, n-pentane; 4, n-hexane; 5, benzene for the temperature range 44.9 to 54.1°C. (a) Series combination of columns (i) 30% methyl stearate in n-octadecanol mixture on Chromosorb P at 20.5% w/w with (ii) 25 % w/w squalane on Chromosorb P. (b) 25% w/w squalane on Chromosorb P; solid support, 60-80 mesh (ASTM).

By courtesy of *Nature*.

resulting increase in sample size (as much as a factor of 10^3 over that used for analytical work), thermal excursions in the 10 to 20°C region have been reported (84, 85). With industrial-scale columns having diameters of 12-in. or more (87, 88), finding a means of dissipating or controlling this thermal energy becomes imperative. Any improvement in radial heat flow would reduce the magnitude of the thermal events and, to this end, workers have utilized internal baffles, metal helices, and annular columns (88). Although these devices also serve to improve the carrier gas flow profile, a major source of reduced performance in large columns, they are very expensive, and a high degree of empiricism is called for in using them.

It was envisaged (78–80) that a temperature-independent column might be successfully applied in the area of preparative gas chromatography. By employing a stationary phase exhibiting an extended range of melting, such a column should undergo localized melting and subsequent solidification of the sorbent with the passage of the solute band. Through this mechanism, it was hypothesized that column performance would be improved considerably over that exhibited by a conventional liquid stationary phase. Accordingly, an experimental study (89, 90) has been made to determine the validity of this hypothesis. During this investigation, thermal transients were monitored with a 0.010-in. diameter wire chromel-constantan thermocouple installed in a tubing union in the 3/8-in. o.d. copper column, as shown in Figure 18. In order to obtain a representative thermal measurement without unnecessarily perturbing the carrier gas flow profile, the thermocouple bead was positioned 3 cm upstream of the feedthrough insulators and concentric with the tube axis. Preliminary experiments indicated a decrease in the magnitude of thermal excursions by a factor of 5.5 when the probe was moved from near the

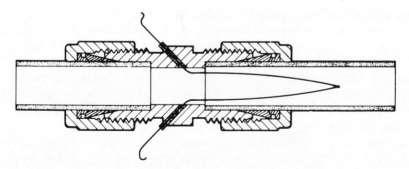

Fig. 18.   Probe for thermal peak measurement (90). Tubing is filled with coated chromatographic support.
     By courtesy of the Foxboro Company, Foxboro, Mass.

column inlet to near the outlet. However, comparision of probe response at the 25 and 50% locations along the length of the column, showed only a 30% drop in the signal. These findings are in accordance with Scott's observation (83) that excess plate temperatures are most noticeable over the first 25% of the column, no doubt because of the initially higher solute concentration at the column inlet.

The column initially studied utilized a "self-compensated" transition sorbent mixture comprising 60 mole % oleic acid in stearic acid (see Figure 14) deposited at a 20% weight loading on 45–60 mesh Chromosorb A. Since the useful transition span of this sorbent was from 43 to 53.5°C, and 100% liquidity occurred at 55°C, operation at higher temperatures would yield "normal" GLC operation. The thermal probe was installed at a location 25% from the inlet of the column.

Figure 19 gives some typical chromatograms and thermal transient records determined at two column temperatures; one corresponding to GLC operation, and the other at a temperature within the transition region of the sorbent. Markedly reduced thermal peak amplitudes can be seen in the transition sorbent for both 5- and 100-μl n-pentane samples, as compared with the substantial temperature excursions when the column is operated in the GLC mode. Note also that significant improvements in peak symmetry, efficiency, and resolution of the minor isopentane peak are displayed by the transition sorbent. A more quantitative presentation of the foregoing parameters as a function of column temperature appears in Figure 20, where exothermal (I_+) and endothermal (I_-) peak intensity, peak asymmetry ratio, and theoretical plate height all increase dramatically as the temperature is raised to the level where the column no longer functions as a transition sorbent and reverts to normal GLC operation. The large samples injected relative to the free volume of the column caused the feed volume limitation (91) to be exceeded for all data presented in Figure 20; therefore, a certain fraction of the total plate height stems from this source. On further examination, this fact merely serves to brighten the potential for the transition sorbent column, which exceeded the feed volume requirement by 32% because of its higher efficiency, whereas the GLC runs only experienced a 12% excess of feed volume. According to these figures, therefore, the observed improvement in efficiency in the transition region is real and may even be improved in the absence of feed volume limitations.

Subsequent examination of higher-boiling solutes on this column resulted in very asymmetric peaks that were attributed at the time to insufficient volatility of the solute in the column. To overcome this difficulty and thus facilitate the examination of a range of solutes, a second column was made utilizing 15 mole % oleic acid in stearic acid as

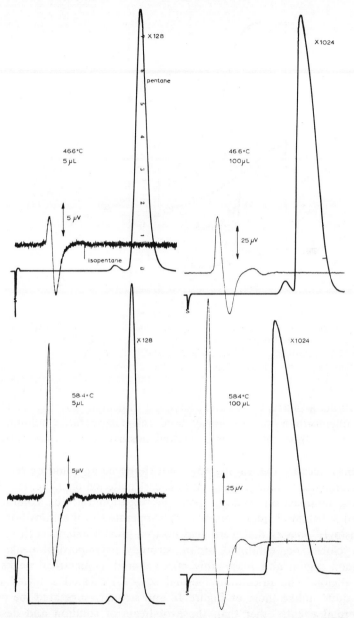

Fig. 19.   Chromatograms (90) of isopentane and *n*-pentane with a 60 mole % oleic acid in stearic acid column. Upper analyses in transition region operation, lower in GLC region; 1°C = 65 μV.

By courtesy of the Foxboro Company, Foxboro, Mass.

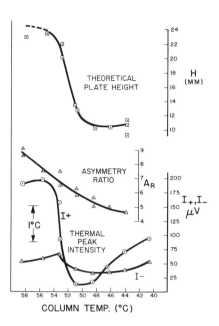

Fig. 20. Thermal peak intensity, asymmetry ratio, and theoretical plate height as a function of column temperature for 100-μl *n*-pentane samples in a 60 mole % oleic acid in stearic acid column (90). Transition to GLC operation at 53.5°C; 1°C = 65 μV.

By courtesy of the Foxboro Company, Foxboro, Mass.

the stationary phase, this mixture having a transition span (55–64°C) about 12°C higher than the previously used self-compensated column. Solute retention on this column has a marked positive temperature dependence throughout the transition region, which should further improve column efficiency because the leading edge of a solute band would be retarded by its proximity to the moving region of excess plate temperature, thus leading to peak compaction. Indeed, pentane retention on this column again showed substantial improvements when operated in the transition region. Surprisingly, solutes that were held up to a greater extent, such as hexane and cyclohexane, exhibited broad, strongly asymmetrical peaks in the transition region and reasonably efficient and symmetrical peaks in the GLC region. The in-column thermal sensor indicated a broad thermal event quite unlike those of Figure 18 and strongly suggested the presence of thermal events other than those of heats of solution and desorption alone.

Since the transition region and normal GLC measurements were taken at 62.2 and 64.6°C, respectively, it is unlikely that insufficient solute volatility could alone account for the substantial worsening of transition region performance with solutes of higher capacity ratio. A probable cause for this observed degradation must be the modification of the binary component sorbent to the extent that it becomes a tertiary system, the third component being the solute band as it passes through the column. Solutes exhibiting low capacity ratios, such as n-pentane at 2.7, are not very soluble in the sorbent; hence the tendency to modify the thermal behavior of the stationary phase is minimal. Longer retained solutes, however, such as hexane and cyclohexane (capacity ratios 10 and 20, respectively) are very soluble in the sorbent and readily form a transient tertiary system with it. Thus the broad thermal probe peaks probably represent the combined result of the heats of sorption, fusion (of the stationary phase), mixing, desorption, and crystallization.

Although the findings just cited limit the range of applicability of transition sorbents to solutes whose capacity ratio lies in the approximate range of 2 to 5, usually we can select for a given solute, a solvent weight loading, transition sorbent, and column temperature (related to the preceding by the transition span) that will result in enhanced column performance. Since, however, in maintaining relatively low solute solubility, we are working in the region where feed volume may constitute a sizeable portion of the band-broadening process, a compromise between solution and feed volume restrictions may prove necessary.

Using binary component carboxylic acid stationary phases, it has been observed that to ensure sufficient volatility of the solutes in the column (and thus to achieve low retentions), the transition sorbent should be selected such that the resulting transition span is about 20°C higher than the boiling point of the solute. It is seen that this rough empirical guideline was held for n-pentane alone in both columns discussed here. This relation has also been upheld by studies involving the injection of large benzene samples into columns containing mixtures of higher-melting dicarboxylic acids as the transition sorbent (transition span 103–112°C). Thermal peaks were reduced from 4°C in GLC operation to near zero at the beginning of the transition span, accompanied by a 100% improvement in column efficiency. The capacity ratio of benzene ranged from 4.8 to 4.0 over the transition region, as compared with a value of about 25 in the earlier oleic-stearic columns. No anomalous thermal probe response was noted, in complete contrast to benzene response on the lower temperature column.

Thus, although the use of transition sorbents in preparative gas chromatography appears to be restricted to samples of relatively low solubility in a given stationary phase under the prevailing conditions of temperature and solvent weight loading, it is to be noted that this constraint results in short elution times, a matter of great importance for maximum throughput in preparative chromatography.

C. Transition Sorbents as Variable Selectivity Stationary Phases

A widely used technique for solute identification has been the determination of the retention volume of the unknown sample in two columns of different selectivity (92, 93). In quantitative analysis, the separation of complex mixtures has been effected at times by combining different stationary phases either as series columns or in a single column (82, 94, 95). All the foregoing methods reflect the need for a sorbent whose selectivity can be varied at will over a large range by manipulating a convenient experimental parameter, such as the column temperature. Several workers have reported a method for peak identification (96, 97) that relies on variations in temperature dependences of the Kovats Indices for various classes of solutes. These changes in selectivity with temperature, which occur as a result of differences in the partial molar heats of vaporization of the solutes in solution, are relatively small for conventional homogeneous stationary phases. The use of liquid crystals (38–47) and, to a lesser extent conventional sorbents (9, 15) operated in the region of their freezing point, brings about a thermally induced change in solute selectivity. In both cases, the alteration of selectivity can be attributed to a change in the physical state of the sorbent (i.e., the formation of an ordered mesophase of swarms of solvent molecules in liquid crystals and to the creation of a solid organic adsorbent for the frozen stationary phase). The latter technique (9, 15) has yielded complete reversals in elution order, but at the expense of greatly reduced column efficiency and peak symmetry. What is required, then, is a stationary phase whose identity or composition can be controlled over a range of temperature such that the retention response of different classes of solutes at different temperatures can serve as a method of identification (98, 99).

Returning, for the moment, to the earlier general discussion of transition sorbents, let us recall that the increase in retention throughout the transition region is a function both of the increasing volume of liquefied sorbent and of any change in the resulting solute partition coefficient. When transition sorbents were applied toward achieving temperature-independent retention (78–80), it was realized that any large variation of this latter retention-determining parameter for various classes of solutes would limit the effectiveness of the sorbent to achieve temperature-independent retention for all these solutes. To avoid this situation, the two components of the transition sorbent were chosen to have very nearly the same sorbing power for a given solute (cf. Figure 14). Thus it was possible to construct a column that was invariant to temperature for alkanes, aromatics, and ketones, as illustrated previously in Figure 13. It therefore became evident that in order to achieve widely varying retention response for these classes of solutes over the transition region of a sorbent, the components of the stationary phase mixture should be chosen to have both a substantial thaw–melt region (i.e., solid solution loop or mutually depressed freezing point) and differing retention behavior of each constituent sorbent toward the various classes of solutes.

To illustrate the mechanism that results in variable solvent composition, Figure 21 depicts a hypothetical temperature-composition phase diagram of substances A and B, in the form of a solid solution loop. Actual phase diagrams are generally more complex than this, but the resulting argument is applicable to most systems where, irrespective of type, miscibility of the two solvents is observed. For a mixture comprising X_2 mole % B in A, temperatures below T_1 will result in complete solidification of the melt. At T_1 the mixture begins to liquefy, yielding liquid with a composition corresponding to X_1; in this region, the melt is very concentrated with respect to substance B. At successively higher temperatures more liquid is generated, and its composition corresponds to the intersection of the T axis with the liquidus curve. When temperature T_2 is attained, all the solid has melted to yield a liquid, which has the original composition X_2. Therefore, if components A and B differ appreciably in solute selectivity, a column that utilized such a mixture as a stationary phase should exhibit a strong temperature dependence of solute selectivity over the transition region.

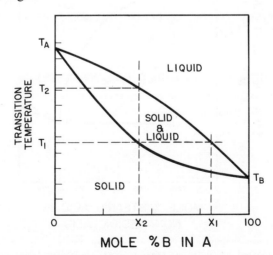

Fig. 21.   Hypothetical solid solution phase diagram of substances A and B (99).
By courtesy of *Analytical Chemistry*.

What is only inferred from Figure 21 is the temperature dependence of the actual volume of liquid that would be available for GLC partitioning. There is no liquid present at temperatures below T_1, and the primary retention mechanism is one of adsorption onto the solid organic surface. For the present application, this region is avoided. With increasing temperature, GLC partitioning rapidly becomes dominant, and a general increase in retention for all solutes is observed. As temperatures approach T_2, retention shifts continue to occur because of the increasing volume of liquid phase present and, most important, because of the changing composition of the liquid sorbent. Thus the main effect of low liquid volume at temperatures near T_1 is a reduction of the effective range of stationary phase compositions that will yield satisfactory GLC column performance. Experience has shown, however, that this is not a serious limitation.

The transition sorbent mixture chosen for experimental verification of the variable selectivity concept comprised stearic acid and nonanedioic acid. The experimentally determined phase diagram for this solvent system is shown in Figure 22. Examination of the phase diagram shows that the eutectic composition at about 80 mole % S in N (S and N denoting stearic

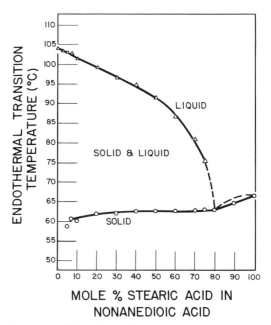

Fig. 22. Temperature–composition phase diagram (99) for mixtures of stearic acid and nonanedioic acid (coated 20% w/w on chromatographic support).

By courtesy of *Analytical Chemistry*.

acid and nonanedioic acid, respectively) would melt over a very short span of temperature to yield that same composition of liquid, whereas according to the discussion associated with Figure 21, a 10 mole % S and N solid mixture would have a transition region from 62 to 102°C, thus yielding liquid state compositions ranging from 80 to 10 mole % S and N. Obviously, chromatographic utilization of the latter sorbent mixture would be preferred, owing to its wide range of available liquid compositions. Figure 23 gives retention data for benzene with each of the pure solvent columns, as well as with columns containing various mixtures of the two. The initial retention increase in the region of 60 to 62°C is predicted by the phase diagram and is a manifestation of the transition from GSC to GLC

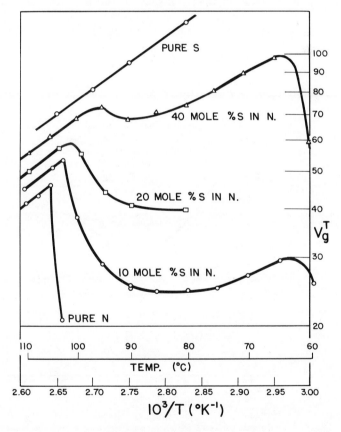

Fig. 23.   Log V_g^T as a function of $1/T$ for elution of benzene from columns containing 0, 10, 20, 40, and 100 mole % stearic acid in nonanedioic acid (99).
    By courtesy of *Analytical Chemistry*.

operation as a result of the onset of melting of the sorbent. At higher
temperatures in the region of 85 to 93°C, the rate of melting of the
stationary phase increases, producing a general shift in retention until, at
an even higher temperature, retention maxima occur at the respective
temperatures of complete liquefaction of each stationary phase. These
maxima also correspond to their respective values on the liquidus curve in
Figure 22. Note that the region from about 87 to 102°C for the 10 mole %
S in N column has the greatest shift in benzene retention. The greatest
shift in liquid composition occurs within the same region on the phase
diagram; hence this column was selected for further investigation.

Figure 24 illustrates the temperature dependence of retention for a
variety of solutes on this column. We observe one type of retention
response for the normal alkanes and a much more active behavior for the

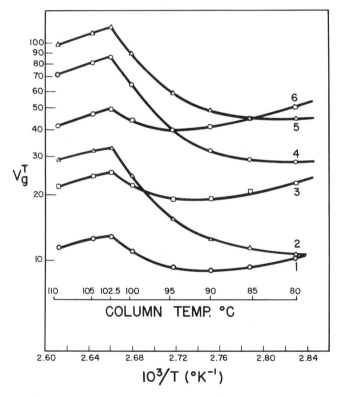

Fig. 24. Log V_g^T as a function of $1/T$ for various solutes with a 10 mole % stearic acid in
nonanedioic acid column (99). Solutes: 1, n-hexane; 2, methyl acetate; 3, n-heptane; 4,
2-propanol; 5, methyl propyl ketone; 6, n-octane.
By courtesy of *Analytical Chemistry*.

more polar acetate, alcohol, and ketone. Toluene, omitted on this plot for clarity, exhibits a retention response intermediate between that of the alkanes and that of the more polar solutes. Specific retention volumes for all solutes at temperatures above 80°C were independent of sample size and solvent weight loading in the range of 9 to 20% w/w on 80–100 mesh Chromosorb P.

Since retention measurements are a function of the volume of stationary phase as well as its identity, both of which change markedly with temperature, a much clearer picture of solute selectivity can be obtained by normalizing all retentions to that of n-heptane, as in Figure 25. Here we see the normal linear convergence of the alkane relative retentions with increasing temperature that occurs on conventional sorbents. The magnitude of the negative shift is about 13% over the 80 to

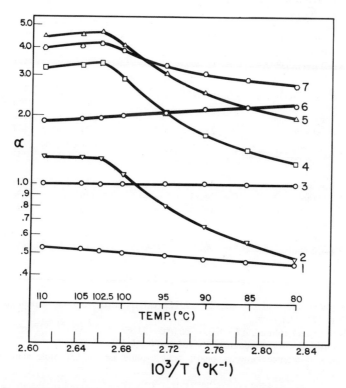

Fig. 25. Log relative retention as a function of $1/T$ for various solutes on a 10 mole % stearic acid in nonanedioic acid column (99). Solutes: 1, n-hexane; 2, methyl acetate; 3, n-heptane; 4, 2-propanol; 5, methyl propyl ketone; 6, n-octane; 7, toluene.
By courtesy of *Analytical Chemistry*.

TABLE II

Index Difference Methods for Solute Identification[a]

Solute	a $\delta I^{N}_{23°C}$	b $\delta I^{S}_{23°C}$	c $\delta\Delta I^{S,N}_{23°C}$	d $\Delta I^{N-S}_{103.4°C}$
Methyl acetate	30	13	132	240
Ethyl acetate	28	9	127	231
Propyl acetate	28	9	126	228
Acetone	37	13	149	289
MEK	35	9	147	273
MPK	35	9	147	266
1 – Hexene	5	1	13	38
1 – Heptene	5	1	15	29
1 – Octene	5	0	15	33
2 – Propyl ether	19	– 1	70	118
n-Propyl ether	23	1	75	122
n-Butyl ether	18	2	72	121
Benzene	24	8	91	153
Toluene	26	8	91	154
Ethyl benzene	26	8	88	152
m-Xylene	25	7	90	157
o-Xylene	27	7	97	170
Propyl benzene	25	7	86	149
Methanol	40	– 12	137	373
Ethanol	24	– 8	156	350
2 – Propanol	19	– 8	155	329
2 – Butanol	22	– 6	155	320
i-Butanol	25	– 6	155	327
n-Butanol	29	– 5	162	346

a   Nonanedioic acid column, $\Delta T = 23°C$ (extrapolated from measurements over the range 107 to 124.5°C).

b   Stearic acid column, $\Delta T = 103.4 - 80.0°C = 23°C$.

c   Transition sorbent column, $\Delta T = 103.0 - 80.0°C = 23°C$.

d   Index difference between stearic acid column and nonanedioic acid column at 103.4°C.

102.7°C portion of the transition region. This shift is greatly exceeded by the more polar solutes, where positive increases over the same temperature span amount to 175% for methyl acetate and 2-propanol, 136% for methyl propyl ketone, and 51% for toluene. Although the numerical values of the relative retention shifts just quoted alter significantly depending on the alkane used for normalization, they nevertheless attest to the dramatic positive increases in selectivity for polar or easily polarizable solutes with increasing temperature, as contrasted with the relatively small decreases in α with increasing temperature, which is observed in most conventional stationary phases (96, 97).

Let us examine the effectiveness of this column as a tool for solute identification. Table II compares three methods of solute identification using δI and ΔI values for the different solutes, where δI is the difference in the retention index for a given solute–solvent pair over a defined temperature interval and ΔI is the retention index difference for a solute on two different columns at the same temperature. The absolute values of the Kovats Indices for the pure solvent and transition sorbent columns are given elsewhere (99). The effectiveness of the two-column method (92, 93) is shown by the ΔI data from the two pure solvent columns at 103.4°C, defined as $\Delta I_{103.4°C}^{N-S}$ in Table II. Since the indices were reproducible to within ± 1 index unit (99), all classes of solutes can be easily identified by the magnitude of the ΔI term. Such is not the case, however, for the δI data on each of the pure solvent columns, namely, $\delta I_{23°C}^{N}$ and $\delta I_{23°C}^{S}$, where it is impossible to distinguish between acetates, ketones, and aromatics on these data alone.

The picture is much more promising in the case of index difference on the transition sorbent column at 80 to 103°C, defined as $\delta \Delta I_{23°C}^{S:N}$ in Table II. These data indicate that the variable selectivity sorbent possesses the specificity of solute identification shown earlier by the dual-column ΔI method, together with the operation advantages of single-column analysis. Unlike the dual-column method, however, the transition sorbent column has the ability to alter its specificity at will to obtain the best resolution of a complex sample. This is clearly depicted in Figure 26, where chromatograms of an eight-component test mixture are shown at temperatures corresponding to the two extremes of sorbent selectivity. It is seen that n-octane has an identical 2-min total retention time at both temperatures, in great contrast to the substantial movement of the more polar solute peaks. The intermediate nature of the toluene retention shift as compared with n-octane and 2-butanol is quite evident. In view of the sensitivity of these novel sorbents to alterations in column temperature, particularly in the region where the composition is changing rapidly,

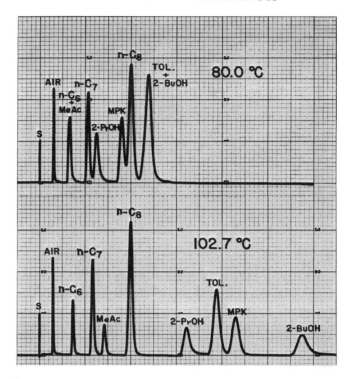

Fig. 26. Chromatograms (99) illustrating variable selectivity stationary phase at temperatures corresponding to the extremes of available polarity.
By courtesy of *Analytical Chemistry*.

column oven temperature stability and spatial uniformity are mandatory for precise solute identification.

It appears that investigations involving mixtures of liquid crystals having significantly different functionality and molecular configuration, at compositions removed from the eutectic, may yield useful spans of variable solute selectivity. Indeed, a recently reported study (100) of nematic liquid crystal mixtures implies a 36°C span wherein the "melt" is a nematic mixture, followed by a 47°C span in which the nematic and isotropic states coexist, the latter component altering in composition from 100 to 30 mole % over that temperature range. Alternately, we could deliberately introduce a nonmesomorphic component that destroys the liquid crystalline structure at certain composition levels and allows it to exist at other levels. Thus it would be possible to control the extent of mesophase formation and, ideally, the selectivity of the column.

Conjugate properties favoring more regular compound packing of two different molecules in a mesophase structure is quite uncommon. Nevertheless, examples of molecules whose melts do not exhibit liquid crystalline behavior in one component systems but which do manifest these properties in a mixture have been reported (101). These unusual systems could have potential in this application if their temperature–composition phase diagrams are favorably oriented.

The use of a thermally unstable solvent–additive complex to achieve variable solute retention is another area in which investigation could be rewarding. When the complex dissociates, the functional groups of the constituent molecules, which may have been shielded or negated by other groups while in the complexed state, are now available to interact with the solute molecules. Great care must be taken to determine that the complex formation is reversible, however; otherwise unreliable retention data will result on subsequent temperature cycles.

Another promising solvent system involves the use of immiscible solvents, with one undergoing a phase transition over a reasonably wide span of temperature. The work of Langer and Sheehan (102) on selective stationary phases should prove helpful in the screening of potential sorbent mixtures.

V. CONCLUSION

It is hoped that the reader now appreciates that there are many ways of utilizing the unique thermal and solubility properties associated with phase transitions in a gas chromatography column. It is hoped that the following summary oi the applications discussed in this chapter will spur new investigations in these and other related areas.

1.   Study of the thermal and solubility characteristics of pure and multicomponent solvent systems.

2.   Investigation of solvent–additive and solute–solvent complexing.

3.   The use of multicomponent liquid crystal sorbents to enhance separation factors.

4.   The study of solid organic adsorbents and their potential for specific interactions.

5.   The characterization of solid-solid transitions in both inorganic and organic compounds.

6.   The investigation of the thermal and solution properties of thin films deposited on active supports.

7.   Chromatographic investigation of pretransformation effects in organic compounds.

8.   Temperature-independent solute retention.

9.   The enhancement of peak parameters in preparative gas chromatography.

10.   Variable selectivity stationary phases.

References

1. J. R. Conder, in *Progress in Gas Chromatography*, J. H. Purnell, ed., Interscience, New York, 1968, p. 243.
2. S. J. Hawkes and E. F. Mooney, *Anal. Chem.*, **36**, 1473 (1964).
3. J. H. Purnell, *Gas Chromatography*, Wiley, New York, 1962, p. 93.
4. J. H. Purnell, *Gas Chromatography*, Wiley, New York, 1962, p. 214.
5. A. R. Ubbelohde, *Melting and Crystal Structure*, Oxford University Press, London, 1965, p. 218.
6. National Bureau of Standards (U.S.), Spec. Publ. 260 1972 Suppl., p. 19.
7. E. T. Rangel, M. S. thesis, Rice University, Houston, Texas, September 1956.
8. F. I. Stalkup and R. Kobayashi, *AIChE J.*, **9**, 121 (1963).
9. C. G. Scott, in *Gas Chromatography 1962*, M. Van Swaay, ed., London, Butterworths, 1962, pp. 39, 176.
10. A. R. Ubbelohde, *in* Ref. 5, p. 227.
11. P. Urone, J. E. Smith, and R. J. Katnik, *Anal. Chem.*, **34**, 476 (1962).
12. S. Dal Nogare, *Anal. Chem.*, **37**, 1450 (1965).
13. W. Fiddler and P. C. Doerr, *J. Chromatogr.*, **21**, 481 (1966).
14. A. G. Altenau, R. E. Kraemer, D. J. McAdoo, and C. Merritt, *J. Gas Chromatogr.*, **4**, 96 (1966).
15. R. R. Claeys and H. Freund, *J. Gas Chromatogr.*, **6**, 421 (1968).
16. R. L. Pecsok, A. de Yllana, and A. Abdul-Karim, *Anal. Chem.*, **36**, 452 (1964).
17. A. R. Ubbelohde, *in* Ref. 5, p. 306.
18. J. H. Purnell, S. P. Wasik, and R. S. Juvet, *Acta Chim. Acad. Sci. Hung.*, **50**, 201 (1966).
19. A. R. Ubbelohde, *in* Ref. 5, p. 76.
20. P. F. McCrea, Ph.D. thesis, University of Wales, 1968, p. 68.
21. G. W. Gray, *Molecular Structure and the Properties of Liquid Crystals*, Academic Press, New York, 1962, p. 115.
22. D. B. Cox and J. F. McGlynn, *Anal. Chem.*, **29**, 960 (1957).
23. F. H. Stross and S. T. Abrams, *J. Am. Chem. Soc.*, **73**, 2825 (1951).
24. J. A. Corbin and L. B. Rogers, *Anal. Chem.*, **42**, 974 (1970).
25. J. Serpinet, *Nature Phys. Sci.*, **232**, 42 (1971).
26. J. Serpinet, *J. Chromatogr.*, **68**, 9 (1972).
27. C. M. Drew and W. M. Bens, in *Gas Chromatography, 1968*, C. L. A. Harbourn, ed., Elsevier, New York, 1969, p. 3.
28. W. J. Baker, E. H. Lee, and R. F. Wall, in *Gas Chromatography*, H. J. Noebels, R. F. Wall, and N. Brenner, eds., Academic Press, New York, 1961, Chapter III.
29. N. C. Saha and J. C. Giddings, *Anal. Chem.*, **37**, 822 (1965).
30. A. R. Ubbelohde, *in* Ref. 5, p. 224.
31. A. R. Ubbelohde, *in* Ref. 5, pp. 26–29.
32. J. Serpinet, C. Daneyrolle, M. Trocaz, and C. Eyraud, *C. R. Acad. Sci. Paris, Ser. C*, **273**, 1290 (1971).
33. G. Untz and J. Serpinet, *C. R. Acad. Sci. Paris, Ser. C*, **273**, 1424 (1971).
34. M. B. Evans, *J. Chromatogr. Sci.*, **10**, 425 (1972).
35. R. K. Iler, *The Colloid Chemistry of Silica and Silicates*, Cornell University Press, Ithaca, N.Y., 1955.
36. I. Halász and I. Sebastian, *Angew. Chem., Int. Ed.*, **8**, 453 (1969).
37. P. F. McCrea, unpublished results, 1970.
38. H. Kelker, *Z. Anal. Chem.*, **198**, 254 (1963).
39. M. J. S. Dewar and J. P. Schroeder, *J. Am. Chem. Soc.*, **86**, 5235 (1964).

40. M. J. S. Dewar and J. P. Schroeder, *J. Org. Chem.*, **30**, 3485 (1965).

41. E. M. Barrall, R. S. Porter, and J. F. Johnson, *J. Chromatogr.*, **21**, 392 (1966).

42. H. Kelker and E. von Schivizhoffen, *Advances in Chromatography*, Vol. VI, J. C. Giddings and R. A. Keller, ed., Dekker, New York, 1968, p. 247.

43. D. E. Martire, P. A. Blasco, P. F. Carone, L. C. Chow, and H. Vicini, *J. Phys. Chem.*, **72**, 3489 (1968).

44. H. Kelker and A. Verhelst, *J. Chromatogr. Sci.*, **7**, 79 (1969).

45. L. C. Chow and D. E. Martire, *J. Phys. Chem.*, **75**, 2005 (1971).

46. D. G. Willey and G. H. Brown, *J. Phys. Chem.*, **76**, 99 (1972).

47. J. P. Schroeder, D. C. Schroeder, and M. Katsikas, in *Liquid Crystals and Ordered Fluids*, J. F. Johnson and R. S. Porter, eds., Plenum Press, New York, 1970, p. 169.

48. L. C. Chow and D. E. Martire, *J. Phys. Chem.*, **73**, 1127 (1969).

49. W. L. Zielinski, D. H. Freeman, D. E. Martire, and L. C. Chow, *Anal. Chem.*, **42**, 176 (1970).

50. A. B. Richmond, *J. Chromatogr. Sci.*, **9**, 571 (1971).

51. A. B. Littlewood, *Gas Chromatography*, 2nd ed., Academic Press, New York, 1970, p. 114.

52. H. Kelker, B. Scheurle, and H. Winterscheidt, *Anal. Chim. Acta*, **38**, 17 (1967).

53. P. J. Taylor, R. A. Culp. C. H. Lockmüller, and L. B. Rogers, *Separation Sci.*, **6**, 841 (1971).

54. A. B. Littlewood, *in* Ref. 51, p. 246.

55. C. Yun and A. G. Fredrickson, in *Liquid Crystals and Ordered Fluids*, J. F. Johnson and R. S. Porter, eds., Plenum Press, New York, 1970, p. 256.

56. E. F. Carr, in *Ordered Fluids and Liquid Crystals*, R. S. Porter and J. F. Johnson, eds., American Chemical Society, Washington, D.C., 1967, p. 76.

57. B. T. Guran and L. B. Rogers, *J. Gas Chromatogr.*, **3**, 269 (1965).

58. C. S. G. Phillips, in *Gas Chromatography, 1970*, R. Stock, ed., The Institute of Petroleum, London, 1971, p. 1.

59. B. T. Guran and L. B. Rogers, *J. Gas Chromatogr.*, **5**, 574 (1967).

60. O. Smidsrød and J. E. Guillet, *Macromolecules*, **2**, 272 (1969).

61. W. W. Hanneman, *J. Gas Chromatogr.*, **1**, 18 (1963).

62. J. A. Favre and L. R. Kallenbach, *Anal. Chem.*, **36**, 63 (1964).

63. R. L. Grob, G. W. Weinert, and J. W. Drelich, *J. Chromatogr.*, **30**, 305 (1967).

64. A. F. Isbell and D. T. Sawyer, *Anal. Chem.*, **41**, 1381 (1969).

65. F. Geiss, B. Versino, and H. Schlitt, *Chromatographia*, **1**, 9 (1968).

66. C. S. G. Phillips and C. G. Scott, in *Progress in Gas Chromatography*, J. H. Purnell, ed., Interscience, New York, 1968, p. 121.

67. J. J. Duffield and L. B. Rogers, *Anal. Chem.*, **34**, 1193 (1962).

68. W. W. Hanneman, C. F. Spencer, and J. F. Johnson, *Anal. Chem.*, **32**, 1386 (1960).

69. R. S. Juvet and F. M. Wachi, *Anal. Chem.*, **32**, 290 (1960).

70. J. Tadmor, in *Chromatographic Reviews*, M. Lederer, ed., vol. 5, Elsevier, London, 1963, p. 233.

71. F. M. Zado and R. S. Juvet, in *Gas Chromatography, 1966*, A. B. Littlewood, ed., Elsevier, New York, 1967, p. 283.

72. R. S. Juvet, V. R. Shaw, and M. Aslam Kham, *J. Am. Chem. Soc.*, **91**, 3788 (1969).

73. P. F. McCrea, Ph.D. thesis, University of Wales, 1968, p. 88.

74. J. H. Purnell, in *Gas Chromatography 1966*, A. B. Littlewood, ed., Institute of Petroleum, London, 1967, p. 3.

75. S. H. Langer, C. Zahn, and M. H. Vial, *J. Org. Chem.*, **24**, 423 (1959).

76. H. B. Knight, L. P. Witnauer, J. E. Coleman, W. R. Noble, and D. Swern, *Anal. Chem.*, **24**, 1331 (1952).

77. H. G. McAdie, *Can. J. Chem.*, **40**, 2195 (1962).
78. P. F. McCrea and J. H. Purnell, *Nature (London)*, **219**, 261 (1968).
79. P. F. McCrea and J. H. Purnell, in *Gas Chromatography 1968*, C. L. A. Harbour, ed., Institute of Petroleum, London, 1969, p. 446.
80. P. F. McCrea and J. H. Purnell, *Anal. Chem.*, **41**, 1922, (1969).
81. J. C. Smith, *J. Chem. Soc.*, 974 (1939).
82. G. P. Hildebrand and C. N. Reilley, *Anal. Chem.*, **36**, 47 (1964).
83. R. P. W. Scott, *Anal. Chem.*, **35**, 481 (1963).
84. J. Peters and C. B. Euston, *Anal. Chem.*, **37**, 657 (1965).
85. A. Rose, D. J. Royer, and R. S. Henly, *Separation Sci.*, **2**, 229 (1967).
86. K. P. Hupe, U. Busch, and K. Winde, in *Advances in Gas Chromatography*, A. Zlatkis, ed., Preston Technical Abstracts , Evanston, Ill., 1969, p. 107.
87. A. B. Carel, R. E. Clement, and G. Perkins, in *Advances in Gas Chromatography*, A. Zlatkis, ed., Preston Technical Abstracts, Evanston, Ill., 1969, p. 113.
88. D. T. Sawyer and G. L. Hargrove, in *Progress in Gas Chromatography*, J. H. Purnell, ed., Interscience, London, 1968, p. 325.
89. P. F. McCrea and J. H. Purnell, Paper 61, presented at the Analytical Division, of the American Chemical Society, Meeting, New York, September 1969.
90. P. F. McCrea, Research Report No. 12731-T97, The Foxboro Company, Foxboro, Mass., November 1969.
91. D. T. Sawyer and J. H. Purnell, *Anal. Chem.*, **36**, 457 (1964).
92. D. A. Leathard and B. C. Shurlock, in *Progress in Gas Chromatography*, J. H. Purnell, ed., Interscience, New York, 1968, pp. 18, 23.
93. G. Schomburg, in *Advances in Chromatography*, vol. 6, J. C. Giddings and R. A. Keller, eds., Dekker, New York 1968, p. 211.
94. R. A. Keller and G. H. Stewart, *Anal. Chem.*, **36**, 1186 (1964).
95. A. B. Littlewood and F. W. Wilmott, *Anal. Chem.*, **38**, 1031 (1966).
96. N. C. Saha and G. D. Mitra, *J. Chromatogr. Sci.*, **8**, 84 (1970).
97. R. A. Hively and R. E. Hinton, *J. Gas Chromatogr.*, **6**, 203 (1968).
98. R. Annino and P. F. McCrea, 1970 Pittsburgh Conference on Analytical Chemistry and Applied Spectroscopy, paper 90, Cleveland, 1970.
99. R. Annino and P. F. McCrea, *Anal. Chem.*, **42**, 1486 (1970).
100. R. A. Bernheim and T. A. Shuhler, *J. Phys. Chem.*, **76**, 925 (1972).
101. J. S. Dave and J. M. Lohar, *Chem. Ind. (London)*, 597 (1959).
102. S. H. Langer and R. J. Sheehan, in *Progress in Gas Chromatography*, J. H. Purnell, ed., Interscience, New York, 1968, p. 289.

Production-Scale Gas Chromatography

JOHN R. CONDER, *Department of Chemical Engineering, University of Wales, University College of Swansea, Swansea, Wales*

I. INTRODUCTION

Production chromatography is a promising new industrial separation technique complementing distillation, extraction, and other established large-scale unit operations for separating gaseous and liquid mixtures. It is a scaled-up form of preparative chromatography.

After the birth of gas chromatography two decades ago, the rapid development of analytical chromatography was soon followed by applications on the preparative side. Early preparative chromatographs were essentially analytical ones into which large samples were injected. With this progeniture, it is not surprising that operating requirements for preparative chromatography are often confused with those for analysis. It is often considered, for example, that the sample size should not be increased beyond a point where resolution begins to suffer or where the width of the injected band exceeds about one-quarter of the emergent bandwidth. These concepts of "overloading" lead to unnecessary restrictions on preparative performance.

On the still larger, production scale, performance becomes all-important, and a better appreciation of the requirements for optimizing performance is essential if the process is to be economically viable. Increased understanding will also provide a basis for the systematic engineering design of the equipment. The object of a large-scale separation process is to separate the desired quantity of material in the minimum time at the lowest possible cost. With the stimulus of increasing commercial interest (1–8) in the large-scale separation possibilities of chromatography, several authors have made progress recently in establishing the design factors for production chromatography. These are described in Section II. Many of these considerations are relevant also to preparative operation, where they should lead to significant improvements in performance.

This chapter is mainly concerned with gas chromatography, but much of the theory is also applicable to liquid chromatography. Although gel permeation, adsorption and ion-exchange chromatography have been scaled up to columns 4 ft in diameter, no experimental work appears to have been reported on liquid-liquid chromatography beyond the preparative scale at the present time.

A.   Chromatographic Separation Techniques

A number of chromatographic and semichromatographic methods have been devised for separating gaseous and liquid mixtures. Many are quite ingenious. Generally, the techniques that have been exploited most successfully are the simpler ones, which do not require a complexity of moving parts and valving.

The simplest technique is the conventional one, whereby an "inert" mobile phase is moved continuously through a fixed bed or stationary phase and the feed mixture to be separated is injected as a discrete batch. This is the conventional way in which gas-liquid (GLC), gas-solid (GSC), and liquid-liquid (LLC) columns are operated for analysis. Objections are sometimes raised to using this approach for production-scale purposes. Since the components being separated occupy only a small part of the column at any moment in time, the packing is not being used as efficiently as in, say, distillation, where the whole of the packing is actively separating all the time. A further argument sometimes advanced is that process engineers prefer continuous techniques to batch procedures, and the method just described is not continuous.

Both these objections are overcome by using a repetitive injection technique under automatic control. Feed batches are injected consecutively and repeatedly, and the repeat interval is set so that the leading edge of one batch just catches up with the trailing edge of the previous one as they reach the column outlet. If, in addition, the feed bands are deliberately made wide, as optimization theory is shown to require, 90% or more of the packing is active in separation at all times. The sequencing of injection and direction of effluent to appropriate traps in the collection manifold is linked to the detector signal and put under automatic control. The operation is then effectively as continuous as, for example, a reciprocating compressor.

The conventional approach has been in use for preparative purposes almost as long as for analysis. It is the basis of most commercial preparative units, with or without repetitive automatic injection. For production chromatographs, repetitive automatic injection is obviously essential. Scale-up is accomplished in two ways—by increasing the batch size and by increasing the column diameter. The boundary between the preparative and production scales can only be somewhat arbitrarily drawn, since throughput depends very much on the type of separation to be carried out. As a rough guide, it may be noted that few commercial preparative GC instruments take columns larger than 2 in. (5 cm) in diameter; a 4-in.-diameter unit usually represents the small pilot-plant scale of operation. The production scale of operation may thus be said to begin at about 2 to 4 in. diameter. The first workers to use columns 4 in. or more in diameter were Huyten, van Beersum, and Rijnders (9), who studied the increase in plate height (HETP) with increasing diameter, and suggested mixing the contents of the gaseous cross section at intervals along the column to promote radial mixing and to keep down the HETP. Bayer and Hupe (10–15) studied the apparatus requirements and found that it was possible to keep plate heights below 4 mm even with columns up to 50 cm (20 in.) diameter. The Abcor Corporation (1–7) of Massachusetts has also concentrated attention on the use of mixing devices to keep down HETP. They manufacture columns with diameters ranging from 4 in. to 1 ft to customers' requirements and have projected the use of columns with diameters up to 20 ft; significantly, this covers the size range of industrial distillation towers. Recently, a French company (8) has produced columns of comparable size, with commercial developments in view.

At present, attention is turned toward more fundamental aspects of design. Economic operation requires the best possible use of the column capacity and carrier gas. The definition of the economically important variables and their relation to the chromatographic parameters has been considered by several authors and is reviewed in Section II.B.

B. Semichromatographic Techniques

The desire to make processing truly continuous has given rise to several separation methods that lie between chromatography and extraction in their principle of operation. They are here collectively described as "semichromatographic" in accordance with a recommendation (16) that the term "chromatography" be restricted to techniques in which one phase is truly stationary. To allow a continuous flow of feed while preserving the freedom to have more than two pure products, some form of multidimensional phase movement is required. Mosier (17), Taramasso (18–20), and others (21, 22) have described apparatus in which both the gas and sorbent phases are moved, but in directions at right angles to each other. No attempts appear to have been made to scale up such techniques, perhaps because of their inherent complexity.

The most extensively investigated semichromatographic techniques are associated with the names of Benedek and Szepesy (23–25), Barker (29–32, 34–36), and Tiley (37–40), among others (26–28, 33, 41–43). In these cases, movement of the phases occurs in one dimension only, with the gas moving countercurrently to the sorbent phase. The material to be separated is fed in continuously to the middle of the column and splits into two products. One component travels in the same direction as the gas; the other, more strongly sorbed, travels with the sorbent phase. For the sorbent material, Barker has used liquid phases coated on a support, as in conventional chromatography, whereas Tiley let the liquid flow over a stationary packing, as in extractive distillation.

Barker's machine was developed in three stages over a period of 10 years (44), and the principle has been further elaborated in a "Sequential Separator" designed by Unidev Ltd. of Crawley, Sussex. These machines are not yet known to have achieved significant commercial success. Earlier (32–36) drawbacks, such as bulk and failure to utilize more than half the total length of column for separation, have been largely overcome, but several objections remain. The machines are mechanically complex, requiring gas seals between moving parts, which does not augur well for scale-up much beyond 1-in. column diameter. Operational flexibility is limited, since the flow rates are dictated within close limits by the thermodynamics of the separation system (32). In common with other countercurrent semichromatographic techniques, there is a price to be paid for injecting the feed continuously. The number of pure product components obtainable is limited to two, and separation efficiencies over all components appear to be very poor in comparison with the conventional technique.

The various semichromatographic approaches may yet find a place in the broad spectrum of separation processes; in the author's view, however, the conventional technique with repetitive automatic injection has the more promising future. Barker has comprehensively reviewed (44) the semichromatographic techniques, which have not yet been exploited beyond the preparative scale and are not discussed further here.

C. The Role of Production Chromatography

Since production gas chromatography provides a new method of separating liquid and gaseous mixtures, we need to consider how it compares with other techniques, such as distillation (in its many forms), liquid-liquid extraction, and fractional crystallization. These are the chief established methods for separating liquid-liquid mixtures, and their technology is highly developed, particularly in the case of distillation.

The market for separating gas mixtures is much smaller than for liquids, and chromatography has not yet been exploited in this area, except on the preparative scale.

The various techniques of separation are compared in Section IV, but an indication of the advantages chromatography offers over distillation is relevant at this stage. Briefly, they are as follows:

1. Components with relative volatilities close to 1.0 can be separated since, by judicious selection of the stationary phase, much more favorable relative volatilities can usually be obtained;

2. Materials having low vapor pressures or subject to heat sensitivity can be processed more easily, without resort to expensive techniques such as vacuum, azeotropic, or extraction distillation;

3. Chromatography is particularly useful where products of high purity are required, since the product specification does not critically affect the production rate.

These virtues have been responsible for the widespread deployment of chromatography on the preparative scale. The technique already provides an indispensable complement to distillation for general-duty laboratory separation; one oil laboratory (46, 47), for example, has used a ratio of about one chromatograph to every five distillation columns for a number of years. On the production scale, the types of products that might be separated advantageously by chromatography cover a wide field. They include pharmaceutical compounds, essential oils, terpenoids, steroids, alkaloids, metal chelates, isotopes, and light hydrocarbons used as solvents and synthetic intermediates.

II. THEORY

The "conventional technique," as described in Section I, involves repetitive injection of discrete batches of feed. If the sequencing is controlled automatically, the process is effectively continuous. The amount of feed injected in each batch and the interval of time between injections determine the rate of processing or throughput.

The feed components separate progressively as they pass along the column, but this is partly offset by spreading of the component bands, which occurs concomitantly. Increasing the column length has a beneficial net effect on resolution of the components because the final separation between band centers is proportional to the column length L, whereas band spreading increases only with $\sqrt{L}$. When chromatography is used for analysis, the object is to achieve a high resolution. Hence long, narrow columns with low theoretical plate heights are employed. For preparative and production purposes, however, it is not resolution that matters; rather, we must pay attention to throughput, or a parameter closely related to throughput, at a specified purity level.

A. Data Reporting

Meaningful comparisons between separation techniques can be made only if all the relevant parameters are taken into account. In a paper distinguished by brevity and emphasis, Perry (48) documented well the inadequate data reporting that frequently appears in the literature. Results are often reported as size of sample separated with a column of specified diameter. However, unless we know the relative volatilities of the components separated, the time required, and the purities obtained, it is not possible to judge whether the performance is good or bad. Whereas 100 ml/hr on a 4-in.-diameter column might be a good processing rate for a mixture of essential oils of relative volatility 1.1 at 99% product purity, it might be a very poor result for a mixture of benzene and toluene at 90% purity. This example also illustrates the importance of the time variable; increasing the column length may benefit the sample size or product purity, but it also increases processing time.

The author suggests that reports of preparative and production separations should always include the following data as the minimum requirement:

1. Separation problem specification
   (a) Relative volatilities of components in the mixture to be separated.
   (b) Product percentage purities.
2. Performance obtained
   (c) Amounts of separated components produced in unit time.
   (d) Column length and cross section.
   (e) Carrier gas flow rate.

To permit more detailed comparison of results, particularly as between different chromatographic techniques, the following parameters should be quoted in addition:

(f) Relative retention of components on the stationary phase used.
(g) Retention time or volume.
(h) Column temperature.
(i) Fractional recovery (if fraction-cutting is used).
(j) Trapping efficiency.

B. Performance Criteria

The first step in considering the theory of production chromatography is to decide how performance should be measured. It is then possible to consider how to choose the design and operating parameters to maximize performance. The question was discussed at the Fourth Gas Chromatography Symposium in 1962. Sakodynsky (49) suggested using the "specific capacity," defined as the number of grams that can be separated per unit of column cross section. It was pointed out that this definition is meaningful only if the separation problem is also specified, via the parameters listed in Section II.A, for example. This is true of other definitions of performance to be described.

The specific capacity is unsatisfactory as a criterion of performance for production applications because it takes no account of the time variable. Sawyer and Purnell (50) made use of the throughput, the amount of material separated in unit time. This parameter, together with others related to it, has since been adopted frequently as a measure of chromatograph performance (48, 51–58).

Perry (48) has proposed a "yield index," defined as the product of the throuput and number of theoretical plates measured on the preparative chromatogram. This definition represents an attempt to take the throughput and product purity simultaneously into account in one parameter. For production as opposed to preparative purposes, it suffers from two disadvantages. First, it does not measure the process economics in any way. Second, if feed-band optimization principles are followed, the number of plates is more meaningful when measured on an elution peak at infinite dilution.

Hupe (57) and Timmins, Mir, and Ryan (55) have pointed out that the basic criterion for economic evaluation is the cost per unit mass of material processed. This parameter is related to the throughput Q by the equation

$$\text{cost/unit mass} = Q/G \qquad (1)$$

where G is the processing cost per unit time. Hupe has further expressed G as the sum of several costs, such as depreciated capital, maintenance, and carrier gas. A similar approach has been taken by Timmins, Mir, and Ryan

(55), who defined two parameters; the column utilization efficiency $Q/$(column volume) and the carrier utilization efficiency $Q/$(carrier flow rate). The authors considered that the economic optimum is bracketed by the maxima in the utilization efficiencies. Maximizing the column utilization efficiency leads to the smallest possible column volume: maximizing the carrier utilization efficiency leads to a minimum cost for product and carrier-handling systems but requires a larger column. Simultaneous use of these two parameters thus gives only a rough guide to the optimum conditions, which are better investigated by way of Q/G.

Another parameter proposed (58) for optimization is Q/H, throughput per unit HETP. This is related to the column utilization efficiency. Neither parameter provides an overall performance criterion.

In conclusion, the only parameter that expresses the overall economic requirements is Q/G, throughput divided by total cost per unit time. Assuming fraction-cutting can be employed, the best criterion of performance for preparative work is probably the throughput itself. Both criteria are meaningful only if the separation problem specification is also cited (Section II. A).

C. Throughput

The throughput of feed is defined by

$$Q_f = \frac{m_f}{t_R} \tag{2}$$

where m_f is the mass of feed injected per batch and t_R is its retention time in the column. Units conversion factors for throughput are listed in Table I. If the component bands overlap at the column outlet and fraction-cutting of the overlapped region is employed to give pure components, the throughput of material recovered is

$$Q_r = rQ_f = \frac{rm_f}{t_R} \tag{3}$$

where r is the fractional recovery, the ratio of pure material recovered after cutting, to feed material.

Although fraction cutting gives $r < 1$, it is always better to overlap the component bands than to avoid the need for cutting by increasing column length and resolution (59, 60). The reason is that t_R increases with length and reduces Q_f faster than r rises (60).

TABLE I
Units Conversion Factors for Throughput

1 lb/hr = 8.40×10^3 lb/350-day year
        = 0.454 kg/hr
        = 3.81×10^3 kg/350-day year
        = 3.75 UK ton/350-day year
1 lb/day = 0.156 UK ton/350-day year

10^5 lb/350-day year = 11.9 lb/hr
                     = 5.42 kg/hr
                     = 44.6 UK tons/350-day year

1 kg/hr = 2.20 lb/hr
        = 1.85×10^4 lb/350-day year
        = 8.40×10^3 kg/350-day year
        = 8.25 UK tons/350-day year

1 kg/day = 0.343 UK ton/350-day year

10^5 kg/350-day year = 26.2 lb/hr
                     = 11.9 kg/hr
                     = 98.4 UK tons/350-day year

1 UK ton/350-day year = 0.267 lb/hr
                      = 0.121 kg/hr

For a liquid of specific gravity 0.8
1 l/hr   = 1.76 lb/hr
        = 1.48×10^4 lb/350-day year
        = 0.8 kg/hr
        = 6.71×10^3 kg/350-day year
        = 6.40 UK tons/350-day year

Equations 2 and 3 apply when each injection is delayed until the previous batch has emerged from the column, as is often done when injections are made by hand. Greater throughput can be obtained if injections are made more frequently. This is best done in such a way that, when the component bands reach the outlet, the fastest-traveling component from one batch has just caught up with the slowest-traveling one from the previous batch. This is illustrated in Figure 1 for a binary mixture. The optimum cycle is such that component 2 overlaps equally with component 1 from the same batch and with component 1 from the next batch. If t_{R1} and t_{R2} are the retention times of the two components,

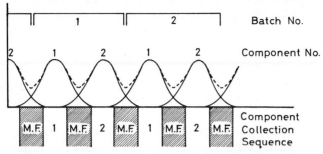

Fig. 1. Chromatogram and collection sequence for repetitive injection, binary feed mixture. M.F. = mixed fraction, which is collected separately and recycled to the feedstock.

the throughput is

$$q_r = r q_f = \frac{r m_f}{2(t_{R2} - t_{R1})} \tag{4}$$

for a binary mixture.

Equations 2 and 4 have been used to derive expressions for throughput in terms of design and operating parameters for the column. An early treatment (50) was subject to the feed bandwidth restriction for Gaussian-shaped bands. This unnecessary restriction (Section II. D) has been lifted by Conder, who has obtained (53, 54) the following expression for throughput:

$$q_r = \frac{r n_f}{n} \left\{ \frac{\alpha}{6(\alpha + 1)} \quad \frac{M_f p_f}{RT} \quad \frac{\pi d^2 \epsilon \bar{u}}{4} \right\} \tag{5}$$

where α is the relative volatility of the pair of components on which the separation is based; M_f is the molecular weight and p_f the partial pressure of the vaporized feed; T is the column temperature; d is the column diameter; ϵ is the porosity of the packing; and $\bar{u}$ is the compressibility-corrected interstitial velocity of the mobile phase. In addition, n_f and n are reduced plate numbers related, respectively, to the number of plates N_f occupied by the feed band after entering the column and to the total number of plates N in the column, by the expressions

$$n_f = \frac{N_f}{6/a} \tag{6}$$

$$n = \frac{N}{(6/a)^2} \tag{7}$$

In these equations, a reflects the thermodynamic nature of the components being separated through the relation

$$a = 2\left(\frac{\alpha-1}{\alpha+1}\right)\left(\frac{k}{1+k}\right) \tag{8}$$

where k is the mole distribution ratio, for the second emerging component, between stationary and mobile phases.

The virtue of expressing throughput in terms of reduced plate numbers is that these are independent of the thermodynamics of the system. Then we can invoke a universal relationship (see, e.g., Figure 2) between n_f and the reduced number of plates required (n) that does not depend on the particular components being separated. The effect of the nature of the components appears only in the factors $\alpha/6(\alpha+1)$ and $M_f p_f / RT$ in the throughput (Eq. 5).

In order to use Eq. 5 it is first necessary to consider the feed-band profile.

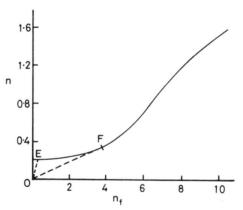

Fig. 2. Plot of n against n_f for a fractional recovery $r = 0.6$, fractional impurity $\eta = 0.01$, and bands of equal mass (60).

D. Feed-Band Width

The key to optimizing performance is the study of feed band profile. The feed band has two dimensions: width (representing the number of plates into which the discrete feed sample is introduced) and height (representing the concentration of sample in the mobile and stationary phases). Preparatively, the two dimensions are often not distinguished, and only the overall sample size is considered (95). No provision is made to control the two variables separately. If such control is attempted, however, a considerably better performance can often be achieved. For production applications, it is essential to control the width and concentration separately and to optimize the two variables. The feed should then be introduced with a rectangular band profile to obtain the maximum throughput from the chosen values of width and concentration.

Consider first low concentrations, where the bands remain symmetrical during passage through the column. Glueckauf (61a) and Said (62, 63) have treated the case of injection bands that are narrow, causing the eluted bands at the outlet to have a Gaussian shape. A small error (61) concerning component bands of unequal height has been corrected (59, 62). These authors obtained plots for evaluating the fractional recovery r on fraction-cutting overlapped bands, at a fractional impurity η, for given values of relative volatility and theoretical plate number. The treatment has been extended by Haarhoff, van Berge, and Pretorius (59), by Glueckauf (61b), and by Conder and Purnell (53) to allow for bands of any width. Conder (53, 60) has shown that if both the bandwidth and the number of plates are allowed to vary freely, there are optimum values of bandwidth, for each η, where the throughput, q_r, and q_r/G go through maxima. This is illustrated in Figure 2, where n is plotted against n_f for incompletely resolved component bands, giving a pure-component fractional recovery r of 0.6. Here n is the reduced number of plates required to separate the components to this extent when the feed band occupies n_f plates. From Eq. 5, the throughput is proportional to rn_f/n, and the maximum throughput is proportional to the reciprocal of the slope of the broken line OF in the figure. Since reduced plate numbers are used, the plot is applicable to any pair of solutes with any stationary and mobile phases.

At any given higher concentrations, where the bands at the outlet are asymmetrical, optimum values of n_f and n should again exist which give maximum values of q_r.

Bands that give maximum throughput are much wider than those often used in preparative work. It is often suggested or implied (64) that feed-band width should not exceed about one-quarter of the outlet bandwidth (elution mode of operation). This criterion merely defines the limit for the bands to have a Gaussian shape, as derived by Glueckauf (61) and van Deemter and co-workers (66). It corresponds to point E in Figure 2 and has no relevance to preparative or production work. The increase in throughput obtained by operating with bands wider than at point E [overload elution or elutofrontal mode (52, 53)], is only reduced by an increase in the column length requirement (n). Hence q_r rises beyond point F, the column length required increases faster than the amount of feed injected (n_f); thus q_r begins to fall. Feed-band width and column length are

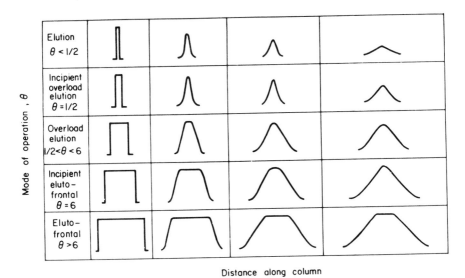

Fig. 3. Change in band profile with distance along column for each mode of operation (53). The mode of operation parameter $\theta = N_f/\sqrt{N} = n_f/\sqrt{n}$ is determined by the feed-band width at a given column length. The bandwidth and height show a large change during passage in the elution mode, none in the elutofrontal mode.

By courtesy of Robert Maxwell and Company.

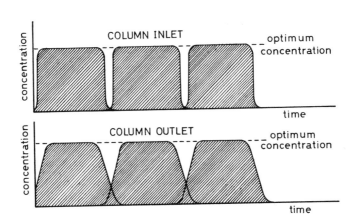

Fig. 4. The elution mode of operation (narrow feed bands). Concentration versus time curves at the two ends of the column for bands injected in succession, to overlap at the outlet.

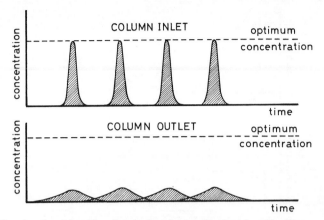

Fig. 5. The elutofrontal mode of operation (very wide feed bands). Concentration versus time curves at the two ends of the column for bands injected in succession, to overlap at the outlet.

thus closely related design variables and should be chosen to give maximum q_r or q_r/G.

Two further reasons for preferring wide bands to narrow, elution bands may be mentioned. The first, which contributes to the form of the plot in Figure 2, is illustrated in Figures 3, 4, and 5. With narrow, elution bands, the feed-band width is much less than the width at the column outlet. Hence, as shown in Figure 4, the major proportion of the packing in the first half of the column lies between successive bands and is not being used for separation at any moment in time, whereas the packing in the other half is not taking its full concentration capacity of solute. With wider feed bands (Figure 5), on the other hand, the band has a more nearly constant width and concentration from inlet to outlet, and all parts of the column are more fully utilized.

The second advantage of using wide bands concerns the additional band broadening associated with band skewing at high solute concentrations. At low concentrations, close to infinite dilution of solute in the stationary phase, the bands at the column outlet are approximately symmetrical. (This is often not true for gas-solid adsorption.) As the concentration is raised, nonlinearity of the partition isotherm, heat of solution (67), and other effects, come into play, as described elsewhere (68). The result appears in Figure 6: the retention volume increases, and the outlet band is skewed and has a greater width. The effect is very marked with elution (narrow) bands (Figure 6a), and it has led many authors to conclude that the concentration must be kept low; liquid-phase mole fractions of 0.01 to 0.1 have been mentioned (50). With wide bands (Figure 6b), the ratio b'/b

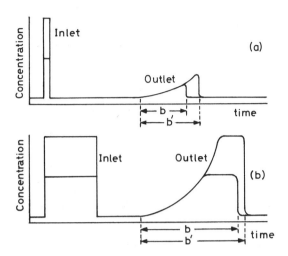

Fig. 6. The effect of doubling the solute concentration on bandwidth at the column outlet when the bands are asymmetrical due to high concentration (54). (*a*) Narrow feed band (elution mode of operation); (*b*) very wide feed band (elutofrontal mode). The ratio b'/b is smaller in (*b*) than in (*a*).
By courtesy of Pergamon Press, Oxford.

is much smaller, so that the effect of concentration-induced band broadening, in requiring increased column length, is not so great.

E.   Feed-Band Concentration

Feed-band width and concentration are closely linked variables. Increase in either tends to increase component bandwidth at the outlet and so requires a longer column to maintain separation. This leads, as already described, to the existence of an optimum bandwidth. Concentration does not necessarily pass through an optimum value. Throughput increases continuously with feed concentration if the relative volatility α of the two components is constant (i.e., independent of concentration). Sometimes, however, α falls with increasing concentration; the required column length increases, and a (high) optimum concentration exists. The optimum concentration for maximum q/G (Section II. B) is lower than that for maximum q. The effect of concentration on band profile has been described by Conder (54).

F.   Column Temperature

The effect of column temperature is related to that of solute con-

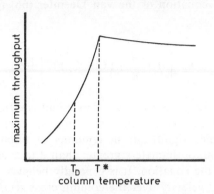

Fig. 7. Variation of maximum possible throughput with column temperature (54). For heat-sensitive materials, the curve does not extend above the degradation temperature T_D.

centration in the stationary and mobile phases (54). At high temperatures, the attainable concentration is limited by the requirement that the solute vapor pressure cannot exceed the operating gas pressure in the column. At low temperatures a limit may be set by a tendency for the stationary phase, greatly swollen by condensed solute, to be stripped from the packing. The two limits set a maximum possible throughput which varies with temperature, as in Figure 7, where T^* is the temperature at which the two concentration limits (vapor pressure and stationary phase stripping) coincide and T_D is the degradation temperature of a thermally unstable solute; T^* is usually close to the solute boiling point. On this basis, Conder (54) developed criteria for selecting the best operating temperature for different solute types.

G. Theoretical Plate Height

A great deal of work has been published on the subject of plate height, particularly with reference to preparative columns of diameter between 0.5 and 10 cm. We shall survey this work first, although, as we see later, there is some doubt about how far conclusions extend to column diameters greater than 10 cm. Several good reviews of preparative efficiency are available (see, e.g., Refs. 69–72).

1. Plate Height Theory

The height equivalent to a theoretical plate (HETP, h) is commonly represented by an equation of the van Deemter (66) form, with an extra term added:

$$h = \left[A + \frac{B}{\bar{u}} + (C_l + C_g) \bar{u} \right] + h_p \qquad (9)$$

where A, B, C_l, and C_g are velocity-independent coefficients, and $\bar{u}$ is the average interstitial gas velocity in the column; h_p is an extra contribution to h which becomes significant in columns of about 0.5 cm or more diameter. With column diameters greater than 2 to 5 cm, h_p is usually the dominant term in the equation. It is generally believed that h_p arises from unevenness in flow velocity over the column cross section.

Random unevenness, different at different points along the column, was postulated by Giddings (69) and investigated by Littlewood (73, 74). This is considered (73, 75) to be the main effect giving rise to the C_g term in the van Deemter equation.

A nonrandom variation of flow over the cross section was observed in early work by Huyten, van Beersum, and Rijnders (9), which has provided the experimental basis of much of the subsequent discussion of preparative efficiencies. These authors divided the cross section of a 7.5-cm-diameter column into annular sections and used a wet analysis technique to study the flow profile. They found that the velocity was least in the center and greatest near the wall. The same bowl-shaped profile was observed by Frisone (76). Giddings and Fuller correlated the velocity profile with an observed increase in particle size from center to wall (77); they suggested that the increase was due to size segregation on packing the column: larger particles roll further than small ones down the outside of the packing cone.

The nonuniform velocity profile was considered by Giddings (69) to carry the solute band further near the outside of the column than at the center. This is illustrated in Figure 8. The resultant band spreading is relaxed by lateral diffusion, described by Littlewood (74) and Sie and Rijnders (75) as being composed of two processes:

1. Molecular diffusion;
2. Anastomosis or "convective" diffusion, arising from the repeated division and combination of fluid streams in the interstices of the packing.

The effective diffusion coefficients are, respectively, γD_g and $\alpha d_p \bar{u}$, where γ is the tortuosity of the packing, D_g is the gas diffusivity, α is a constant for the packing geometry, and d_p is the particle diameter. The band spreading caused by the velocity profile interacts with the two lateral

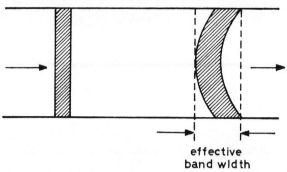

effective
band width

Fig. 8. The effect of a nonuniform velocity profile in increasing bandwidth.

diffusion mechanisms to give a plate height term (74, 75) as follows:

$$h_p = \frac{0.5\kappa d^2 \overline{u}}{\gamma D_g + \alpha d_p \overline{u}} \tag{10}$$

In Eq. 10, d is the column diameter and κ is a dimensionless flow profile factor proportional to $(\Delta u/\overline{u})^2$, where Δu is the amplitude of the cross-sectional velocity variation; the value of κ depends on the shape of the velocity profile (69, 73, 74). According to this interpretation of h_p, the plate height in large-diameter columns depends on the diameter, flow velocity, and κ.

The value of κ depends on the homogeneity of the packing: the more irregular the packing, the larger κ. On correlating published data, Sie and Rijnders (75) found that κ decreases with increasing ratio of column to particle diameter, as represented in Figure 9. For $d/d_p > 40$, as is usual in production-scale columns, the figure shows no definite trend in κ but does indicate that κ will depend very strongly on packing uniformity, perhaps by as much as two orders of magnitude. This suggests, first, that theoretical prediction of the size of h_p is very difficult, if not impossible, and second, that packing technique is very important.

Frisone (76) found that filter paper "washers," soaked in stationary liquid phase and incorporated in the packing, reduced the HETP. It is not clear whether this resulted from the retarding effect of added stationary phase on band velocity near the column wall, as the author supposed, or from improved lateral mixing.

a. **Contrary Views.**   The body of experimental work to which Eq. 10 relates is largely based on columns of less than 10 cm diameter. There is little experimental evidence to support the idea that the explanation so far described, of loss of efficiency in large-diameter columns, applies to

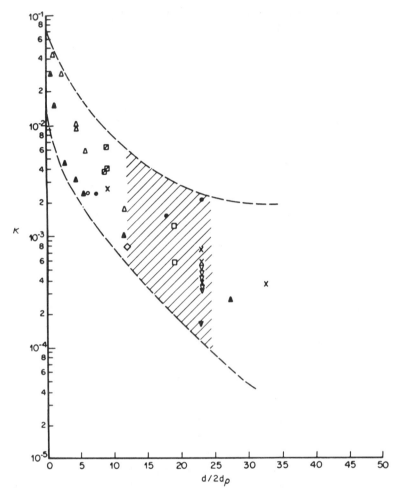

Fig. 9. Envelope plot of flow profile factor κ against ratio of column radius to particle diameter. Points plotted by Sie and Rijnders (75) from experimental data in the literature. Hatched area, region of usual analytical-scale columns, or preparative columns with coarser particles; d_p = particle diameter.
By courtesy of Elsevier Publishing Company.

production-scale columns. Indeed, Dixmier, Roz, and Guiochon (51) have doubted whether the explanation is fully correct on any scale of operation. They pointed out that when Huyten, van Beersum, and Rijnders (9) compared different packing techniques, increased radial variation in velocity was accompanied by improved efficiency—the opposite of what would have been expected.

Doubts about the widely held interpretation of h_p are supported by results recently obtained by Hupe, Busch, and Winde (78). When columns packed by fluidization and by filling and bouncing were studied by providing thermocouples at different radii, solute bands traveled faster at the center than at points closer to the wall. This conflicts with the finding of Huyten et al. (9) that the carrier gas traveled faster nearer the wall if it is assumed, as is usually done, that the ratio of carrier velocity to band migration velocity is constant over the column cross section. Hupe and co-workers failed in their initial attempt (78) to explain the results by postulating that the density of the packing, and therefore the concentration of stationary phase, increased toward the periphery. Thus, when the packing density near the periphery was deliberately increased with a tubular plunger, the solute band moved faster through the compacted material. This gave a more uniform cross-sectional band profile, and a higher number of theoretical plates was obtained overall. To reconcile the conflicting evidence, the authors suggested that smaller spaces between particles, permitting faster solute exchange, would give faster rates of travel of solute bands. This explanation is difficult to accept because the speed of the band is determined by the equilibrium partition coefficient in normal circumstances, not by mass-transfer exchange rates.

b.   Temperature Effects.   It is possible that some of the band profile phenomena observed are due to transient heat of solution effects in the bed. Since, as a rule, absorption is exothermic and desorption endothermic, a "temperature wave" travels with the solute band. The height of the wave may be as little as a degree or two, or 50°C, depending on conditions (78–80). In 1-in.-diameter columns, the fall in temperature on desorption extends to a large overshoot (79, 80), indicating that the column is far from adiabatic and that the temperature change is partly relaxed by heat transfer through the column walls. With a 4-in.-diameter column, the overshoot appears much smaller, being less than 1°C at 0.1 in. from the wall and negligible in the body of the packing (78). Thus 4-in. columns appear to be nearly adiabatic in operation, implying that there is a significant radial temperature gradient associated with the migrating band due to the non-uniform migration velocity profile.

If the column were nonadiabatic, the temperature wave traveling with the band would be relaxed by heat transfer out of the column. The resulting radial temperature gradient would produce a higher band velocity at the column axis than at the periphery, increasing the HETP. The net result, however, would be to reduce the HETP if the effect were opposed by another influence that tended to raise the band velocity at the periphery. Such an opposing influence is provided by the increase in porosity from axis to periphery, as previously described. This compensation between temperature and velocity gradients provides a possible explanation both for the paradox pointed out by Dixmier (51) in Huyten's (9) results and for the conflict between the observations of Huyten and Hupe (78). On the other hand, the radial temperature gradient in columns of 3 to 4 in. diameter, as used by Huyten and Hupe, appears to be very small.

c. **Wall Effect.** Schwartz and Smith (81) have studied the wall effect and have demonstrated that when wide-diameter columns are packed with uniform cylindrical particles, the porosity has a maximum one particle diameter from the wall. Golay (82) drew attention to the effect of the wall on HETP, but Giddings (77) considered that its influence was much less than that of the radial size segregation of particles occurring on packing the column. Assuming that the wall effect dominates the flow profile, Hupe (83) obtained the equation

$$h_p = \frac{2.83 d^{0.58}}{\bar{u}^{1.886}} \tag{11}$$

where the units are: h_p, cm; d, cm; $\bar{u}$, cm/sec. This equation predicts a dependence of h_p on $\bar{u}$ very different from that of Eq. 10. The available experimental evidence (7, 65) favors Eq. 10 in that h_p increases with increasing $\bar{u}$. A second comparison between the equations, on the basis of the variation of h_p with column diameter, cannot be made satisfactorily because of the large masking effect of nonuniformity of packing, already pointed out. Both equations predict, and experiment confirms, a general increase in h_p with diameter.

d. **Summary.** Despite the great amount of attention devoted to wide-column plate height over the last 15 years, its origin remains far from clear. This uncertainty is partly due to lack of reproducibility of conditions between one laboratory and another and partly to the large number of factors, not all appreciated, which influence plate height and band profile measurements. Most important among these are injector, detector, and other extracolumnar contributions to plate height. The theoretical bases of Eqs. 10 and 11 are insecure. Regarded purely as empirical relations, however, Eq. 10 appears to be more satisfactory than Eq. 11.

2. Production-Scale Plate Height

Bayer, Hupe, and Mack (14) observed that the rate of increase of h with d fell off as the diameter increased, as shown in Table II. Spencer and Kucharski (84) observed a similar trend, with an actual fall in plate height above 2.5 in. diameter, as in Figure 10. The fall in HETP is not predicted

TABLE II
Plate Height Variation with Column
Diameter—Results Obtained (14) with
Particles of Uniform Size Using a Standardized
Packing Procedure

Column diameter, cm	Plate height h, mm
1	2.0
4	2.8
10	3.1
50	4.0

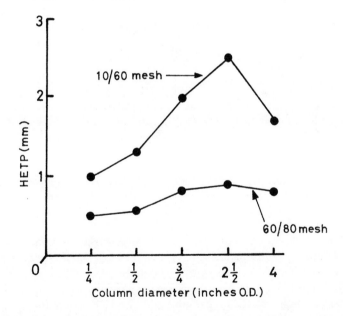

Fig. 10.   Plot of the height equivalent to theoretical plate h as a function of Chromosorb P mesh size and column diameter (84). Stationary phase was silicone rubber UCCW98. Method of packing was vertical vibrating or gentle bouncing, adding a scoopful at a time. By courtesy of Hewlett-Packard Company.

by either Eq. 10 or Eq. 11. Equation 10 is applicable only when the column diameter is small enough or the length great enough for lateral diffusion to relax the radial nonequilibrium caused by the velocity profile. In sufficiently wide columns, radial equilibrium is not achieved at all. In this case, Giddings has shown (69, 85) that h_p increases with length and is independent of diameter. It has been suggested (51, 69) that, in principle, wide columns of diameter exceeding, say, 4 in. should be more efficient than preparative columns. However, the limited experimental evidence quoted in Table II and Figure 10 is not conclusive on this point. Nevertheless, it seems that with a very careful packing technique, the efficiency of a production column can be kept down to little more than that of a preparative column.

3. Packing Uniformity

It is clear from the preceding discussion that packing uniformity is perhaps the most important single factor influencing plate height in preparative and production columns. Several workers have investigated the technique of packing the column (14, 65, 78, 83, 86, 87), but opinions differ regarding the best technique. Higgins and Smith (65) prefer "mountain packing," whereas Littlewood (88) opts for "bulk packing," and Guillemin (87) favors fluidization. One way to achieve a uniform packing is to use an extremely narrow particle size range (14, 84, 86), as indicated in Figure 10, but the advantage is somewhat offset for production purposes by the higher cost of narrow-mesh packing. Another way is to use a conical-shaped plunger to control the variation of packing density over the cross section (78).

4. Mixing Devices and Column Design

Several methods, apart from controlling packing technique, have been used to maintain column efficiency with wide columns.

The most common technique is to incorporate mixing devices at intervals along the column. These may be either insertions, such as washers and baffles (7, 76, 82), or thin mixing chambers (46, 47, 89) provided by constructing the column in a number of short lengths, each supplied with a porous disc at each end of the packing and connected by narrow diameter tubing. Giddings (69) has pointed out that the volume of the connecting tubing should be less than 0.01 times the volume of the column sections, and this is very easily achieved in practice. The law of diminishing returns applies to the number of mixing devices used in the column: a small

number can be of great help in overcoming the effects of flow maldistribu-
tion (90). With n mixing devices, the plate height h_n is given (90) by

$$\frac{h_n - h_{vD}}{h_p} = \frac{1}{n+1} \tag{12}$$

where, in a column without mixing devices, h_{vD} and h_p are, respectively, the
intrinsic plate height of the packing, as given by the van Deemter equation
(Eq. 9 without the h_p term), and the additional contribution from the
velocity profile.

A different approach to maintaining efficiency is to use an array of
small-diameter columns in parallel. This requires close matching of the
packing densities and weights of stationary phase (45). Although successful
on the preparative scale (91, 92), the method is unlikely to be feasible for
the production scale because of the need to match accurately a very large
number of columns. The objective of minimizing the radial velocity
gradient might also be attained with a single annular column (69), provided a uniform angular distribution of packing could be achieved. Finned
columns have been evaluated by Wright (93) and Reiser (94). The plate
height of a 3-in.-diameter finned column was 3.0 mm, compared with 1.3
mm for a 1.75-in. unfinned column, suggesting that finned columns are less
effective than, for example, matched parallel columns as a way of reducing
HETP.

5. Other Influences on HETP

Effects other than velocity profile contribute to the plate height in
production chromatography. The effects of "overloading" (i.e., increasing
the size of batch injected) are sometimes considered under this heading. A
more productive approach is to treat the width and concentration of the
feed band separately, as described in Sections II.D and II.E. The width
then becomes an influence on the number of plates in the column, rather
than on HETP. The effect of feed concentration was described in Section
II.E.

Changes in cross section at the ends of the column or column sections
cause nonuniformities in velocity, or, in the extreme case, dead gas pockets
(69). The effects are usually minimized either by using inlet and exit cones
(14) or by incorporating porous discs to enclose a thin gas distribution
chamber at each end of the column (46, 47, 89). This is further discussed in
Section III.C.3.

6. Conclusions

Of the approaches to reducing HETP in wide columns so far studied, those most suitable for production-scale application appear to be (a) careful packing technique, (b) use of narrow mesh-range packing, (c) modification of packing density with a shaped plunger, and (d) provision for periodic gas mixing. A combination of these methods, implying both reduction of h_p and increase in n in Eq. 12, may in practice give better results than either method alone. In support of this view, we may note that, at large diameter-to-length ratios, h_p depends on both packing uniformity and the length of column between mixing points (69, 85). The consensus of opinion is that plate heights can be obtained (at infinite dilution of solute) of the order of several millimeters for columns of 10 to 50 cm diameter. This compares with values of 1 cm or more when no special precautions are taken.

H. Design and Operating Conditions

For optimum economic operation, a production chromatograph must be designed to maximize the ratio of throughput to total cost per unit time; that is, q_r/G (Section II.B). Equation 5 leads to the following expression for q_r/G:

$$\frac{q_r}{G} = \frac{1}{G} \frac{rn_f}{n} \left[\frac{\alpha}{6(\alpha+1)} \frac{M_f p_f}{RT} \frac{\pi d^2 \epsilon \bar{u}}{4} \right] \tag{13}$$

where G is the sum of the operating costs and the capital cost amortized over a period of time and is an increasing function of column diameter, column length, and carrier gas flow rate. The form of the function has not yet been determined, although sufficient experimental and theoretical information probably exists in the literature to make an approximate formulation possible. Hupe (57) and Ryan and Dienes (4) have made predictions of the effect of column diameter on G.

In the absence of knowledge of G, complete optimization procedures remain to be developed, although partial methods have recently been described (53–55, 60). Another recent procedure (57) is restricted to the elution mode of operation (i.e., to narrow bands) and also assumes that the cycling time is equal to the retention time. For these reasons, it is not discussed further. The other procedures serve to identify the important factors for design calculations. These are as follows (expressed in terms of Eq. 13).

1. Feed-Band Width and Column Length

The feed-band width is proportional to n_f, whereas column length is proportional to n:

$$L = \frac{9nh}{\left[\dfrac{\alpha - 1}{\alpha + 1} \dfrac{k}{1+k} \right]^2} \tag{14}$$

In Eq. 14, q_r depends on the ratio n_f/n, whereas G depends only on n through L; the n_f and n are closely related and should be chosen jointly at the design stage to maximize q_r/G. This might be done using plots such as that in Figure 2 in conjunction with Eq. 13 and an expression for G derived from a costing analysis. An analogous procedure, using the carrier utilization efficiency (Section II.B), has been outlined by Timmins, Mir, and Ryan (55).

2. Fractional Recovery

The column need not be made so long that the component bands are completely resolved at the outlet. Higher throughputs are achieved by employing fraction-cutting (Section II.C). This gives a shorter column and allows more frequent injection of batches.

3. HETP

The increase in height equivalent to a theoretical plate (HETP) with increasing column diameter has often been considered to be the main bar to performance with large diameter columns. This has tended to divert attention from the relation between performance and injection band profile. In the author's view, the performance of many units would be improved even more by attention to feed band profile (Sections II.D and II.E) than by efforts to reduce HETP. Thus the reduction in HETP obtained by taking steps such as those mentioned in Section II.G.6 is often a factor of 2 or 3. This will lead to a proportional reduction in column length, but the resulting improvement in q_r/G will not be as great because G is not strongly dependent on length and q_r is independent of HETP. The improvement to be expected by designing for optimum bandwidth, concentration, and column temperature may be up to tenfold, depending on how the system has previously been operated (e.g., to the left of point E in Figure 2).

Although the decrease in efficiency with increase in column diameter is well known, it is not always realized that increase in plate height can be partly beneficial. Thus a twofold increase in HETP requires a doubling of column length but also permits a compensating doubling of the length of column initially occupied by the feed band. The net effect is that throughput is unaffected by HETP, as consideration of Eq. 5 shows. In preparative chromatography, where throughput is the best criterion of performance, HETP is not necessarily of any significance at all. This is so if the column length is treated as a design variable and not fixed. Only if the oven is small or the relative volatility very close to unity does column length, and hence HETP, become important. Dixmier, Roz, and Guiochon (51) have also pointed out that the performance of a preparative column is limited more by the loadability of the column than by loss of efficiency with increasing diameter.

It is concluded that: (a) in preparative chromatography, HETP has no influence on performance except under imposed limiting conditions; and (b) in production applications, there is some advantage in taking steps to reduce the HETP, but less than is often supposed.

4. Relative Volatility

Equation 13 shows that the value of α has only a small effect on q_r, provided the column length is not fixed. If a separation were made twice as difficult by reducing α from 1.2 to 1.1, q_r would fall by only 5%. This applies to repetitive injection with the cycle time equal to twice the difference in component retention times. If the cycle time were set to equal the retention time, α would have a greater effect on throughput (see Eq. 38, Ref. 53). Although q_r is little affected by α, Eq. 14 shows that α strongly influences L. Since column length enters into the calculation of G, the stationary phase should be chosen to give the highest possible relative volatility for the pair of solutes that are the most difficult to separate.

5. Feed Vapor Pressure

Since an increase in concentration of solute in the band requires a concomitant increase in column length to maintain the chosen degree of separation, q_r is directly proportional to p_f, and G is also affected by p_f.

6. Column Temperature

The column temperature influences q_r/G in three ways: through α, through the factor $1/T$, and through controlling p_f at temperatures below the boiling point of the solute. The result of combining these factors was described in Section II.F.

TABLE III
Costs Projected by Abcor (4) for Separation of α- and β- Pinenes, for Columns of Varying Diameter[a]

Costs	Column diameter, ft					
	1	2	3	4	10	20
Throughput, lb/yr	115,000	460,000	1,030,000	1,840,000	11,500,000	46,200,000
Equipment cost, $	50,000	89,000	112,000	158,000	474,000	1,090,000
Annual operating cost, $						
Maintenance and taxes	5,000	8,900	11,200	15,800	33,200	76,400
Operating labor	17,000	17,000	17,000	17,000	34,000	34,000
Utilities and supplies	2,000	2,500	4,500	7,000	43,800	175,000
Packing replacement	1,000	1,500	2,500	3,000	18,700	75,000
	25,000	29,900	35,200	42,800	129,700	360,400
Depreciation	5,000	8,900	11,200	15,800	47,400	109,000
Total annual cost, $	30,000	38,800	46,400	58,600	177,100	469,400
Total cost, cents/lb	26	8.5	4.5	3.2	1.5	1.0

[a]These are future costs as projected for the period after production gas chromatography is reasonably established (i.e., following installation and operating experience gained from a modest number of commercial units in operation processing a range of materials).

165

7. Carrier Velocity

An increase in the carrier velocity $\bar{u}$ both raises the HETP and allows faster cycling. Since q_r depends on the cycle time but is not affectd by the HETP, q_r is proportional to $\bar{u}$, as Eq. 13 confirms. However, G depends on HETP by way of column length. Since n is fixed quite independently of $\bar{u}$, by considerations described in Section II.H-1, we can write

$$L \propto h \propto (\bar{u})^m \tag{15}$$

The theoretical and experimental evidence reviewed in Section II.G suggests that $0 < m < 1$, m being close to 1 at low or moderate velocities and falling at high velocities. How strongly G depends on $\bar{u}$ varies with velocity and scale of operation. The velocity thus has a large effect on both throughput and cost per unit time, but a smaller influence on q_r/G, depending on circumstances.

8. Column Diameter

The effect of diameter is rather different from that of the other parameters considered. Instead of being a parameter to be optimized, it determines the overall scale of operation. The diameter is selected to give a specified throughput.

It is of interest to consider how the cost of the product G/q_r varies with the scale of operation. This can be worked out from Eq. 13 if an analytical expression for G is available. Abcor have given a projection of costs (4) for the separation of α- and β-pinene reproduced as Table III. On fitting the total annual costs to a proportionality of the form

$$G \propto d^{m'}$$

it is found that m' increases from 0.3 when $d = 1$ ft to 1.5 when $d = 20$ ft. Hence we have

$$\frac{q_r}{G} \propto d^{m''} \propto q_r^{m''/2}$$

where, since $q_r \propto d^2$, m'' falls from 1.7 when $d = 1$ ft to 0.5 when $d = 20$ ft. The cost of the product therefore falls off quite rapidly at first with increasing scale of operation and more slowly when the scale is very large, as illustrated by the figures in the bottom row of Table III.

III. EQUIPMENT

In principle, production-scale equipment requirements are the same as for preparative gas chromatography, except that provision must be made to clean up and recycle the expensive carrier gas. Because of the difference of scale, however, and the importance of economic considerations, greater attention is required to the design of individual units of the chromatograph. Reviews of preparative apparatus are available elsewhere (96).

A schematic diagram of a production gas chromatograph appears in Figure 11. A number of authors have given detailed descriptions of apparatus, and a short list of these (selected either because the carrier was recycled or because columns 4 in. or more in diameter were used) is given in Table IV. Some features of the equipment requirements, and solutions adopted by these and other authors, are now described.

A. Carrier Gas

Air, the cheapest carrier gas, should be used when feasible; however, many materials to be separated are not completely stable chemically in the

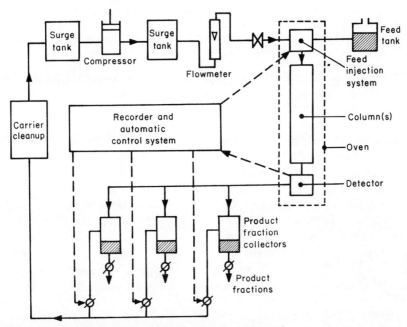

Fig. 11.   Block diagram of a production gas chromatograph.

TABLE IV
Selected Detailed Descriptions of
Large-Scale Chromatography Equipment

Work	Column diameter, in.	References
Carrier gas recycled		
Bayer, Hupe, et al., 1959–1964	0.2–4	10–15
Frisone, 1961	2	76
Abcor, 1967	4, 6, 12	6
ELF, 1972	5	8
Conder and Shingari, 1972	4	98
Carrier gas not recycled:		
Carel and Perkins, 1966–1969	4, 6, 12	46, 47
Debbrecht et al., 1965	4	89

presence of air at column temperatures. Oxygen may also attack the stationary phase, the support sometimes acting as catalyst (97, 98). For most separations, nitrogen, the next cheapest gas, is most suitable, although helium is sometimes used in the United States.

B.   Feed Introduction

As nearly as possible, the injection system should give an ideal rectangular band profile, without tailing (Section II. D). Controlled injection periods usually range from several seconds to several minutes. Liquid feeds have to be vaporized, even though most of the vapor subsequently condenses on entering the column; on-column (liquid-state) injection is considered not feasible at high solute concentrations.

The problem of providing the necessary heat capacity and heat transfer rates for vaporization is solved in some designs by using a bypass injector (15). This allows the whole of the cycle time between injections to be used for evaporation. It is not usually permissible for heat-sensitive compounds because of the extra time of exposure to high temperature. In-stream vaporization may be accomplished with a high-surface-area metal or glass surface, such as that provided by small beads (103, 107a), helices (76), or steel wool (15). The recommended temperatures range from 10°C to as high as possible above the boiling point of the highest boiling material (see, e.g., Refs. 15, 50, 70). In practice, the value must depend on the thermal stability of the materials being separated. Lower temperatures might be used if the carrier gas flow were reduced during the period of injection (99, 100); but this is difficult when the gas is recycled. A spray evaporator has been used (101). Albrecht and Verzele (102), who tried other evaporator designs, advocated a long, narrow, empty tube. They noted that the relative merits of different designs depend on the size of batch being injected, and this should be borne in mind when considering adapting preparative equipment for production-scale designs. Hupe has recently reviewed preparative inlet systems (103).

C. Columns

1. Stationary Phase

Besides giving good relative volatility for the components being separated, the stationary phase should have a very low vapor pressure at column temperature, since packing replacement or recoating is expensive. Selection of the stationary phase is facilitated by the large amount of data available in the literature of analytical chromatography (104).

2. Support

Chromatographic supports are expensive. Fairly coarse supports of the firebrick or Chromosorb P types have been used (6, 76), although Carel and Perkins (46) advocated raw Celite 545, which has the advantage of being a few hundred times cheaper than the usual supports. Use of a cheap packing, however, is not necessarily an advantage, particularly if precautions are taken so that the packing does not have to be replaced periodically. It was pointed out in Section II. C. 3 that overall costs may be reduced by using a narrow-mesh support, in spite of the expense of the support itself.

3. Column Construction

The technique of packing the columns has been described by several authors (14, 46, 65, 76, 78), and the design of columns for high efficiency was considered in Section II. G. 4. To minimize end effects and obtain a uniform distribution of gas over the cross section, cones (14, 101) or thin distributor sections closed by a porous disc (46, 47, 89) are normally employed. Cones have usually been studied by observing the overall performance of the system of column plus cones (9, 102), but Musser and Sparks (105) recently investigated the details of velocity distribution and axial dispersion in the inlet cone by itself. There is agreement (9, 105) that wide-angle (60–90°) inlet cones do not degrade performance and that they perform best if packed to about 80% of their volume. The outlet cone should probably be packed completely (9, 102). It is sometimes considered that solute overloading occurs in the narrow tip of the inlet cone (9, 102), which, accordingly, should be packed with uncoated material (105). In the author's view, however, provided wide feed bands are employed, no overloading phenomenon should occur because the volume of packing occupied by the feed band does not vary with column cross section.

D. Detectors

A detector is required to monitor the separation and to provide a signal for controlling the injection and collection sequence. For preparative work, it is possible to connect a high-capacity thermal conductivity detector (katharometer) directly in-stream between the column and the fraction collection system (89). This is not usually possible with the high gas flow rates used on the production scale, and it is easier to take off a bleed stream to a katharometer (46) or flame ionization detector (8). The katharometer offers some advantages when studying the performance of the unit, but precautions must be taken to deal with peak inversion and other effects resulting from use of high solute concentrations (98). A katharometer cannot be used if the carrier gas is air.

E. Fraction Collection

On leaving the column the solute vapors are condensed from the carrier gas by some form of cold trapping. The design of cooler condensers for mixtures of vapors with noncondensing gases was the subject of a classic paper written nearly 40 years ago by Colburn and Hougen (106), who assumed that condensation occurred entirely at the interface between gas and bulk liquid. It is characteristic of the solutes and concentrations involved in large-scale chromatography that aerosols are formed. Thus Colburn's design treatment cannot be used; designs must be tested by experiment.

The collection efficiency of a trap is improved by increasing the heat transfer area, and consequently many designs have included some form of packing (15, 76). The chief obstacle to attaining high collection efficiency with a packed trap is the formation of aerosols. The approaches to solving the problem are of two types: either the aerosol is allowed to form and is then removed, or formation is hindered.

Removal of an aerosol is accomplished in various ways: with an electrostatic precipitator (107), with a porous medium (108), or by scrubbing the gas with a suitable solvent. Electrostatic precipitators usually give efficiencies of at least 90% and as high as 98%, but they entail some hazard because of the high voltages used (6–100 MV). Carel and Perkins (46) have patented a trap that has a porous stainless steel plate and a large metal condensing surface of good thermal conductivity. Condensation of components was stated to be best accomplished at the highest (subboiling) temperature possible. As used by the F and M Company (89), this trap routinely provided 90 to 98% collection for a variety of types of organic compounds. Scrubbing with a solvent is capable of giving high collection efficiency but is not suitable for production purposes if a further operation is required to separate the solvent. In a modified form of this type of trap, the problem of solvent recovery has been solved, and collection efficiencies of 95 to 98% have been attained (109).

The second approach to dealing with aerosols is to prevent their formation, usually by employing a thermal gradient across the trap. Stevens and Mold (110) have suggested that the thermal gradient increases collection efficiency by promoting turbulence in the gas stream. The main purpose of the thermal gradient, however, is to prevent the concentration–temperature curve from crossing the saturation curve by supplying a small amount of heat to the gas during the condensation process (111). The method and its theory were originally developed by Colburn and Edison (112), and various forms of it have been proposed or adopted chromatographically (110, 113, 116). Collection efficiencies are usually greater than 90%.

Other collection methods that have been used include co-condensation with an added vapor and total trapping of vapor and carrier gas. Although successful for preparative purposes, these methods are unsuitable for large-scale application. A more appropriate, and completely different, collection device is the cyclone-type separator described by Bayer, Hupe, and Witsch (12), which uses centrifugal forces to trap the aerosol. Separators of this type usually trap fog particles larger than those (up to 10^{-2} mm) separated by an electrostatic precipitator (117).

Trapping efficiency varies with both the nature and the concentration of the vapor to be collected. Colburn and Edison have shown (112) that aerosols form most easily with vapors of high molecular weight. It is therefore particularly important to take account of aerosol formation when designing traps to collect such materials as essential oils, steroids, and pharmaceutical chemicals.

Since there are at least two components and a mixed fraction to be trapped, a manifold is required to direct the column effluent to the appropriate trap in turn. The sequence is controlled by a signal from the detector. Bayer (15) inserted sintered metal discs at the entrance to each collector to prevent diffusion of vapor into the wrong traps. The same object is achieved in the Hupe-Corning APG402 preparative chromatograph (118) by having a separate reverse purge through the manifold.

F. Carrier Recycle System

Chromatographs with columns diameters up to 12 in. (47) have been operated without a gas recycling system. This saves on capital cost but makes operation extremely expensive because of the large quantities of nitrogen or other carrier gas consumed. Even on the scale of 2-in. column diameter (76), it is preferable to recirculate the carrier gas. Recycle systems usually consist of a diaphragm or nonlubricated compressor, surge tanks, and a carrier gas clean-up unit (6, 8, 14, 15, 76, 98). A carrier clean-up unit is necessary to remove small quantities of solute remaining in the gas stream from the fraction collectors. Frisone (76) used four cold traps, one of molecular sieve cooled in liquid nitrogen and three immersed in a mixture of dry ice and acetone containing, respectively, activated charcoal, dehydrite, and ascorite. Bayer employed a cooled tube of activated charcoal (15), and Abcor (6) used molecular sieve or activated charcoal.

G. Automatic Control

In order to achieve effectively continuous operation, the feed introduction and fraction collection systems are put under automatic control. It is desirable to actuate the sequencing with the detector signal. If a preset time-based program is provided (107a), it is difficult to make fraction cuts at the right point every time, owing to small changes in the column temperature and other operating conditions. By employing contact switches or photocells, valve operation can be triggered at preset percentages of full-scale recorder deflection on the chosen peaks (118, 119). Time delays are also incorporated in the program. In a more sophisticated control system, electronic discrimination of concentration, time, and slope can be incorporated in a flexible logic unit (6).

H. Equipment Costs

Table V presents an analysis of capital equipment costs projected by Abcor (3) for a system with a column 4 ft in diameter and 15 ft long. These figures were projected from pilot plant data and refer to an assumed throughput of 400,000 lb/yr, but the costs do not vary greatly with throughput or nature of the components being separated, unless the column diameter is changed. (For the effect of diameter on costs, see Table III.)

TABLE V
Projected Equipment Costs for a 4-ft
Diameter Column Processing 400,000 lb/year (3)[a]

Equipment	Cost, $\$ \times 10^3$
Feed preparation and injection	25
Column	35
Detection and control system	7
Fraction collection and heat exchange	50
Carrier recycle system	30
Process piping and buildings	44
Engineering and construction	76
Total	267

[a]By courtesy of *The American Chemical Society*.

IV. COMPARISON OF SEPARATION PROCESSES

It is evident from the variety of techniques described in earlier sections that a complete spectrum of separation methods now exists, ranging from fractional distillation, through semichromatographic methods and extraction, to conventional chromatography. The technology of some of these processes, especially distillation, is highly developed, whereas that of large-scale chromatography is still comparatively primitive, and some of the semichromatographic techniques have been tried on the analytical scale only.

If any of the newer methods are to be widely used, comparative performance data are essential. It is difficult to make such comparisons now since, as pointed out in Section II, performance data are usually inadequately reported. There are, however, a number of important differences between chromatographic and other separation techniques. The relative merits and demerits are now surveyed.

A. Chromatography and Distillation

1. Difficult Separations

The basic difference between chromatography and distillation is the presence of a fixed sorbent phase with a very high surface area per unit volume of packing. Virtually instantaneous equilibrium is achieved between the two phases, and a very large number of transfer units, or theoretical plates, is obtained per unit column length. This means that separations too difficult for distillation can be carried out by chromatography. An example will make this clear.

Suppose that a binary mixture of components with a relative volatility (α) of 1.2 is to be separated to give products of 99% purity. For distillation, assume that an economic optimum reflux ratio is selected which requires twice the minimum number of theoretical plates and that the overall plate efficiency is 0.5. The number of plates required is given by Fenske's (120) equation

$$N_{req} = \frac{2}{0.5} \left\{ 1 + \frac{\log \left[99 / (100 - 99) \right]^2}{\log 1.2} \right\}$$

$$\approx 200$$

If the column has a diameter of 2 ft, the plate height is of the order of 1 ft., and the total length of column required is about 200 ft. For chromatographic separation, it is usually possible, by suitable choice of stationary phase, to enhance $\alpha - 1$ about 2.5 or more times (3), so that $\alpha = 1.5$. From Eq. 8, $a \approx 0.4$. Optimization considerations require a reduced number of plates (n) in the column somewhat less than 0.35 (Section II. D). Equation 7 then gives the number of plates required

$$N_{req} = 36 n_{req} / a^2 \approx 80$$

Allowing a factor of 2 for band broadening caused by use of high concentrations, and assuming, conservatively, that the plate height is about 1 in. a 2-ft-diameter column, the total length of column required is about 13 ft.

This example of a moderately difficult separation problem shows that distillation necessitates too long a column to be feasible, whereas chromatography is a practical proposition, requiring a column of the order of one-fifteenth as long.

The calculations also demonstrate that the separation by chromatography calls for rather more plates that distillation (although the definition of a plate is quite different in the two cases). This disadvantage, however, is completely reversed by the much lower plate height obtained in gas chromatography, which accordingly requires shorter columns. It has been assumed that the column diameter is the same in both cases. It may well be that a greater chromatographic diameter would be required to achieve the same throughput as in distillation, but this would lead to only a small increase in HETP (Section II. G. 2) and column length. The balance of advantage regarding column volume, therefore, still lies with chromatography, unless the throughput per unit column cross section is of the order of 30 or more times less than for a distillation column. The relative throughputs cannot be calculated as easily from theory as can the relative column lengths. Much experimental work remains to be done in comparing throughputs and costs for distillation and chromatography.

Repetition of the calculation with different assumed figures leads to the conclusion that chromatography is preferable to distillation (on grounds of feasible column length) when the relative volatility is less than about 1.3 to 1.5, depending on the throughput and degree of purity

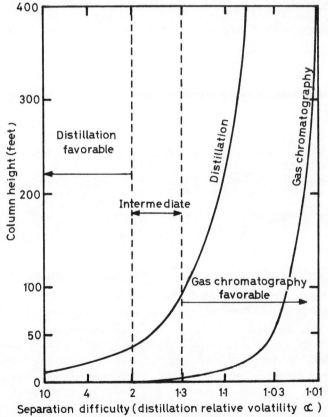

Fig. 12. Required column height as a function of separation difficulty, for distillation and gas chromatography (3). The numerical figures are only indicative, since column height depends on percentage purity, column diameter, and other factors.
By courtesy of *Chemical Engineering News*.

required. Similar conclusions have been reached by Abcor (3), as shown in Figure 12. A general relation between the numbers of plates required for chromatography and distillation has been deduced by Conder (121).

2.   Purity and Throughput

The throughput of a distillation column of given size is very sensitive to the product purity required. The relation between the two parameters is such that, even at zero throughput (total reflux), a definite minimum number of plates is required for a given purity specification. A chromatographic column always gives some throughput at the desired purity, even when it contains only a few plates and hence gives a low fractional recovery.

A calculation with Gilliland's correlation (122), for the same example as in the previous section, shows that an increase in purity from 99 to 99.9% would entail a reduction in distillation throughput by a factor of about 2.5. The change in chromatographic column throughput (at a fractional recovery of 0.6) would be of the order of 20% (see Figure 13).

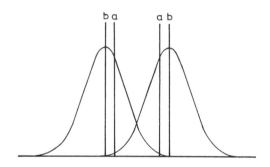

Fig. 13.   Effect of change of product purity specification on fractional recovery. *aa*, cut points for 99% purity; *bb*, cut points for 99.9% purity.

Chromatography is thus particularly suitable where high product purities are required. It also offers an unusually large degree of flexibility in operating conditions. The purity specification can be drastically changed with only a small effect on throughput. Therefore, considerable error can be tolerated in the design parameters and can be compensated readily by adjustment of the operating conditions without serious loss of performance.

3.   Heat-Sensitive Materials

Processing of heat-sensitive materials by distillation requires vacuum techniques, with their associated high equipment costs. In addition, the lower vapor pressure leads to low throughput.

No vacuum equipment is required for chromatography because the carrier gas pressurizes the system to atmospheric pressure or above. Moreover, leakage problems are less serious at pressures above ambient than below. Economically, the cost of recycling the carrier can be set against the cost of maintaining the vacuum.

A more important virtue of chromatography for thermally unstable materials is that its much shorter residence times permit higher vapor pressures and throughputs. Several factors are responsible for shorter residence times. First, operation is single-pass and there is no reboiler where the material is held at a high temperature for an indefinite period of time. Flash vaporization can be used to effect the change of state. Second, the columns are short, as explained in Section IV.A.1. Third, high carrier flow rates are used, whereas in distillation the flow rates are limited by gravity, flooding, or mechanical entrainment. Finally, there is no permanent holdup in the column; in distillation some of the material is required to prime the column and much more is lost during the long time required for equilibration after startup.

4.   Azeotropes

It is often necessary to separate mixtures showing moderately large deviations (usually positive) from Raoult's law. Such mixtures form an azeotrope, for which $\alpha = 1$, at some composition. The products of distillation are one pure component and the azeotrope, which cannot be separated further with a single tower. Complete separation requires other separation techniques or use of supplementary towers, as in extractive ar azeotropic distillation systems. Such systems have a higher capital cost than a single distillation tower, as well as a higher heat requirement and running cost.

The stationary phase in a chromatographic column changes α and thus performs the same function as the extracting agent in azeotropic or extractive distillation. Hence only one chromatographic column is required to separate azeotrope-forming mixtures. These can be separated as easily as more ideal mixtures.

5.   Multicomponent Mixtures

An array of multiple distillation towers is also required when the feed mixture is to be separated into several pure components. In general, $n-1$ towers are required to yield n pure, nonazeotrope-forming components from an n-component mixture.

In principle, a single chromatographic column can separate any number of pure components. This is usually easily achieved for three or four components. Separation of more components requires careful selection of the stationary phase, or combination of phases; but the designer is greatly assisted by the extremely large body of analytical chromatographic literature on this subject.

6. Disadvantages of Chromatography

For easy separations, normally carried out by distillation, chromatography would be expensive, for several reasons. The inert carrier gas used in gas chromatography has to be cleaned up and recycled continuously. Even if air is permissible as carrier, inefficient cold trapping makes a further clean-up stage both economic and necessary to prevent pollution. The need for fairly good temperature control requires provision of a constant-temperature oven or, at least, of a thermostatted jacket, rather than just insulation as used on a distillation column. Finally, column packing, particularly the support, is much more expensive than that in a packed distillation column, and it may require recoating two or three times a year. As commercial experience is gained, however, changes of packing may become unnecessary (see Table VI).

B. Liquid-Liquid Extraction and Fractional Crystallization

Liquid-liquid extraction shares with chromatography the advantage that ease of separation is a variable that is under the designer's control, because of the presence of a solvent. Continuous liquid-liquid extraction is the analog of liquid-liquid chromatography, just as extractive distillation is the analog of gas-liquid chromatography.

Choice of the extracting solvents is much more limited than that of a GLC stationary phase, because the two phases must be mutually immiscible, and this requirement is most difficult to meet at high solute concentrations. After extraction, a further separation process, often distillation, is needed to separate solute and solvent. Separation of solute from the gas phase in GLC is a simpler and less expensive process. In addition, GLC offers far greater separating efficiency because mass transfer in the gas phase and in thin liquid layers is much faster than in extraction.

The process commonly described as fractional crystallization is a relatively expensive technique. Solids handling, the need to allow for drainage and separation of occluded mother liquor, the large heat load, and the time required for mass and heat transfer make this a more cumbersome and costly process than fluid-fluid contacting processes. It is

TABLE VI

Current and Long-Range Projected Costs
for a 4-ft Diameter Column Separating
α - and β - Pinenes[a]

	Current	Long-Range
Throughput, lb/year	1,840,000	1,840,000
Equipment cost, $	246,000	158,000
Annual operating cost, $		
Maintenance and taxes	24,600	15,800
Operating labor	17,000	17,000
Utilities and supplies	7,000	7,000
Packing replacement	50,600	3,000
Depreciation (equipment written off over 10 years), $	24,600	15,800
Total annual cost, $	123,200	58,600
Cost of product, cents/lb	6.7	3.2

[a]Data projected by Abcor (4) from pilot-plant data with 4-in.-diameter column.

used mainly for separating the xylenes, since the relative volatilities of these compounds are so close to unity that distillation is not feasible. This is one separation for which chromatography would appear to offer many advantages.

C. Conclusions

By far the most commonly used separation technique at present is fractional distillation at atmospheric pressure. This technique offers long-established virtues of simplicity, low cost, and highly developed technology. It is likely to remain the method of choice for straightforward separations where the materials have good relative volatilities, are thermally stable, and are not required at a high level of purity.

For easy separations, chromatography would be more expensive than distillation. Chromatographic costs, however, may be expected to come down once the technique is established and experience has been gained in the installation and operation of commercial units. Table VI shows current and long-range projected costs for a 4-ft-diameter column separating α- and β-pinenes (4). The biggest factor in the halving of the envisaged product cost is an anticipated change in packing replacement technique.

For many separations, ordinary distillation at atmospheric pressure is expensive. Frequently it is not even practical, and other, more expensive, methods are currently used. These include extractive, azeotropic, and vacuum distillation, liquid-liquid extraction, and fractional crystallization. In these cases, chromatography may well offer a cheaper and more flexible alternative. As previous sections have indicated, the technique is particularly attractive when the separation involves relative volatilities close to unity, a high purity requirement, heat-sensitive materials, or azeotropes, or when several pure components are required. Such separations are encountered in most chemical industries, and many types of product might be reduced in cost by using chromatographic separation. These include essential oils, terpenoids, steroids, alkaloids, pharmaceutical compounds, metal chelates, isotopes, close-boiling isomers, and light hydrocarbons used as solvents and synthetic intermediates.

The most well-documented chromatographic separation is that of α- and β-pinenes. Hupe (57) has projected a cost of roughly 20 cents/lb for a throughput of 925,000 lb/yr at 98.5% purity on a 4-ft-diameter column. He assumed separation at 165°C on Carbowax 20M ($\alpha = 1.4$) with an HETP of 0.3 cm. A design for this separation on a 4-ft-diameter column has also been made by Abcor (7); the throughput was 1,840,000 lb/yr; the total operating cost, including depreciation, was about 7 cents/lb at 1966 price levels. Comparative separation costs published by Abcor (4) for gas chromatography and the corresponding conventional method are quoted in Table VII. The figures show considerable cost reduction for chromatographic separation. Typical throughputs of very pure product were quoted by Valentin (8) as being in the range 200 to 1500 ml/hr on a 4-in.-diameter column (450,000–3,400,000 lb/yr on a 4-ft column), capacity varying according to relative volatility and other conditions. Heat-sensitive materials will give lower throughputs, depending on the available vapor pressure. With the present emphasis on optimizing performance, many new applications can be expected for production gas chromatography in the next few years.

TABLE VII
Processing Cost Comparison
(Pilot-Scale Data) for Production Gas Chromatography
versus Current Conventional Method (4)

Product Material	Current Processing Method	Current Cost, $/lb	Projected GC Cost, $/lb
Fragrance material	Vacuum distillation	1.10	0.50
Vitamin intermediate	Fractional crystallization	0.70	0.50
Essential oil	None, except by GC	—	0.50
Vitamin	Extraction and crystallization	10.00	5.00
Fragrance	Vacuum distillation	0.50	0.25
Vitamin intermediate	Vacuum distillation	0.50	0.35
Solvent	Distillation	0.40	0.20
Essential oil components	Vacuum distillation	0.60	0.30
Hydrocarbon isomers	Distillation	0.70	0.30
Terpene chemicals	Vacuum distillation	0.50	0.20
Steroid	None, except by GC	—	10.00

Notation

A, B, C_l, C_g velocity-independent coefficients of the van Deemter equation

a $2\left(\dfrac{\alpha-1}{\alpha+1}\right)\left(\dfrac{k}{1+k}\right)$

D_g diffusivity of solute in gas phase

d internal diameter of column

d_p average diameter of packing particles

G total processing cost per unit time (i.e., the sum of depreciation and operating costs)

h height equivalent to a theoretical plate (HETP), measured with an elution (narrow) band; h is a function of solute concentration, but is often specified for infinite dilution

h_p excess of HETP for large-diameter column, over HETP for a similar analytical column

k equilibrium mass distribution ratio (moles of solute in stationary phase)/(moles of solute in mobile phase), at infinite dilution. Usually, $k \gg 1$ in production GLC columns.

L column length

m, m' indices

m_f mass of feed (total or component) injected

M_f molecular weight (average or component) of feed

N number of theoretical plates in column, measured with elution (narrow) band at infinite dilution

N_f number of theoretical plates occupied by feed band at column inlet

n (a) reduced value of N, $= Na^2/36$; (b) an index

n_f reduced value of N_f, $= N_f a/6$

p_f partial pressure of feed (total or component) in gas stream immediately before entering column

Q_f throughput of feed (i.e., mass passed in per unit time) for repetitive injection with cycle time equal to t_R

Q_r throughput of material recovered after fraction cutting of adjacent overlapping components, with cycle time equal to t_R

q_f throughput of feed (i.e., mass passed in per unit time) for repetitive injection with minimum cycle time

q_r throughput of material recovered after fraction cutting of adjacent overlapping components, with minimum cycle time

R gas constant

r ratio of mass of material recovered, after fraction cutting of adjacent overlapping components, to mass of feed material (fractional recovery)

T column temperature

T_D temperature above which a thermolabile solute degrades in the column

t_R retention time of solute in column

$\bar{u}$ compressibility-corrected interstitial velocity of mobile phase

α relative volatility (>1) of two components to be separated

γ tortuosity of packing

ϵ porosity of packing

η fractional impurity ($0 < \eta < 1$) of one component in another from which it has been separated

θ mode of operation parameter, $= N_f/\sqrt{N}$

κ flow-profile factor determining h_p (Section II.G.1)

References

1. Anon., *Chem. Eng. News*, June 28, 1965, p. 46.
2. Anon., *Chem. Eng. News*, May 16, 1966, p. 18.
3. Anon., *Chem. Eng. News*, May 23, 1966, p. 52.
4. J. M. Ryan and G. L. Dienes, *Drug Cosmet. Ind.*, **99**, 60 (1966).
5. Anon., *Oil Gas J.*, January 1, 1968.
6. Abcor, Inc., Cambridge, Mass., *Chem. Eng.*, February 12, 1968; advertising literature of Abcor, Inc., 341 Vassar Street, Cambridge, Mass. 02139.
7. J. M. Ryan, R. S. Timmins, and J. F. O'Donnell, *Chem. Eng. Prog.*, **64**, 53 (1968).
8. Technical information, "Separations by Industrial Chromatography," ELF Centre de Recherches (P. Valentin), Boite Postale 1, 69-Solaize, France.
9. F. H. Huyten, W. van Beersum, and G. W. A. Rijnders, in *Gas Chromatography 1960*, R. P. W. Scott ed., Butterworths London, 1960, p. 224.
10. E. Bayer, in *Gas Chromatography 1960*, R. P. W. Scott, ed., Butterworth's, London, 1960, p. 236.
11. E. Bayer and H. G. Witsch, *Z. Anal. Chem.*, **170**, 278 (1959).
12. E. Bayer, K. P. Hupe, and H. G. Witsch, *Angew. Chem.*, **73**, 525 (1961).
13. E. Bayer, G. Wahl, and H. G. Witsch, *Z. Anal. Chem.*, **181**, 384 (1961).
14. E. Bayer, K. P. Hupe, and H. Mack, *Anal. Chem.*, **35**, 492 (1963).
15. E. Bayer, *J. Chem. Educ.*, **41**, A755 (1964).
16. D. H. Desty et al. in *Vapour Phase Chromatography*, D. H. Desty, ed., Butterworths, London, 1957, p. xi; A. Klinkenberg, *ibid.*, p. 375.
17. Cities Service Research and Development Co. L.C. (Mosier), U.S. Patent 3,078,647 (1963); quoted in Ref. 44.
18. D. Dinelli, S. Polezzo, and M. Taramasso, *J. Chromatogr.*, **7**, 447 (1962).
19. (a) S. Polezzo and M. Taramasso, *J. Chromatogr.*, **11**, 19 (1963); (b) M. Taramasso and D. Dinelli, *J. Gas Chromatogr.*, **2**, 150 (1964).
20. M. Taramasso, *J. Chromatogr.*, **49**, 27 (1970).
21. A. J. P. Martin, *Disc. Faraday Soc.*, **7**, 332 (1949).
22. S. Turina, V. Krajovan, and T. Kostomaj, *Z. Anal. Chem.*, **189**, 100 (1962).
23. M. Freund, P. Benedek, and L. Szepesy, in *Vapour Phase Chromatography*, D. H. Desty, ed., Butterworths, London, 1957, p. 359.
24. P. Benedek, L. Szepesy, et al., *Acta Chim. Acad. Sci. Hung.*, **14**, 3, 19, 31, 339, 353, 359 (1958); quoted in Ref. 45.
25. P. Benedek, L. Szepesy, and S. Szepe, *Gas Chromatography*, V. J. Coates, H. Noebels, and I. Fagerson, eds., Academic Press, New York, 1963, p. 410.
26. H. Pichler and H. Schulz, *Brennst. Chemie*, **39**, 148 (1958).
27. H. Schulz, in *Gas Chromatography 1962*, M. van Swaay, ed., Butterworths, London, 1962, p. 225.
28. R. P. W. Scott, in *Gas Chromatography 1958*, D. H. Desty, ed., Butterworths, London, 1958, p. 189.
29. P. E. Barker and D. Critcher, *Chem. Eng. Sci.*, **13**, 82 (1960).
30. P. E. Barker and D. I. Lloyd, *J. Inst. Petrol.*, **49**, 73 (1963).
31. W. H. Husband, P. E. Barker, and K. D. Kimi, *Trans. Inst. Chem. Eng.*, **42**, 287 (1964).
32. P. E. Barker and D. H. Huntington, *Gas Chromatography 1966*, A. B. Littlewood, ed., Institute of Petroleum, London, 1966, p. 135.
33. D. Glasser, *Gas Chromatography 1966*, A. B. Littlewood, ed., Institute of Petroleum, London, 1966, p. 119.

34. M. R. Hinchcliffe and P. E. Barker, *New Sci.*, June 20, 1968, p. 640.
35. P. E. Barker and S. Al Madfai, *J. Chromatogr. Sci.*, **7**, 425 (1969).
36. P. E. Barker and Universal Fisher Engineering Co. Ltd., Crawley, Sussex, Patent Appl. 5764/68 and 44375/68.
37. G. R. Fitch, M. E. Probert, and P. F. Tiley, *J. Chem. Soc.*, 4875 (1962).
38. B. J. Bradley and P. F. Tiley, *Chem. Ind. London* 743 (1963).
39. P. F. Tiley, *J. Appl. Chem.*, **17**, 131 (1967).
40. D. W. Pritchard, M. E. Probert, and P. F. Tiley, *Chem. Eng. Sci.*, **26**, 2063 (1971).
41. G. Natta, *Chim. Ind. Milan*, **24**, 43 (1942).
42. W. Kuhn, E. Narten, and M. Thurkauf, *Helv. Chim. Acta*, **41**, 2135 (1958).
43. O. Grubner and E. Kucera, *Collect. Czech. Commun.*, **29**, 722 (1964); quoted in Ref. 44.
44. P. E. Barker, "Continuous Chromatographic Techniques," in *Preparative Gas Chromatography*, A. Zlatkis and V. Pretorius, eds., Wiley-Interscience, New York, 1971, p. 325.
45. J. H. Purnell, *Gas Chromatoghaphy*, Wiley, New York, 1962, p. 410.
46. A. B. Carel and G. Perkins, Jr., *Anal. Chim. Acta*, **34**, 83 (1966).
47. A. B. Carel and G. Perkins, Jr., *J. Chromatogr. Sci.*, **7**, 218 (1969).
48. J. A. Perry, *Chem Ind. London*, 576 (1966).
49. C. A. Sakodynsky, *Gas Chromatography 1962*, M. van Swaay, ed., Butterworths, London, 1962, p. 64.
50. D. T. Sawyer and J. H. Purnell, *Anal. Chem*, **36**, 457, (1964).
51. M. Dixmier, B. Roz, and G. Guiochon, *Anal. Chim. Acta*, **38**, 73 (1967).
52. J. R. Conder and J. H. Purnell, *Chem. Eng. Progr. Symp. Ser.* **65** (No. 91), 1 (1969).
53. J. R. Conder and J. H. Purnell, *Chem. Eng. Sci.*, **25**, 353 (1970).
54. J. R. Conder, *Chem. Eng. Sci.*, **28**, 173 (1973).
55. R. S. Timmins, L. Mir, and J. M. Ryan, *Chem. Eng.*, **1969**, 170.
56. K. Sakodynsky and S. Volkov, *J. Chromatogr.*, **49**, 76 (1970).
57. K. P. Hupe, *J. Chromatogr. Sci.*, **9**, 11 (1971).
58. D. A. Craven, *J. Chromatogr. Sci.*, **8**, 540 (1970).
59. P. C. Haarhoff, P. C. van Berge, and V. Pretorius, *Trans. Faraday Soc.*, **57**, 1838 (1961).
60. J. R. Conder and M. K. Shingari, *J. Chromatgr. Sci.*, in press.
61. (a) E. Glueckauf, *Trans. Faraday Soc.*, **51**, 34 (1955); (b) *ibid.*, **60**, 729 (1964).
62. A. S. Said, *J. Gas Chromatogr.*, **1**, 20 (1963).
63. A. S. Said, *J. Gas Chromatogr.*, **2**, 60 (1964).
64. See, for example, Refs. 57, 65, and 103.
65. G. M. C. Higgins and J. F. Smith, *Gas Chromatography 1964*, A. Goldup, ed., Institute of Petroleum, London, p. 94.
66. J. J. Van Deemter, F. J. Zuiderweg, and A. Klinkenberg, *Chem. Eng. Sci.*, **5**, 271 (1956).
67. R. P. W. Scott, *Anal. Chem.*, **35**, 481 (1963).
68. See, for example, J. R. Conder, "Physicochemical Measurement by Gas Chromatography," in *Progress in Gas Chromatography*, vol. 6, *Advances in Analytical Chemistry Instrumentation*, J. H. Purnell, Wiley, ed., New York, 1968, pp. 209–270; J. R. Conder and J. H. Purnell, *Trans. Faraday Soc.*, **65**, 824 (1969).
69. J. C. Giddings, *J. Gas Chromatogr.*, **1**, 12 (1963).
70. D. T. Sawyer and G. L. Hargrove, "Preparative Gas Chromatography" in *Progress in Gas Chromatography*, vol. 6, of *Advances in Analytical Chemistry Instrumentation*, J. H. Purnell, ed., Wiley, New York, 1968, pp. 325–359.
71. V. Pretorius and K. de Clerk, in *Preparative Gas Chromatography*, A. Zlatkis and V. Pretorius, eds., Wiley, New York, 1971, p. 1–53.
72. R. E. Pecsar, in *Preparative Gas Chromatography*, A. Zlatkis and V. Pretorius, eds., Wiley, New York, 1971, pp. 73–141.

73. A. B. Littlewood, in *Gas Chromatography 1964*, A. Goldup, ed., Institute of Petroleum, London, p. 77.
74. A. B. Littlewood, *Anal. Chem.*, **38**, 2 (1966).
75. S. T. Sie and G. W. A. Rijnders, *Anal. Chem. Acta*, **38**, 3 (1967).
76. G. J. Frisone, *J. Chromatogr.*, **6**, 97 (1961).
77. J. C. Giddings and E. N. Fuller, *J. Chromatogr.*, **7**, 255 (1962).
78. K. P. Hupe, U. Busch, and K. Winde, *J. Chromatogr. Sci.*, **7**, 1 (1969).
79. J. Peters and Euston, *Anal. Chem.*, **37**, 657 (1965).
80. A. Rose, D. J. Royer, and R. S. Henly, *Separation Sci.*, **2**, 229 (1967).
81. C. E. Schwartz and J. M. Smith, *Ind. Eng. Chem.*, **45**, 1209 (1953).
82. M. J. E. Golay, in *Gas Chromatography (Second International Symposium 1959)*, H. J. Noebels, R. F. Wall, and N. Brenner, eds., Academic Press, New York, 1961, Chapter 2.
83. K. P. Hupe, described in Ref. 14.
84. S. F. Spencer and P. Kucharski, *Facts and Methods*, **7** (4), 8 (1966); Hewlett-Packard, 1966.
85. J. C. Giddings and G. E. Jensen, *J. Gas Chromatogr.*, **2**, 290 (1964).
86. J. Pypker, *Gas Chromatography 1960*, R. P. W. Scott, ed., Butterworths, London, p. 240.
87. C. L. Guillemin, *J. Chromatogr.*, **12**, 163 (1963); *ibid.*, **30**, 222 (1967); *J. Chromatogr.*, **4**, 104 (1966).
88. A. B. Littlewood, *Gas Chromatography 1964*, A. Goldup, ed., Institute of Petroleum, London, p. 321.
89. F. J. Debbrecht, R. H. Kolloff, L. Mikkelsen, A. J. Martin, G. R. Umbreit, and C. E. Bennett, F and M Technical Paper 33, 1965.
90. L. Mir, *J. Chromatogr. Sci.*, **9**, 436 (1971).
91. T. Johns, M. R. Burnett, and D. W. Carle, *Gas Chromatography (Second International Symposium, 1965)*, H. J. Noebels, R. F. Wall, and N. Brenner, eds., Academic Press, New York, 1961, Chapter 20.
92. J. Bohemen, Ph.D. thesis, Cambridge, 1960.
93. J. L. Wright, *J. Gas Chromatogr.*, **1**, 10 (November 1963).
94. R. W. Reiser, *J. Gas Chromatogr.*, **4**, 390 (1966).
95. See, for example, R. M. Bethea and J. M. Bray, *J. Chromatogr. Sci.*, **9**, 758 (1971).
96. See, for example: (a) A. Zlatkis and V. Pretorius, eds., *Preparative Gas Chromatography*, Wiley-Interscience, New York, 1971; (b) M. Verzele, in *Progress in Separation and Purification*, vol. I, E. S. Perry, ed., Wiley-Interscience, New York, 1968, p. 83.
97. M. B. Evans and J. F. Smith, *J. Chromatogr.*, **28**, 277 (1967).
98. J. R. Conder and M. K. Shingari, unpublished work.
99. S. J. Hawkes, in *Gas Chromatography 1962*, M. van Swaay, ed., Butterworths, London, p. 65.
100. D. Jentzsch, *J. Gas Chromatogr.*, **5**, 226 (1967).
101. A. Rose, D. J. Royer, and R. S. Henley, *Separation Sci.*, **2**, 211 (1967).
102. J. Albrecht and M. Verzele, *J. Chromatogr. Sci.*, **9**, 745 (1971).
103. K. P. Hupe, Ref. 96a, p. 55.
104. See, for example, A. B. Littlewood, *Gas Chromatography*, Academic Press, London, 1962, p. 89 ff.
105. W. N. Musser and R. E. Sparks, *J. Chromatogr. Sci.*, **9**, 116 (1971).
106. A. P. Colburn and O. A. Hougen, *Ind. Eng. Chem.*, **26**, 1178 (1934).
107. See, for example: (a) E. P. Atkinson and G. A. Tuey, *Gas Chromatography 1958*, D. H. Desty, ed., Butterworths, London, p. 281; (b) A. E. Thompson, *J. Chromatogr.*, **6**, 454 (1961); (c) D. W. Fish and D. G. Crosby, *J. Chromatogr.*, **37**, 307 (1968).

108. (a) E. C. Schluter, Jr., *Anal. Chem.*, **41**, 1360 (1969); (b) K. D. Kilburn, in *Gas Chromatography 1962*, M. van Swaay, ed., Butterworths, London, p. 66.
109. P. Barrett and J. R. Conder, unpublished work.
110. R. K. Stevens and J. D. Mold, *J. Chromatogr.*, **10**, 398 (1963).
111. H. F. Johnstone, M. D. Kelley, and D. L. McKinley, *Ind. Eng. Chem.*, **42**, 2298 (1950).
112. A. P. Colburn and A. G. Edison, *Ind. Eng. Chem.*, **33**, 457 (1941).
113. A. Klinkenberg, *Gas Chromatography 1958*, D. H. Desty, ed., Butterworths, London, p. 211.
114. R. Teranishi, J. W. Corse, J. C. Day, and W. G. Jennings, *J. Chromatogr.*, **9**, 245 (1962).
115. H. Schlenck and D. M. Sand, *Anal. Chem.*, **34**, 1676 (1962).
116. R. C. Hoffman and A. Silveira, *J. Gas Chromatogr.*, **2**, 107 (1964).
117. J. M. Coulson and J. F. Richardson, *Chemical Engineering*, 2nd ed., vol. 2, Pergamon Press, Oxford, 1968, Chapter 8.
118. Advertising literature for K. P. Hupe APG 402 preparative chromatograph: Dr. Hupe Apparatebau, D-75, Karlsruhe, Germany; Corning Glass Works, Corning, N.Y.
119. C. Nebel and W. A. Mosher, *J. Gas Chromatogr.*, **6**, 483 (1968).
120. M. R. Fenske, *Ind. Eng. Chem.*, **24**, 482 (1932).
121. J. R. Conder, to be published.
122. E. R. Gilliland, *Ind. Eng. Chem.*, **32**, 1220 (1940).

Study of Polymer Structure and Interactions by Inverse Gas Chromatography

JAMES E. GUILLET, *Department of Chemistry, University of Toronto Toronto, Canada*

I. INTRODUCTION

Although polymeric materials, particularly the silicone rubbers, and polyethylene oxides have been used extensively as stationary phases since the earliest studies of gas chromatography, the application of chromatography to the study of polymers per se is much more recent.

Polymeric materials of commercial interest generally are of such high molecular weight that they have negligible vapor pressures at temperatures below their decomposition threshold, and hence they are not suitable for analysis by conventional gas chromatography. This was considered to be an insuperable obstacle and, as a result, most of the earlier research effort on chromatography of polymers was restricted to liquid phase partition, absorption, or exclusion chromatography.

A partial solution to the problem of polymer analysis by GLC was developed independently in four different laboratories about 15 years ago

(1–4). This involved pyrolysis of the polymer supported on a heated filament or plate situated in the entrance to the gas chromatographic column. The "pyrograms" so obtained have been used extensively to provide data on the composition and structure of macromolecules. The method has the disadvantage that the investigator is not observing directly the polymer itself but only its decomposition products, and the nature of these products depends very much on the conditions of the pyrolysis. For this reason the data obtained are often difficult to interpret. In spite of these defects, however, the method has proved specially valuable for the analysis of complex copolymers (5) and block copolymers (6). More recently, the method has been applied to the study of sequence distribution in copolymers by Michajlov et al. (7). Commercial apparatus for utilizing the technique have been available for more than a decade and there is now an extensive literature on the subject.

II. INVERSE CHROMATOGRAPHY

In principle, the most satisfactory method of studying polymers by GLC involves the use of "inverse chromatography." Whereas in conventional GLC some property of an "unknown" sample in the moving phase is determined by its interaction with a "known" stationary phase, in inverse chromatography, the properties of an "unknown" stationary phase are determined by its interaction with a "known" moving phase.

In essence, the experiment involves sending a pulse of vaporizable molecules along a narrow tube whose inner wall is coated with a thin film of the polymer to be investigated (Figure 1). These molecules undergo random diffusional motion in all directions, and upon this motion is superimposed a velocity U in the forward direction, maintained by a flow of inert carrier gas. In general, because of diffusion, each of the probe

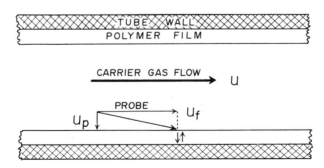

Fig. 1.   Schematic diagram of molecular probe experiment.

molecules has a velocity component U_p perpendicular to the flow direction which causes these molecules to impinge on the polymer surface at the wall. If collision with the wall involves no interaction, there will be no alteration in the component of velocity U_f in the forward direction. On the other hand, any interaction with the polymer will result in a retardation of the net translational velocity of the probe molecules along the tube direction. The strength and nature of the interaction can be deduced from this change in velocity by application of rather simple physicochemical considerations. The vaporizable molecules in the gas phase have been designated "probe" molecules by Guillet (8), who refers to the experiment as a "molecular probe" experiment.

The necessary apparatus is simple, consisting of a tube coated with a thin film of polymer, or alternatively, a short column packed with an inactive material containing the polymer dispersed as a thin film on the surface. In some cases the column is also packed directly with the polymer in either film, fiber, or powder form. A uniform flow of inert gas is maintained through the column, and a pulse of probe molecules is injected at one end and detected at the other by a suitable detector. A small pulse of noninteracting gas can also be injected with the probe molecules to aid in detection of the carrier gas front.

As in conventional gas chromatography, it is convenient to express the results of the experiment in terms of the retention volume V_r, defined as the volume of carrier gas required to elute the probe molecules, rather than the net translational velocity U_f. The fundamental quantity from which the various interactions can be deduced is the specific retention volume V_g, which is defined by the expression:

$$V_g = \frac{273}{T} \frac{V_r}{w}$$

where T (°K) is the temperature of the column at which the flow rate is measured, V_r is the retention volume corrected for pressure drop along the column, and w is the weight of the polymer film.

In a typical experiment, the value of V_g for a particular probe molecule (e.g., hexadecane) is determined for a specified polymer at a series of temperatures. The data are plotted in a generalized retention diagram in the form of log V_g as a function of the reciprocal of absolute temperature $(1/T)$. A typical curve for a semicrystalline polymer with probe molecules having weak molecular interaction is presented in Figure 2. Such a curve would be expected for polypropylene and crystalline polystryene using hydrocarbon probe molecules such as the linear alkanes; for experimental reasons, however, it is unlikely that a single linear alkane

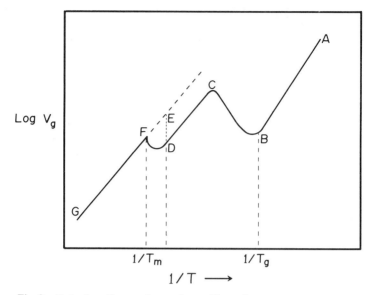

Fig. 2.   Retention diagram for semicrystalline polymer.

could be used over such a wide temperature range. If the vapor–pressure–temperature relationship is accurately known for the probe molecules, a great deal of valuable information can be obtained. The correct explanation for the shape of such curves is due to the work of Guillet and co-workers (8–11).

In the temperature region corresponding to the segment of the curve $A-B$, the polymer is below its glass transition temperature, and penetration of the probe molecule into the bulk of the polymer phase is precluded. Retention in this case is attributable to the sum of terms due to condensation and adsorption on the polymer surface. The curve approximates a straight line, whose slope at any point is equal to $(\Delta H_v - \Delta H_a)/2.3R$, where ΔH_v is the latent heat of vaporization of the probe molecule and ΔH_a is the enthalpy of adsorption. If the value ΔH_v is known, the latter quantity can be determined with some degree of accuracy.

The deviation from linearity represented by point B can be related to the glass transition temperature of the polymer (9, 10). The region BC corresponds to nonequilibrium absorption of the probe molecules in the polymer phase. The molecules begin to penetrate the bulk phase, but the rate of diffusion is still low; until point C is reached, therefore, absorption through the complete film thickness is not attained during the time of the pulse. From this region of the curve it is possible in principle to obtain information regarding the diffusion coefficient D and its temperature coefficient for the probe molecule in the polymer.

Equilibrium absorption of the probe molecules in the amorphous phase of the polymer occurs in the region CD. Retention in the region is caused by three processes: (1) condensation on the surface, (2) surface adsorption, and (3) solution in the bulk polymer phase.

There are now at least two adsorption terms relating to the two surfaces accessible to the probe molecules (i.e., the polymer–gas interface and the polymer–support interface). Additional terms may come from interfaces between crystalline and amorphous regions in the polymer. The slope of the linear portion of this curve is given by $(\Delta H_v - \sum_i b_i \Delta H_{a_i} - h_m)$ $/2.3R$. The b_i coefficients relate to the proportion of retention due to adsorption at each interface; these factors which can be evaluated experimentally (12), are related to the surface-to-volume ratio for the polymer film. For thick films it is sometimes possible to ignore these terms.

In the region DF of the curve, the polymer is undergoing a melting transition. If the crystalline phase is impermeable, as seems likely for most noninteracting probe molecules, this curve gives information about the size and distribution of crystalline regions and the percentage of crystallinity at any point during the melting process. The data are obtained by extrapolating the linear portion of the curve GF to lower temperatures. This represents the curve for 100% amorphous polymer. Since the quantity V_g is proportional to the weight of the adsorbing phase, the percentage of crystallinity corresponding to, say, point D, is given by

$$\% \text{ crystallinity} = \left(1 - \frac{(V_g)_D}{(V_g)_E}\right) \times 100$$

Experiments with polyethylene show that the results obtained using decane as the probe molecule compare quite well with those estimated from density measurements (11).

It is only in this region that the polymer behaves as a truly amorphous liquid, and therefore GLC theory can be applied to obtain thermodynamic quantities relating to the interactions of the polymer bulk phase with the moving probe molecules. In noncrystalline polymers, normal liquidlike behavior is attained at point C, which is a function of the film thickness. Under the usual GLC conditions of column loading and carrier gas flow rates, this is usually about 40 to 50°C above the glass transition temperature T_g of the polymer phase. This important limitation on the region accessible to thermodynamic treatments for large macromolecules has not been adequately understood in the past and may be a considerable source of error in early studies of this type.

III. THERMODYNAMICS OF INVERSE CHROMATOGRAPHY

By application of thermodynamic theory to GLC, we can derive a relation between the solute (probe) activity coefficient at infinite dilution (γ_1^∞) and the specific retention volume. The relation usually employed with conventional liquid stationary phases is

$$\ln\gamma_1^\infty = \ln\left(\frac{273.2R}{V_g^{\,0}M_2P_1^{\,0}}\right) - \frac{P_1^{\,0}}{RT}(B_{11} - V_1^{\,0})$$

$$+ \frac{\bar{P}(B_{12} - V_1^{\,0})}{RT} \tag{1}$$

where M_2 is the solvent molecular weight; $P_1^{\,0}$, $V_1^{\,0}$ and $\bar{P}$ are, respectively, the vapor pressure and molar volume of the solute in the pure liquid state and the average pressure; B_{11} is the gas state second virial coefficient and B_{12} is the cross coefficient, all these quantities being at the operating temperature T (°K). In principle, it is possible to use a single determination of the retention volume to determine the activity coefficient γ_1^∞, since the necessary vapor pressure data are usually available for a variety of solutes. For low-molecular-weight stationary phases, the inclusion of the solvent molecular weight M_2 in the denominator of the first term in the expression causes no particular difficulty; in fact, this relation has been combined with the Flory–Huggins relation by Martire and Purnell (13) and used to estimate the molecular weight of the stationary liquid phase. However, as these authors pointed out, the method becomes inaccurate for molecular weights exceeding about 1500.

Since most high polymers have molecular weights exceeding about 10,000, the inclusion of the M_2 term in Eq. 1 presents a serious difficulty. In the first place, the measured retention volume becomes substantially independent of molecular weight for high-molecular-weight stationary phases, and yet the logarithm of the activity coefficient varies with M_2, reaching a limit of $-\infty$ for large molecular weights. This problem has been resolved by a treatment due to Patterson et al. (14), which is outlined briefly here.

It is noted that the Raoult law activity coefficient γ_1 is the ratio of an activity a_1 to x_1, the mole fraction of component 1 in solution. However, for systems involving very large molecules, the mole fraction is not a convenient variable, and it is not generally used in polymer solution thermodynamics. Instead, it is more useful to consider the ratio of the activity a_1 to either a weight or a volume fraction of the polymeric component. If we choose weight fraction, Eq. 1 becomes

$$\ln\left(\frac{a_1}{w_1}\right)^{\infty} = \ln\left(\frac{273.2R}{P_1^0 V_g^0 M_1}\right) - \frac{P_1^0}{RT}(B_{11} - V_1) \tag{2}$$

where the denominator now contains M_1, the molecular weight of the probe, not of the polymer molecule. This important relation permits the unambiguous experimental determination of thermodynamic properties of polymer solutions at infinite dilution of the solute in the polymer phase, a region of particular importance both from theoretical and practical considerations.

In statistical theories of solution thermodynamics, the solute activity is often expressed in terms of two contributions: (a) a combinatorial (or athermal) entropy and (b) a noncombinatorial (or thermal) free energy of mixing. The latter contribution is characteristic of the polymer–solvent pair. The value of the combinatorial entropy must be calculated from a theory. When the familiar Flory–Huggins approximation is used, we have

$$\ln a_1 = (\ln a_1)_{\text{comb}} + (\ln a_1)_{\text{noncomb}} \tag{3}$$

$$= \left[\ln\phi_1 + \left(1 - \frac{1}{r}\right)\phi_2\right] + \chi\phi_2^2 \tag{4}$$

Neglecting the volume of mixing, the volume fraction is given by

$$\phi_1 = \frac{w_1 v_1}{w_1 v_1 + w_2 v_2} \tag{5}$$

where v_1 and v_2 are the specific volumes of the probe and the polymer, respectively. For polydisperse stationary phases, r is given by

$$r = \frac{(\overline{V}_2^0)_n}{V_1^0} = \frac{(\overline{M}_n)v_2}{V_1^0} \tag{6}$$

where $(\overline{V}_2^0)_n$ is the number-average of the molar volumes in the polydisperse polymer sample, V_1^0 is the molar volume of the monodisperse solute, and v_2 is the specific volume of the polymer, which may be calculated from published $v(M)$ data for sharp fractions. Since v is usually of the form $v(M) = v(\infty) - (\text{const}/M)$, the specific volume of the polydisperse sample should be approximately equal to that of a sharp fraction having $M = \overline{M}_n$, the number average molecular weight.

For the limiting case in which $\phi_2 \to 1$, Eq. 4 becomes

$$\ln\left(\frac{a_1}{w_1}\right)^{\infty} = \ln\frac{v_1}{v_2} + \left(1 - \frac{V_1{}^0}{\overline{M}_2 v_2}\right) + \chi \tag{7}$$

and from Eq. 2 we have

$$\chi = \ln\frac{273.2 R v_2}{V_g{}^0 V_1{}^0 P_1{}^0} - \left(1 - \frac{V_1{}^0}{\overline{M}_2 v_2}\right) - \frac{P_1{}^0}{RT}(B_{11} - V_1{}^0) \tag{8}$$

Equation 8 now gives the relation between the Flory–Huggins thermodynamic interaction parameter and the experimental GLC datum $V_g{}^0$.

Flory and collaborators have recently suggested (15) that a better relation is found by replacing the volume fractions in Eq. 5 by "segment fractions" defined by

$$\phi_1^* = \frac{w_1 v_1^*}{w_1 v_1^* + w_2 v_2^*} \tag{9}$$

where the v^* are specific "core" volumes, rather than specific volumes that also reflect the thermal expansions of the liquids. Generally v^* is also referred to as a reduction parameter and corresponds to the close-packed specific volume of the hypothetical liquid at $0°K$. This substitution now gives a new value for the interaction parameter denoted χ^*, as follows:

$$\chi^* = \ln\frac{273.2 R v_2^*}{V_g{}^0 V_1^* P_1{}^0} - \left(1 - \frac{V_1^*}{\overline{M}_2 v_2^*}\right) - \frac{P_1{}^0}{RT}(B_{11} - V_1{}^0) \tag{10}$$

The role of the term B_{11} in determining values of thermodynamic quantities can be very important. This correction for gas phase imperfections makes only a relatively small difference in the calculations of χ, but the same is not true when thermodynamic data such as enthalpies are to be calculated from the temperature dependence of $V_g{}^0$. From Eq. 2, the partial molar heat of the solute at infinite dilution in the polymer $\bar{h}_1^\infty$, relative to that in the pure liquid solute state h_1^0, is given by

$$\Delta\bar{h}_1^\infty = \bar{h}_1^\infty - h_1^{\,0} = \frac{R\partial\ln a_1}{\partial(1/T)}$$

$$= R\frac{\partial}{\partial(1/T)}\left[-\ln V_g^{\,0}P_1^{\,0} - \frac{P_1^{\,0}}{RT}(B_{11} - V_1^{\,0})\right] \qquad (11)$$

Data of Guillet and Stein (11) on n-decane and n-dodecane in polyethylene showed that the partial derivative with respect to $1/T$ of the logarithmic term is essentially zero, the sign and absolute value of the calculated $\Delta\bar{h}_1^\infty$ being determined largely by the reciprocal temperature derivative of the correction term containing the B_{11} coefficient. Accurate values of B_{11} and the inclusion of the correction term in Eq. 11 are prerequisites, therefore, for meaningful evaluation of such thermodynamic parameters as $\Delta\bar{h}_1^\infty$.

In general the χ values calculated from Eqs. 8 and 10 differ, and it is perhaps too soon to decide which is the more useful relation. Equation 8 gives values that should correlate with the conventional χ parameter data in the literature. However, as pointed out by Patterson et al. (14), the χ^* value has the virtue that the partial molar heat of solution of the probe in the polymer $\Delta\bar{h}_1^\infty$ is given directly by the temperature coefficient of χ^* by the relation

$$\Delta\bar{h}_1^\infty = R\left(\frac{\partial\chi^*}{\partial(1/T)}\right) \qquad (12)$$

The equations derived by Patterson et al. are limited by the approximations used to infinite dilution values of the χ parameter. Recently, however, Conder and Purnell (16–19) extended the theory to obtain thermodynamic data at finite concentrations by the GLC method. The extension to polymer systems at finite concentrations was first proposed by Guillet (8), and such studies have very recently been recorded by Brockmeier, McCoy, and Meyer (20). According to their treatment, the χ parameter at finite concentration is related to the activity a_1 by the relation

$$\chi = \frac{1}{\phi_2^{\,2}}\left(\ln\frac{a_1}{\phi_1} - \phi_2\right) \qquad (13)$$

This gives the χ parameter in terms of volume fractions, which are less convenient than weight fractions for this experiment. Converting to weight fractions, the analogous relation to Eq. 7 for finite concentrations is

$$
\ln\left(\frac{a_1}{w_1}\right) = -\ln[(1-w_1)r + w_1] + \frac{1-w_1}{(1-w_1)+w_1/r} + \chi\left[\frac{1-w_1}{(1-w_1)+w_1/r}\right]^2
$$

(14)

Provided adequate care is taken with the experimental technique, the GLC method now appears to be the method of choice for determining thermodynamic interactions between polymers and small molecules (probes). Precautions necessary to obtain good results are outlined more completely in a recent review by Young (21).

In this section, solution theories involving the Flory–Huggins treatment have been emphasized because these are more generally accepted and useful in polymer chemistry. However, the technique does not depend on the use of any particular solution theory. Indeed, its most important use may well be in testing various models of solution behavior of polymer–solvent systems.

One treatment that is extensively used in practical applications of polymers—the "solubility parameter" concept—serves in predicting the solubility of polymers in various solvents and solvent mixtures. It has proved to be a valuable empirical tool in the paint and lacquer industry, and it is also useful in predicting adhesive interactions.

The solubility parameter δ for any compound is defined from Hildebrand–Scatchard solution theory as

$$
\delta = \left(\frac{\Delta H_v - RT}{v^0}\right)^{1/2}
$$

(15)

where ΔH_v is the heat of vaporization of the compound and v^0 is its molar volume. Since high polymers have no vapor pressure and their molar volume is usually not well defined, there can be no direct experimental determination of δ for a polymer. Therefore, this value must be deduced from other measurements of swelling or solubility, or calculated from an approximate theory.

In principle (8), the GLC experiment is ideal for determining solubility parameters directly for polymer substrates. Combinationof the Flory treatment with the Hildebrand–Scatchard theory gives

$$
\chi = \frac{v_1}{RT}(\delta_1 - \delta_2)^2
$$

(16)

where v_1 and δ_1 are the molar volume and solubility parameter, respectively, of the probe molecule, and δ_2 is the solubility parameter of the polymer. Since the solubility parameter is known or easily calculable for most probe molecules, this represents a direct method for determining such values for the polymer. Once the column is set up, it is an easy matter to inject a wide variety of probe molecules, permitting the rapid accumulation of solubility data on any polymer.

One of the earliest studies involving the use of the GLC method to obtain thermodynamic data on high polymer systems was that of Smidsrød and Guillet (9), who studied solvent interactions in poly N-isopropyl acrylamide by observing the chromatographic behavior of acetic acid, butyl alcohol, naphthalene, α-chloronaphthalene, and n-hexadecane on Chromosorb columns coated with polymer. They showed that normal elution occurred over a wide temperature range only with the first two probe molecules, which were true solvents for the polymer. The latter three probes gave normal behavior only at temperatures 30 to 50°C above the glass transition ($T_g = 130°C$) for the polymer. They suggested the use of this phenomenon as a means of detecting glass transitions in polymers.

The thermodynamic parameters listed in Table I were obtained from activity coefficients calculated using Eq. 1, and these correlated with the known solution behavior of the polymer–solvent systems. However, no data were available to compare the results with those of more conventional methods. A more critical test of the method was given by Patterson et al. (14), who reported χ parameters for n-dodecane–branched polyethylene and n-decane–linear polyethylene that agree well with extrapolations from literature data on mixtures of linear alkanes.

More recently the same systems have been reevaluated by Schreiber et al. (22), who obtained the data given in Table II for the interaction

TABLE I

Thermodynamic Data at 200°C for Poly N-isopropyl Acrylamide,
$\overline{M}_n = 50,100 \pm 2000$

	ΔG_m, cal/mole	ΔH_m, cal/mole	ΔS_m, cal/(deg) (mole)$^{-1}$
Acetic acid	-6740	-2400	$+9.2$
Butyl alcohol	-5390	$+1100$	$+13.7$
Naphthalene	-4360	$+3900$	$+17.5$
α-Chloronaphthalene	-4010	$+3600$	$+11.3$
n-Hexadecane	-1760	$+2200$	$+8.4$

TABLE II

Interaction Parameters, χ

Compound	LPE 145.4°C	152.6°C Average χ̄	BPE 120.0°C	145.1°C	Average χ̄	Δχ̄
3-Me-hexane	0.421	0.385 0.403	0.328	0.335	0.332	.071
n-Octane	0.366	0.350 0.358	0.307	0.300	0.304	.054
2-Me-heptane	0.393	0.391 0.392	0.331	0.340	0.336	.056
3-Me-heptane	0.373	0.364 0.369	0.308	0.300	0.304	.065
2,4-Di Me-hexane	0.391	0.360 0.376	0.336	0.323	0.330	.046
2,5-Di Me-hexane	0.425	0.379 0.402	0.354	0.352	0.353	.044
3,4-Di Me-hexane	0.319	0.296 0.308	0.247	0.249	0.248	.060
2,2,4 Tri-Me-pentane	0.411	0.392 0.402	0.341	0.336	0.339	.063
n-Nonane	0.347	0.330 0.339	0.275	0.282	0.279	.060
2,2,4 Tri-Me-hexane	0.368	0.327 0.348	0.288	0.284	0.286	.062
n-Decane	0.317	0.306 0.312	0.254	0.255	0.255	.057
n-Dodecane	0.293	0.277 0.285	0.227	0.240	0.234	.051
Toluene	0.393	0.395 0.394	0.336	0.338	0.337	.057
Ethylbenzene	0.368	0.372 0.370	0.326	0.328	0.327	.043
p-Xylene	0.319	0.322 0.320	0.270	0.277	0.274	.046
m-Xylene	0.340	0.340 0.340	0.288	0.294	0.291	.049
Mesitylene	0.290	0.274 0.282	0.244	0.249	0.247	.035
Tetralin	0.325	0.318 0.322	0.289	0.279	0.284	.038
cis-Decalin	0.080	0.063 0.072	0.028	0.027	0.027	.045
trans-Decalin	0.061	0.045 0.053	0.017	0.003	0.010	.043
						mean .053

parameters of a variety of hydrocarbons in both branched (BPE) and linear (LPE) polyethylenes. The χ parameters show very little temperature dependence in these systems, but there is a consistent difference between the values for branched and the linear polymer, which the authors attribute to differences of the thermal expansion of the two polymer homologs. The relatively large values of the χ parameter in polymer–solvent systems with such weak interactions are due to differences in the expansion coefficients of the small probe and the large polymer molecule. A linear plot was obtained when χ was plotted as a function of α_1, the expansion coefficient of the probe molecule. The variation in χ between branched and normal alkanes seems related to differing free volumes rather than to special steric effects or differing force fields around the molecules. These results have contributed substantial evidence to support the contentions of Patterson (23) on the importance of the free volume in polymer solution thermodynamics.

Further experimental tests of the method were reported by Summers, Tewari, and Schreiber (24), who studied the interaction of a variety of hydrocarbon solvents with a sample of poly(dimethyl siloxane) having a viscosity molecular weight $\overline{M}_v = 5 \times 10^5$. The χ parameters were deter-

TABLE III

Summary of χ and χ^* Data for Hydrocarbons on Poly Dimethylsiloxane at 25°C

| Compound | GLC | | Vapor sorption |
	χ	χ^*	χ^*
n-Pentane	0.409	0.513	0.51
n-Hexane	0.448	0.524	—
n-Heptane	0.497	0.556	0.53
n-Octane	0.556	0.600	0.57
2-Methylpentane	0.449	0.534	—
2-Methylbutane	0.392	0.440	0.48
2-Methylhexane	0.461	0.520	—
2-Methylheptane	0.521	0.566	—
2,2,4-Trimethylpentane	0.446	0.499	0.47
Benzene	0.814	0.864	0.87
Toluene	0.802	0.833	0.82
p-Xylene	0.800	0.822	0.80
Ethylbenzene	0.828	0.833	0.80

mined by the GLC method and compared with those obtained on the same polymer sample with the same solvents by the classical equilibrium vapor sorption method. The results appearing in Table III indicate excellent agreement between the two methods. The same authors investigated the effect of ignoring the B_{11} term in Eqs. 8 and 10 by calculating the χ parameters with and without the correction for nonideality of the gas phase. The results at two different temperatures are listed in Table IV. It is clear that the virial coefficient term is required if an accuracy of better than 5% in χ is desired and the saturation vapor pressure of the probe molecule P_1^0 exceeds 200 mm Hg under the conditions of the experiment. It is perhaps worth pointing out that under GLC conditions the probe molecules are always at very low partial pressures and hence behave "ideally." The correction term is required because of the non-ideality of the saturated vapor in the vapor pressure measurement used to determine the value of P_1^0 which is used in Eqs. 8 and 10.

It is also interesting to note that in both the polyethylene and poly(dimethyl siloxane) polymers, the χ and χ^* values are independent of temperature in the range of 25 to 70°C, within the precision of the

TABLE IV

Importance of Nonideality Vapor Pressure Correction in
Calculation of Thermodynamic Parameters[a]

| | 25°C | | | 75°C | | |
Compound	P_1^0, mm	χ_{ratio}	χ^*_{ratio}	P_1^0, mm	χ_{ratio}	χ^*_{ratio}
n-Pentane	512.5	1.096	1.075	2123.00	1.287	1.208
n-Hexane	151.2	1.037	1.031	790.5	1.134	1.108
n-Heptane	45.72	1.014	1.013	303.6	1.063	1.054
n-Octane	13.98	1.005	1.007	119.0	1.031	1.029
2-Methylpentane	221.7	1.051	1.043	1025.	1.177	1.137
2-Methylhexane	65.88	1.022	1.019	402.1	1.088	1.075
2-Methylheptane	20.61	1.009	1.009	159.5	1.043	1.039
2,2,4-Trimethylpentane	49.34	1.023	1.018	305.8	1.091	1.078
Benzene	95.18	1.012	1.012	550.9	1.046	1.042
Toluene	28.44	1.006	1.006	203.8	1.026	1.024
p-Xylene	8.85	1.002	1.002	78.75	1.014	1.014
Ethyl benzene	9.89	1.002	1.001	84.75	1.013	1.013

[a]Ratios are corrected/uncorrected values as calculated from Eqs. 8 and 10.

experiments. This implies that the heat of mixing term $\Delta \bar{h}_1^\infty$ is very small for these systems.

Tewari and Schreiber (25) have recently published a study on rubber–hydrocarbon systems that confirms the validity of the GLC method for obtaining χ parameters. A summary of their results is given in Table V. The agreement between the GLC values and those calculated from static vapor pressure measurements is reasonably good, and the consistency of the GLC-derived χ data for the normal alkanes suggests that where discrepancies occur, the GLC method may be the more reliable.

Newman and Prausnitz (27) have recently studied polymer–solvent interactions in polystyrene at temperatures above T_g. Since they found that asymmetric peaks were obtained with polar solutes when Chromosorb W was used as a support, they used Fluoropak, which gave symmetrical peaks, for this purpose. The disadvantage of the latter support is that now the retention volume is flow-rate dependent and must be extrapolated to

TABLE V

Thermodynamic Parameters for Natural
Rubber–Hydrocarbons

	χ^*			χ Tewari and Schreiber (25)	Bristow and Watson (26)
	25°C	40°C	55°C		
n-Pentane	0.758	0.714	0.717	0.580	0.565
n-Hexane	0.658	0.651	0.651	0.534	0.474
n-Heptane	0.607	0.602	0.609	0.504	0.433
n-Octane	0.575	0.564	0.567	0.478	0.475
2-Methylpentane	0.686	0.671	0.674	0.558	
2-Methylhexane	0.626	0.616	0.625	0.524	
2-Methylheptane	0.586	0.577	0.577	0.498	
2,2,4-Trimethylpentane	0.589	0.571	0.580	0.483	0.513
Benzene	0.55[a], 0.560	0.550	0.543	0.456	0.437
Toluene	0.433	0.418	0.422	0.350	0.391
Ethylbenzene	0.409	0.391	0.382	0.346	
p-Xylene	0.345	0.338	0.348	0.270	

[a] Benzene–rubber value from Eichinger and Flory (15).

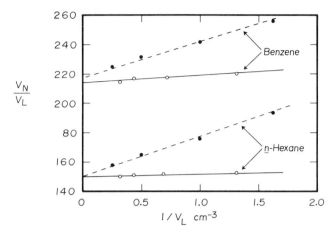

Fig. 3.   Flow rate dependence of retention volumes: open circles, on Chromosorb P; solid circles, on Fluoropak 80.

zero flow rate. This was done by the method of Conder et al., and Figure 3 plots the results obtained with squalane on the two supports, indicating that the extrapolation gives the same value of V_N/V_L, where V_N is the observed retention volume and V_L is the volume of the stationary phase.

Activity coefficients calculated according to Eq. 2 for a number of solutes in polystyrene ($\overline{M}_n = 97,000$) are presented in Table VI. Similar experimental results were obtained for polystyrene having $\overline{M}_n = 37,000$, as would be expected from theory. The activity coefficients for the strongly polar solutes were larger than those for the nonpolar solutes. As temperature increased, the activity coefficients of the nonpolar solutes increase, except for cyclohexane. The activity coefficients of the polar solutes were not significantly affected as the temperature increased from 150 to 200°C.

Newman and Prausnitz also reported χ^* values calculated from Eq. 10 for poly isobutylene with benzene, cyclohexane, and n-pentane. The values are given in Table VII. The values obtained for benzene and n-pentane are in good agreement with the results of Hammers and De Ligny (28) on the same systems, but consistently 10 to 20% lower than the value obtained by extrapolating the data of Eichinger and Flory (15) to infinite dilution. On the other hand, the cyclohexane data are consistent by both the GLC and static measurements. The reason for these discrepancies is not yet determined.

Hammers and De Ligny (28) have also published experimental data on thermodynamic interactions of high-molecular-weight poly isobutylene ($\overline{M}_v = 40,000$) with a variety of normal and branched alkanes, as well as benzene, cyclohexane, and tetrachloromethane. They reported their results

TABLE VI

Activity Coefficients (Weight Fraction) γ_1^∞ for Solvents at Infinite Dilution in Polystyrene, $\overline{M}_n = 97{,}000$

Solvent	150°C	175°C	200°C
Cyclohexane	12.2	10.5	8.92
Benzene	5.44	5.67	5.81
Toluene	5.22	5.29	5.34
Ethylbenzene	4.96	5.47	5.67
Styrene	4.11	4.57	4.69
Dioxane	5.17	5.41	5.71
1,2-Dichloroethane	4.55	4.83	5.10
Chlorobenzene	3.72	3.85	4.04
Nitroethane	11.0	11.6	12.6
Methyl ethyl ketone	9.44	9.30	9.12
Cyclohexanone	5.81	6.14	6.26
Dimethylformamide	10.6	10.8	11.0
Acetonitrile	19.8	19.8	19.7

TABLE VII

Flory χ^* Values (Based on Segment Fraction) for Poly Isobutylene Systems from Chromatographic Measurements

Solvent	25°C	40°C	50°C
Benzene	1.03	0.96	0.92
Cyclohexane	0.55	0.54	0.53
n-Pentane	0.87	0.801	0.797

in terms of the residual partial molar free enthalpy of mixing at infinite dilution, defined by

$$\overline{\Delta G}_M^{0,\text{res}} = -RT\ln p - \frac{p(B_{11} - v_1^0)}{41.303} - \overline{\Delta G}_v^0 - RT\left[\ln\frac{\bar{v}_1}{\bar{v}_2} + \left(1 - \frac{\bar{v}_1}{\bar{v}_2}\right)\right]$$

$$(17)$$

where R is the gas constant $(cal/(mole)(°C))$, T the column temperature $(°K)$, p the solute vapor pressure (atm), B_{11} the second virial coefficient of the solute $(cm^3/mole)$ and the $\bar{v}$ are the partial molar volumes $(cm^3/mole)$ of the solute (1) and the polymer (2) [which can be approximated by the molar volumes v^0 $(cm^3/mole)$]. The first two terms give the standard molar free enthalpy of vaporization of the pure solute.

The standard partial molar free enthalpy of vaporization of the solute from infinitely dilute solution, $\overline{\Delta G}_v^0$, is given by

$$\overline{\Delta G}_v^0 = RT\left(\ln V_N - \ln\frac{41.303\,RTw_2}{M_2}\right) \tag{18}$$

where V_N is the net retention volume (cm^3), and w and M are the weight and the molecular weight, respectively, of the polymer in the column. The last term in Eq. 17 is the Flory–Huggins correction, which accounts for the difference in size of the solute and solvent molecules. The corresponding residual entropy and enthalpy of mixing are obtained by differentiation of Eq. 17 with regard to T.

The results of Hammers and DeLigny are summarized in Table VIII in the form of the ratio of $\overline{\Delta G}_M^{0,res}$ to r_1, the number of segments of the solute (probe) being given by

$$r_1 = \frac{n+1}{2} \tag{19}$$

where n is the number of carbon atoms of the normal alkane. For branched alkanes, the segment number was used as an adjustable parameter that was approximately 4 for the compounds used in this study. The value listed first in Table VIII is the experimental value; beneath it is the value calculated from a simplified version of Prigogine solution theory, which gives at infinite dilution

$$\frac{\overline{\Delta G}_M^{0,res}}{r_1} = A + \left(\frac{BT^2}{r_1^2}\right) \tag{20}$$

The A term represents the interchange energy per segment mole at $0°K$; it is positive and independent of temperature. The remaining term accounts for the difference in thermal expansion of the mixture components and predicts the temperature dependence of the residual entropy and enthalpy of mixing. The B is positive, independent of temperature and not strongly dependent on the type of (apolar) solvent.

TABLE VIII
Experimental $\Delta G_M^{0,\text{res}}$ Data

| Solvent | $\Delta G_M^{0,\text{res}} r_2^{-1}$, cal/mole | | | | | | |
	313.16 °K	333.16 °K	353.16 °K	373.16 °K	393.16 °K	413.16 °K	433.16 °K
n-Pentane	181 ± 4 175	201 ± 2 195	222 ± 1 216	243 ± 2 238	–	–	–
n-Hexane	134 ± 3 136	150 ± 2 150	166 ± 1 165	182 ± 1 181	199 ± 1 198	–	–
n-Heptane	110 ± 3 110	120 ± 2 121	132 ± 1 133	144 ± 1 145	158 ± 1 158	–	–
n-Octane	96 ± 3 93	103 ± 2 101	112 ± 2 111	121 ± 1 120	131 ± 1 131	142 ± 1 142	–
n-Nonane	–	88 ± 3 87	94 ± 2 95	102 ± 1 103	109 ± 1 111	118 ± 1 120	127 ± 1 129
n-Decane	–	80 ± 5 77	84 ± 3 83	89 ± 2 89	95 ± 1 96	102 ± 1 104	109 ± 1 111
2,3-Dimethylbutane	128 ± 3 132	144 ± 2 146	159 ± 2 160	177 ± 3 176	194 ± 5 192	–	–
2,2-Dimethylpentane	131 ± 3 128	143 ± 2 142	156 ± 2 156	170 ± 2 171	186 ± 2 187	–	–
2,2,4-Trimethylpentane	107 ± 3 109	119 ± 2 119	131 ± 2 131	143 ± 2 143	155 ± 2 156	167 ± 2 169	–
2,5-Dimethylhexane	111 ± 3 112	124 ± 2 ¹24	135 ± 2 136	148 ± 2 149	161 ± 2 162	176 ± 2 176	–
Cyclohexane	61 ± 2	68 ± 2	75 ± 1	81 ± 1	88 ± 1	–	–
Benzene	132 ± 2	128 ± 2	134 ± 1	140 ± 1	148 ± 1	–	–
Tetrachloromethane	81 ± 2	89 ± 1	97 ± 1	106 ± 1	117 ± 2	–	–

The experimental data were fitted statistically to Eq. 20 yielding values for $A = 27.1 \pm 1.4$ cal (segment/mole)$^{-1}$ and $B = 0.0136 \pm 0.0002$ cal(segment mole)$^{-1}$(°C)$^{-2}$. The agreement between the calculated and observed values is excellent. On this basis, Hammers and De Ligny concluded that the experimental data on the n-alkanes examined in poly isobutylene could be interpreted quantitavely over a large range of temperatures with the (simplified) Prigogine theory, with two temperature independent parameters, A and B, and that the interchange enthalpy of hydrocarbons in poly isobutylene is larger than that in Apiezon and n-alkane solvents. They also examined the application of Flory–Huggins theory to their systems and found that the data were most consistent if analyzed in terms of the principle of corresponding states.

This type of analysis illustrates the power of GLC method for obtaining thermodynamic information to test various solution theories; but the treatment implied in the use of Eq. 17 is considered to be less satisfactory for high polymer stationary phases than the formulation of Patterson et al. (Eq. 2) because of the uncertainty involved in determining the partial molar volume of the polymer ($\bar{V}_2^0$).

IV.   SOLUBILITY PARAMETERS

Attempts have been made to derive solubility parameters and other more or less empirical estimates of polymer–solvent interactions from GLC data. A number of quite useful studies have yielded results of considerable practical value, particularly in relation to paint and lacquer technology. A recent survey of some of this work has been published by Reichert (29). The method also has value in estimating the solvent resistance of experimental elastomers (30). The difficulty with much of the early work was inherent in the inapplicability of regular solution theory (on which the solubility parameter concept is based) to macromolecular systems. However, recent developments in polymer solution theory, and particularly the work of Patterson and his co-workers, may lead to much better agreement between theory and experiment.

Various authors have proposed schemes relating to solvent–solvent interactions with GLC data. For example, Kovats (31) and also Kaiser (32) have introduced empirical parameters based on gas chromatographic measurements, representing, on the basis of their definition, characteristic values for the polar interactions of volatile substances with a stationary phase. Kovats defined a retention index I based on the behavior of linear aliphatic hydrocarbons having n and $n+2$ carbon atoms.

$$I = 200 \frac{\log V_r - \log V_{rn}}{\log V_{r(n+2)} - \log V_{rn}} + 100n \tag{21}$$

These index units were used by Rohrschneider (33) to characterize the polarity and selectivity of a variety of stationary liquid phases in order to predict column performance in analytical applications. In these experiments, the retention behavior of a polar liquid phase was compared with that of a standard nonpolar liquid phase (squalane) using the "retention index increment" ΔI defined by

$$\Delta I^{sq} = I_{polar} - I_{squalane} \tag{22}$$

The results were useful in correlating the separation efficiency of a variety of stationary phases.

A more direct application to polymeric materials is due to the recent work of Reichert (29, 34), who used GLC to characterize the interactions between solvents and polymeric binders used in paints and lacquers. Reichert used the logarithm of the retention volume at a temperature τ, $\log V_{r\tau}$, where τ corresponds to the boiling point of the solvent at the mean column pressure, determined graphically from the temperature dependence of the retention volumes, as an empirical parameter to estimate solvent–binder interactions. Table IX illustrates typical data for cyclized rubber.

Reichert showed that the $\log V_{r\tau}$ values correlate better with Hansen's polar solvent solubility parameter δ_a than with the Hildebrand solubility parameter δ, particularly for alcohols. The Hansen relation divides the Hildebrand solubility parameter into two terms, as follows:

$$\delta^2 = \delta_d^{\,2} + \delta_a^{\,2} \tag{23}$$

where δ_d is the portion of the solubility parameter due to dispersion forces between molecules and δ_a is the portion due to polar interaction forces, such as hydrogen bonds and dipole–dipole interactions. Similar data were obtained on polyvinyl acetate columns and compared with the empirical parameters suggested by Kovats and Kaiser; but it could not be stated definitely which approach gave the most satisfactory correlations. In general it appears that the more fundamental approach through the calculation of Flory–Huggins χ parameters would ultimately yield more useful data than the various empirical correlations used by Kovats, Kaiser, and Reichert.

TABLE IX

Gas Chromatographic Data for Solvent Retention
on 0.68 g of Cyclized Rubber

Solvent	Boiling Temperature		$\log V_r$ (90°C)	$\log V_{rr}$	δ	δ_a (Hansen)
	T,°C	τ,°K				
MeOH	64.7	360.5	4.20	4.31	14.28	12.40
EtOH	78.3	374.0	4.34	3.97	12.92	10.45
n-PrOH	97.2	393.0	4.39	3.49	11.97	9.10
MeF	31.5	329.5	3.30	4.08	10.05	6.66
EtF	54.0	353.0	3.63	3.92	9.55	6.10
n-PrF	81.2	383.0	3.88	3.51	9.56	6.09
Benzene	80.2	394.0	2.78	2.39	9.15	1.48
Toluene	110.8	412.0	3.15	2.17	8.91	2.18
p-Xylene	138.4	459.7	3.52	1.66	8.75	2.30
n-Hexane	69.0	372.0	2.08	1.97	7.24	(0)
n-Heptane	98.4	402.2	2.40	1.79	7.40	(0)
n-Octane	125.8	431.7	2.75	1.68	7.60	(0)

V. STUDY OF GLASS TRANSITION PHENOMENA IN POLYMERS

The use of GLC to determine glass transitions in polymers is due to the work of Smidsrød and Guillet (9), who demonstrated the effect shown in Figure 2, where a strong reversal of normal chromatographic behavior occurs at T_g. The glass transition represents the temperature at which an amorphous polymer, or the amorphous portion of a semicrystalline polymer, changes from a hard rigid glass to a rubberlike state. This change, which is accompanied by a rapid increase in the diffusibility of small molecules in the polymeric matrix, is associated with the availability of free volume for the motion of polymer segments of about 20 to 40 carbon atoms. Many of the physical and other properties of polymers show discontinuities or changes in slope at this temperature, and hence its determination is of considerable theoretical and practical importance. For a more detailed review of glass transition phenomena, the reader is referred to the review by Boyer (35).

In the GLC experiment, a glass transition can be considered to occur at the temperature at which a "nonsolvent" probe begins to penetrate the bulk of the polymer stationary phase. At lower temperatures, retention is almost exclusively by adsorption on the polymer surface. The discontinuity at T_g is caused by a bulk retention term, which increases rapidly with temperature and reverses the normal trend toward shorter retention times as the temperature is increased. Because of the mechanism of this second term (i.e., the occurrence of bulk diffusion through accessible free volume) the GLC method can be considered to be diagnostic for T_g, when other solid phase transitions relating to group motion also occur in the polymer. Since the size of the discontinuity in the GLC experiment is directly proportional to the thickness of the coating, it may not be observable in very thin coatings.

Working with poly N-isopropyl acrylamide, Smidsrød and Guillet (9) observed the effect only with compounds that were not solvents for the polymer at room temperature. Their results are shown in Figure 4. Acetic acid and n-butane exhibit "normal" linear relationships when log V_g is plotted as a function of $1/T$, whereas hexadecane, α-chloronaphthalene,

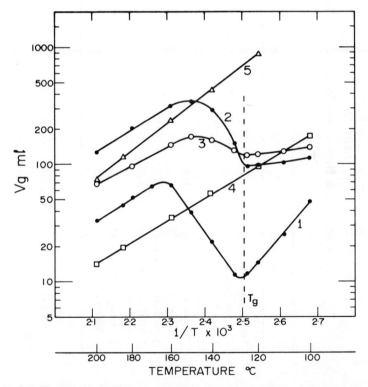

Fig. 4.   Specific retention volume as a function of reciprocal absolute temperature: curve 1, hexadecane; curve 2, α-chloronaphthalene; curve 3, naphthalene; curve 4, butyl alcohol; curve 5, acetic acid.

TABLE X

Summary of Data Derived from Retention Diagrams

Polymer substrate	PVC (A)	PVC (B)	Polystyrene	PMMA
Molecular weight (M_n)	68,000	35,500	51,000	48,000
Column loading, g	0.227	0.116	0.088	0.357
Solute	n-Dodecane	n-Dodecane	n-Dodecane	n-Hexadecane
DSC glass transition, °C	75±2	75±2	95±2	100±2
Temperature of first deviation from linear relation T_1, °C	81	81	88	97
Temperature of curve min T_2, °C	91	91	100	105
Enthalpy of adsorption ΔH_a, kcal/mole	−1.2	−1.3	−2.7	—[a]
Enthalpy of mixing ΔH_m, kcal/mole	+3.8	+3.8	+2.5	+4.4

[a] Vapor pressure data on n-hexadecane not available in this temperature range.

and naphthalene display discontinuities at T_g. The most pronounced effect is with the "poorest" solvent, hexadecane. Lavoie and Guillet (10) extended this work to a variety of other amorphous polymers and demonstrated that the effect was quite general when aliphatic compounds such as dodecane and hexadecane served as probe molecules. They also used the linear portions of the retention diagram below and above T_g to calculate enthalpies of adsorption and mixing for the polymer–probe system. The data of Table X indicate that the temperature at which deviation occurs from normal chromatographic behavior (T_1) correlates better with T_g, as determined calorimetrically, than does T_2, the tempera-

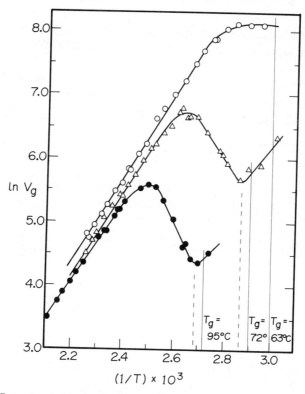

Fig. 5.   Retention diagrams for polystyrenes in the region of the glass transition:

open circles, $\overline{M}_w = 2,100$;

triangles, $\overline{M}_w = 4,000$;

solid circles, $\overline{M}_w = 51,000$.

ture of the minimum of the retention diagram. However, it is known that estimates of T_g by various methods can vary by as much as $\pm 5°C$; thus the precise value to use for the GLC method must await further study.

The glass transition temperature is nearly independent of polymer molecular weight if the $\overline{M}_w$ exceeds about 20,000. However, there is a considerable variation at lower molecular weights. This is evident in Figure 5, obtained by Lavoie and Guillet (36), which shows retention diagrams for polystyrene of three different molecular weights. In this case the T_g values determined calorimetrically agree quite well with the retention curve minima. Other experiments indicated that there was no difference in the position of the retention curve minima between probes of different size, from decane to octadecane.

It should be pointed out that it is not necessary to have the polymer coated on a support to show this effect. For example, Wallace et al. (37) reported the use of short columns packed with poly vinyl chloride (PVC) powder. The retention diagrams obtained are similar to those observed by Lavoie and Guillet (10) for PVC coated on Chromosorb. In this case, the curve minima represent the temperature required for plasticizer uptake ("drying temperature"); the values of V_R in the surface adsorption region, being directly proportional to the surface area of the powder, hence can be used to determine surface areas for PVC in a very simple experiment.

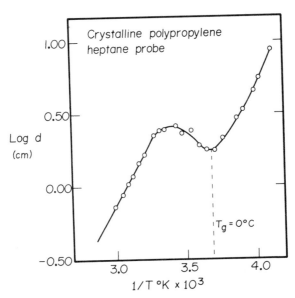

Fig. 6. Retention diagram for crystalline polypropylene using heptane probe (d = chart distance between peaks for methane and heptane).

The GLC method of detecting glass transitions is not restricted to glassy polymers. Semicrystalline polymers containing appreciable amorphous regions display similar behavior. Figure 6 shows the glass transition of crystalline polypropylene (ca. 60% crystallinity) observed using heptane as a probe (38). The value of $T_g = 0°C$ is in agreement with that determined by other physical tests.

Recently the method has been extended to study glass transitions in polyethylene terephthalate. Ateya et al. (39) studied columns made by packing polyethylene terephthalate in powder form, using methanol as a probe. No transition was observed with saturated hydrocarbon probes, but methanol gave values of T_g comparable to those obtained by DTA. After a series of heat treatments to increase crystallinity, the transition became undetectable by DTA, although it was still observable by GLC. Their data are summarized in Table XI.

TABLE XI

Glass Transitions of Polyethylene Terephthalate

Glass transitions	Temperature of heat treatment, °C					
	None	90	130	160	200	230
By GLC	76	78	93	98	100	100
By DTA	78	79	87	89	—[a]	—[a]

[a] Transition not detectable by DTA

Similar studies have been carried out by Galin and Guillet (40) using polyacrylonitrile coated on Chromosorb W or as the unsupported powder. The results for two different column loadings and acetonitrile as the probe appear in Figure 7. Polyacrylonitrile shows two different solid phase transitions, one at 85°C and the other variously reported at between 105 and 140°C. There has been some controversy in the literature regarding which of these is the glass transition temperature. Since both columns give the same retention volumes, although one contains four times the amount of polymer, it is clear that retention at temperatures below 110°C is due to surface adsorption only. A change in slope occurs at about 80°C, but there is no change in accessibility until about 110°C, which clearly identifies the upper transition at around 110°C as the glass transition for polyacrylonitrile. The transition at 85°C denoted T_p has been detected by a change in the linear expansion coefficient by Hayakawa et al. (41), who attributed

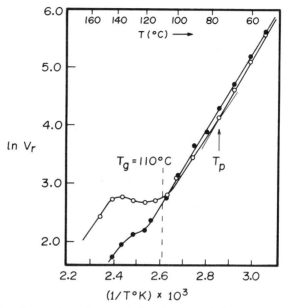

Fig. 7.   Retention diagram for polyacrylonitrile using acetonitrile as probe.

it to the onset of motion in the paracrystalline phase. Although the GLC evidence for this transition from Figure 7 is marginal, it is easily detected from observations of peak shape. Below T_p the GLC peak is strongly skewed, and the peak maxima are very dependent on probe concentration. Above T_p the peaks are symmetrical and give evidence of very little concentration dependence. As in the case of polyethylene terephthalate, the retention curve obtained for dodecane is completely linear; it is proportional to surface area, showing no evidence of transitions. Apparently, probes used for highly polar polymers must themselves be polar molecules.

Similar observations were made by Guillet and Pinchin (41) using columns filled with polyacrylonitrile fiber (Courtelle). The retention diagram for octadecane was linear from 80 to 150°C, as Figure 8 reveals. Therefore, such data can be used to determine heats of adsorption on the fiber surface and the surface area of the fibers by procedures discussed in a later section of this chapter. A similar change in slope of the log V_g versus $1/T$ plot was used by Nakamura et al. (43) to identify a new solid phase transition in cellulose triacetate at 83°C, well below the glass transition at 153°C. They used styrene, n-butyl acrylate, and methanol as probes and confirmed their hypothesis by demonstrating a change in slope of the specific volume–temperature curve at 91°C. They attributed this transition

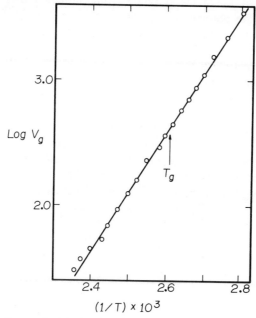

Fig. 8.   Retention diagram for Courtelle fiber (0.210 g) using octadecane probe.

to the onset of side-chain motion in the ester group.

Since the slope of the retention diagram in the surface adsorption region is related to the enthalpy of adsorption on the polymer surface, slope changes must be related to abrupt changes in the surface interaction between probe and polymer above and below the transition. It is not immediately clear why the enthalpy of adsorption should change in this way as a result of segmental or group motion in the polymer; however, the phenomenon appears to be general in highly polar polymers and may indeed be a fruitful area for future research.

VI. DETERMINATION OF CRYSTALLINITY IN MACROMOLECULES

One of the most promising applications of gas chromatography to the study of polymers is in the detection and estimation of crystalline order. In 1965 Alishoev, Berezkin, and Mel'nikova (44) reported studies of mixtures of solid polyethylene or polypropylene powders with glass beads, as stationary phases in a gas chromatograph. They showed that drastic changes in retention volume and peak width occurred in the region of the melting transition T_m of the polymers, but indicated that the effect disappeared when the polymer was in the form of a thin film. Later Altenau et al. (45), in a study of GC columns at cryogenic temperatures, observed abrupt discontinuities in the log V_g versus $1/T$ curve on Carbowax 20M at 52°C. They correctly attributed this condition to changes in the accessibility of the polymer phase due to a melting transition. A similar discontinuity was observed on squalane at $-60°C$. Altenau also assumed this to be due to a melting transition, but it is more probably a glass transition.

In 1970 Stein and Guillet (11) reexamined the polyethylene system using hydrocarbon probes. Contrary to the results of Alishoev et al., they found that discontinuities did indeed appear at the melting transition T_m when the polymer was spread as a thin film on a support such as Chromosorb W. Furthermore, the peaks were narrow and symmetrical both below and above T_m, which suggested that equilibrium sorption was occurring both in the solid and molten liquid phases. Above T_m the log V_g versus $1/T$ curve was linear and reproducible, as represented in Figure 9. The reproducibility of slope and intercept of the least-squares straight line

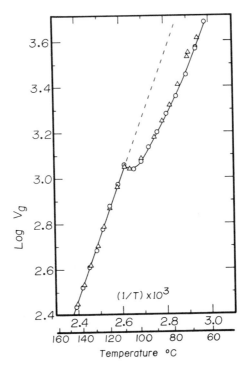

Fig. 9.   Generalized retention curve for polyethylene using dodecane probe: circles, melting curve, first series; triangles, melting curve, 24 hr later.

TABLE XII

Least-Squares Analysis of Retention Data on Polyethylene above T_m

Series	Slope	Standard deviation	%	Intercept	Standard deviation	%	Correlation coefficient
1	2776.6	20.3	0.73	−4.2073	0.0015	0.04	.9998
2	2751.2	12.4	0.45	−4.1449	0.0010	0.02	1.0000
3	2720.6	73.9	2.72	−4.0496	0.0020	0.05	.9972
Composite	2794.6	15.7	0.56	−4.2515	0.0011	0.03	.9996

(Table XII) permits an accurate extrapolation to give $V_g{}'$, the theoretical retention volume for a 100% amorphous polymer at any desired temperature. If we assume that the crystalline regions are impermeable to the probe molecules during the time of passage of a GC peak, then since V_g is proportional to the mass of the absorbing phase, the weight fraction of amorphous material in the polymer at any temperature T less than T_m is given by

$$w_a = \frac{(V_g)_T}{(V_g{}')_T} \tag{24}$$

and hence the percentage of crystallinity can be given by

$$\% \text{ crystallinity } = 100\left(1 - \frac{(V_g)_T}{(V_g{}')_T}\right) \tag{25}$$

This simple relation does not depend on the column loading, and it requires no independent method of calibration. For the polyolefins at least, the relation seems to apply with astonishing reliability; but it has not yet been tested adequately with other types of crystalline polymers. Its greatest application would seem to be in the evaluation of crystallinity in copolymer systems, where other methods of crystallinity determination are difficult to apply.

In applying the GLC method, it is important to be assured that the probe molecule is not dissolving or melting any of the crystalline regions. This is done by observing whether there is a concentration dependence of the retention time in the semi crystalline polymer.[11] Data on a typical experiment are presented in Figure 10. It is clear that for probe concentrations below a certain critical value, the retention time is independent of

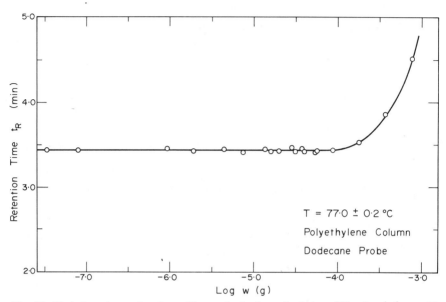

Fig. 10. Variation of retention time with sample size for polyethylene (11) using dodecane as probe molecule ($T = 77.0 \pm 0.2°C$).

probe concentration and is essentially the same as at infinite dilution. The w_a values are also independent of the particular hydrocarbon probe used. Data on polypropylene with hexadecane and tetradecane are plotted in Figure 11. The successful application of the GLC technique to crystallinity in polyethylene prompted the design and construction of an automatic apparatus for collection of such data. Gray and Guillet (46) reported data obtained on an "Automatic Molecular Probe Apparatus," consisting of an automatic system for sample injection and for measurement of gas chromatographic retention time. The apparatus was based on a Varian Aerograph model 1720 gas chromatograph equipped with a thermal conductivity detector. Helium was used as a carrier gas. At a preset cycle time, a mixture of nitrogen and decane vapor was introduced into the carrier gas stream by way of an electropneumatic injection system. The sample size, which was approximately constant for every injection in a series, was as small as practicable. (All samples contained less than 1.5×10^{-6} mole of decane.)

The net gas chromatographic retention time for decane at a given temperature was measured by feeding the output from the thermal conductivity detector into an electronic peak detection system, which measured the time between the peak maxima for nitrogen (noninteracting) and decane (interacting). The corresponding temperature was measured using

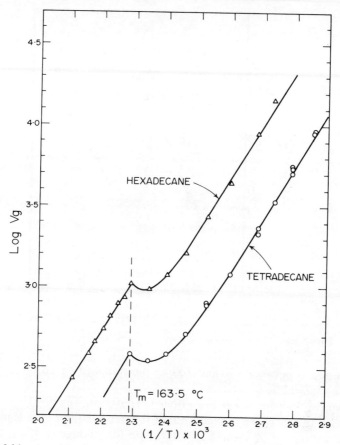

Fig. 11.   Melting transition in crystalline polypropylene using hexadecane and tetradecane probes.

an iron–constantan thermocouple attached to the outside of the gas chromatograph column. A digital printer served for recording the net retention time and the temperature. The carrier gas flow rate, measured with a soap-bubble flowmeter, was adjusted to give retention times of from 10 to 500 sec; retention times were reproducible to ± 0.2 sec at temperatures above the polymer melting points.

Experiments to determine the generalized retention diagram were performed at a heating rate of $0.5°/\text{min}$, using the column oven linear temperature programmer. At temperatures above the polymer melting point, a straight-line relationship was obtained. Using the automatic injection–detection system, the linearity was excellent. In a typical case for

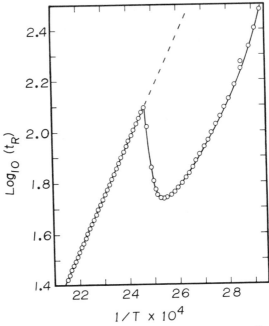

Fig. 12.   Plot of $\log_{10}$ (retention time, sec) versus $1/T\,°K$ for decane on linear polyethylene.

34 data points between 140 and 200°C, the standard deviation in the slope was less than 0.2%.

Typical data for linear polyethylene (Eastman Tenite 3310) appear in Figure 12. In this case the logarithm of the net retention time is plotted, as defined by $t_R \equiv t_p - t_m$, where t_p is the time for passage of the probe peak maxima and t_m that of the marker gas (nitrogen in this case). From the data so obtained, the percentage of crystallinity at any temperature below T_m is calculated by

$$\% \text{ crystallinity } = 100\left(1 - \frac{(t_R)_T}{(t_a)_T}\right) \qquad (26)$$

where t_a is the hypothetical retention time of 100% amorphous polymer calculated from the extrapolated melt curve. This can be done at a series of temperatures with a small computer program to generate the data in Figure 13.

Curve 3 in Figure 13 was obtained for comparison using a Perkin-Elmer differential scanning calorimeter to determine the heat of fusion of a sample of the same polyethylene. The sample was subjected to a heating

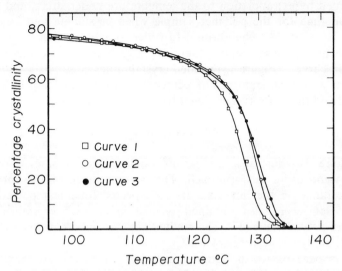

Fig. 13   Crystallinity of high-density polyethylene; melting curves using molecular probe and calorimetric data. Solid circles, DSC data, annealed at 1.25°C/min, heated at 1.25°C/min; open circles, molecular probe data, annealed at 1.0°C/min, heated at 0.5°C/min; squares, molecular probe data, annealed ballistically, heated at 0.5°C/min.

and cooling cycle similar to that used with the gas chromatograph, and its fusion curve was then obtained using the DSC. By integration of this curve, assuming a value of 68.5 cal/g for the latent heat of crystallization of perfectly crystalline polyethylene, the percentage crystallinity of the sample was determined as a function of temperature. The agreement between the calorimetric and the gas chromatographic data is very good when the sample is annealed under comparable conditions (curve 2). Curve 1 illustrates the effect of annealing conditions on the shape of the curve.

   When the polymer on the gas chromatographic column is cooled from the melt to a temperature below the polymer melting point T_m, the retention times for the probe molecule decrease with time. This reflects the decrease in the available amorphous material as crystallization proceeds. The decrease in retention times can be used to follow the crystallization kinetics.

   The maximum possible degree of crystallinity at a given temperature is related to gas chromatographic retention times by the equation

$$\% \text{ crystallinity }_{\text{max}} = 100\left(1 - \frac{t_e}{t_a}\right) \qquad (27)$$

where t_a is the retention time expected for a totally amorphous sample, determined by extrapolation to the temperature under study, and t_e is the retention time for the polymer sample when crystalline and amorphous regions have reached equilibrium. In order to avoid extremely long equilibration times, it was assumed that the pretreatment and heating rate used to obtain curve 2 (Figure 13) yielded crystallinities close to the equilibrium values at each temperature. The percentage of crystallization of a sample at any fixed flow rate is then given by

$$100\% \text{ crystallinity}/\% \text{ crystallinity }_{max} = \left(\frac{t_a - t_m}{t_a - t_e}\right)100 \qquad (28)$$

Figure 14 illustrates the results obtained for a sample of linear polyethylene using this approach. The general shape of the curves and their positions on the time axis are as expected for linear polyethylenes. Similar data were also obtained on the kinetics of crystallization of branched polyethylene (47). The method appears to be sensitive to changes in crystallinity of the order of 0.1%.

It is perhaps worth pointing out that all other determinations of crystallinity ultimately involve X-ray determinations of unit cell dimensions to obtain the theoretical properties of 100% crystalline polymer. Even the calorimetric method must be calibrated by the determination of heat of fusion of the hypothetical 100% crystalline polymer. Thus the GLC method has many advantages, particularly in the study of new polymer or copolymer compositions for which X-ray data are unavailable or inapplicable.

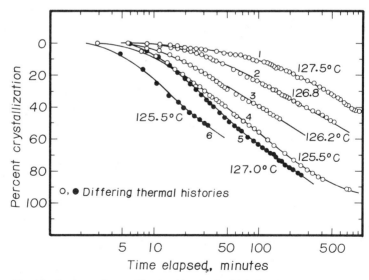

Fig. 14.   Isothermal crystallization kinetics for high-density polyethylene.

VII. SURFACE PROPERTIES OF POLYMERS

At temperatures below the glass transition temperature T_g, polymers become impermeable to a wide variety of probe molecules unless there is a strong interaction resulting in solution (9). In this temperature region it is possible to use gas chromatography to investigate various properties of the polymeric surface.

In previous work, the interaction of the probe molecule with the polymer was measured by its specific retention volume, V_g, a fundamental gas chromatographic parameter that is independent of experimental conditions, depending only on the column temperature and the chemical nature of the polymer and probe molecules. In particular, meaningful retention volumes should be independent of the size of the injected probe sample. This was found, for example, with dodecane dissolved in polyethylene (11), the implication being that the equilibrium concentrations in the polymer and in the gas phase are linearly related for sufficiently dilute solutions. However, when retention of the probe is due to adsorption on the surface of a solid rather than in solution, the isotherms relating concentrations in the gas phase and on the surface are often curved because of surface heterogeneity and incipient saturation of the available sites. Furthermore, the sample sizes normally injected in gas chromatography often span a considerable portion of the isotherm; thus it is not always possible to work at concentrations of probe low enough to ensure effective linearity. Since retention volumes in this case depend on sample size, the shape of the isotherm must also be determined in order to characterize the adsorbate–adsorbent interaction.

A. Adsorption Isotherms

Gas chromatographic techniques have been widely used to determine adsorption isotherms (48, 49). Although most work has involved the use of "frontal" techniques (50), it is also possible to use the conventional elution technique and to determine the shape of the isotherm from a single unsymmetrical peak (51). Starting from the first-order conservation equation of chromatography (52), it can be shown that for each gas phase concentration c of eluted probe vapor, there exists a corresponding retention volume V_c given by

$$a = \frac{1}{m} \int_0^c V_c \, \mathrm{d}c \tag{29}$$

where a is the amount of probe vapor adsorbed on a mass m of adsorbent.

In the absence of all peak broadening factors other than the nonlinearity of the isotherm, one side of the elution peak should be vertical; V_c can be determined experimentally from the other diffuse side of the peak (48, 52). Hence the isotherm relating a and c may be found using Eq. 29.

This method is less accurate than frontal analysis, since all kinetic factors and gas phase volume changes due to vapor adsorption are ignored. However, it has the advantage of being much faster and simpler experimentally; and if conditions are chosen to minimize errors, the resultant adsorption isotherms agree with those obtained by static measurements.

Gray and Guillet (53) have recently published studies of the surface properties of poly(methyl methacrylate) (PMMA) using the GLC method. When large injections of decane are made on a column of PMMA beads at 25°C, the peak shapes in Figure 15 are obtained.

The shape of the experimental elution peaks changed as the amount of n-decane was increased; the front profiles became increasingly diffuse but

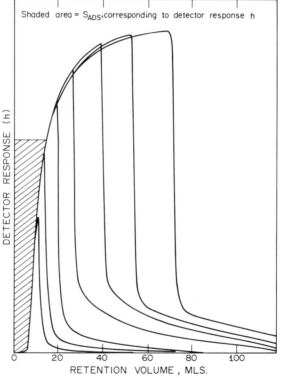

Fig. 15. Chromatographic peak shapes for large injections of n-decane on PMMA beads at 25° C. Injection sizes: 0.03, 0.06, 0.15, 0.3, 0.5, 0.7, and 1.0μl.

fell on a common curve, whereas the rear profiles remained almost vertical. The conditions required for the calculation of isotherms from elution chromatograms were thus obeyed for this system, and Eq. 29 can be applied to obtain the amount of n-decane sorbed on the column as a function of its partial pressure in the following simple manner (54). The partial pressure of the n-decane vapor entering the gas chromatographic detector can be related to the experimental chromatogram by

$$p = \frac{m_{cal}\,qRT}{S_{cal}\,V} \times h \qquad (30)$$

where p is the partial pressure of eluting n-decane at column temperature $T°K$ corresponding to recorder pen deflection h; S_{cal} is the calibration peak area on recorder chart for an injection of m_{cal} moles of n-decane; q is the recorder chart speed; R is the gas constant; and V is the carrier gas volume flow rate at column temperature. The amount of decane sorbed corresponding to this partial pressure p is given by

$$a = \frac{m_{cal}\,S_{ads}}{m S_{cal}} \qquad (31)$$

where a is the amount of decane sorbed (per unit weight of sorbent); m is the mass of sorbent; and S_{ads} is the chart area bounded by diffuse profile of chromatogram, carrier gas front, and height h. By measuring areas S_{ads} corresponding to different values of h, the isotherm relating a and p may be calculated using Eqs. 30 and 31.

A series of isotherms of n-decane on PMMA at a number of temperatures obtained by this method is presented in Figure 16. The isotherms correspond to type III in Brunauer's classification (55), which has been ascribed to a multilayer adsorption where the heat of adsorption is equal to or less than the heat of liquefaction of the adsorbate. As the amount of adsorbate on the surface is increased, the partial pressure of its vapor, as a limit, should approach the vapor pressure of the pure adsorbate at the temperature in question. This was found to be the case for n-decane on PMMA and was confirmed by determining the sorption isotherms at 25°C for n-decane on polystyrene-coated glass beads and poly(vinyl acetate)-coated glass beads. The adsorption per unit geometric area was very similar for a given partial pressure of decane, and the vapor pressure of pure decane was approached asymptotically. Thus for the sorption of n-decane on these glassy polymers, the shape of the isotherm at high coverages is governed primarily by surface saturation.

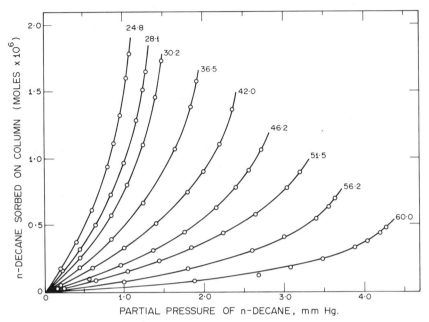

Fig. 16.   Isotherms at the indicated temperatures for *n*-decane on PMMA beads.

When small injections of decane were made (i.e., $< 0.02\mu l$) the peak retention volumes on PMMA remained almost constant for injected sample sizes ranging over four orders of magnitude. The peaks were sharp, with only slight tailing. It is therefore reasonable to assume that in this region the distribution isotherm is effectively linear and that thermodynamic information can be properly derived from peak retention data.

B.   Heats of Adsorption

When the gas chromatographic retention volumes for a polymer–probe system are independent of the size of the probe sample at low sample sizes, the heat of sorption can be readily determined from the change in retention volume with temperature (48, 49).

$$\frac{\partial \ln V_g}{\partial (1/T)} = \frac{\partial \ln (V_s/T)}{\partial (1/T)} = \frac{\Delta H_s}{R} = \frac{\Delta H_v - h_e}{R} \tag{32}$$

where V_g is the specific retention volume measured at 0°C; V_s is the retention volume per unit surface area, measured at column temperature T°K; ΔH_s is the overall heat of desorption from the surface into the vapor phase; ΔH_v is the latent heat of evaporation of the probe molecule; and h_e is the excess enthalpy of sorption of the liquid probe onto the polymer. The retention volumes for small injections of n-decane on PMMA were measured between 25 and 60°C. A plot of the logarithm of the retention volume against reciprocal temperature (Figure 17) was linear, enabling ΔH_s to be determined at effectively zero surface coverage. For this system ΔH_s was found to be the same (12.0 ± 0.2 kcal mole^{-1}), within experimental error, as ΔH_v for n-decane. Thus any specific interaction between the hydrocarbon probe and the polymer surface must be very small. Values of ΔH_s at various surface coverages can also be calculated from the isotherms in Figure 16 and similar values are obtained. For n-hexanol the value of ΔH_s was 12.8 ± 0.2 kcal/mole.

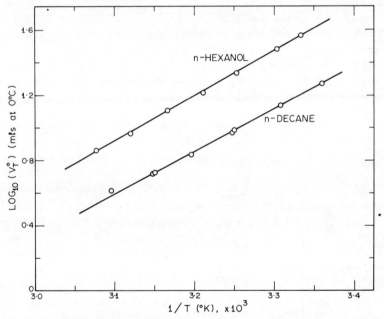

Fig. 17. Arrhenius plots of retention volumes for n-decane and n-hexanol on PMMA beads.

C. Measurement of Surface Areas

Gas chromatographic measurements of the specific surface area of solids have been performed by three different methods. Although not strictly a chromatographic procedure, the thermal desorption method (56) has been widely applied, in certain cases to polymeric substrates (57). The area, which is determined from the amount of nitrogen sorbed at liquid nitrogen temperatures, is usually applied to materials with relatively high specific surface areas.

Secondly, when the adsorption isotherm is linear and surface adsorption is the only retention mechanism, the measured retention volume is linearly related to the surface area for a given adsorbate, adsorbent, and temperature. Thus the retention volume per unit surface area at a given temperature can be calculated by measuring the retention volume for small injections of n-decane on a PMMA column of known surface area. For example, by interpolation of the data in Figure 17, the retention volume is 19.9 ml measured at 25°C, giving a surface retention volume of 80 ml/m^2 for n-decane on PMMA. Hence the surface area of other samples of PMMA can be obtained readily by measuring the retention volume of n-decane on a suitable column of the material. When applicable, this is the most convenient method for determining surface areas, but it requires a linear distribution isotherm for the probe on the polymer surface and the absence of bulk sorption effects.

The third method involves calculation of the isotherm shape from chromatographic data for relatively high coverage of a suitable probe, and applying the Brunauer–Emmett–Teller (BET) approach (55) to deduce the specific surface area from the isotherm. A two-parameter BET equation was used in the form

$$\frac{x}{v(1-x)} = \frac{1}{v_m C} + \frac{C-1}{v_m C} x \qquad (33)$$

where $x = p/p_0$, p is the vapor pressure of the adsorbate over the surface, p_0 is the saturation vapor pressure of pure liquid adsorbate, v is the volume of adsorbate on the surface, v_m is the volume of adsorbate on the surface corresponding to monolayer coverage, and C is a constant (for $0.05 < x < 0.35$). Using values for v and x from the experimental isotherms, $x/v(1-x)$ was plotted as a function of x, and v_m was calculated from the slope and intercept of the resultant straight line. Thus, if the surface area covered by the probe molecule is known, the area of the surface in the chromatographic column can be obtained. The surface area occupied by an n-decane molecule on a polymer surface, where only weak interactions are expected, can be extrapolated approximately from Kiselev's data for a series of n-alkanes on graphitized carbon black (58), where somewhat similar surface species are to be expected. Taking 70 Å^2 as this area and estimating the point of monolayer formation from the slope and intercept of BET plots for values of p/p_0 between 0.05 and 0.35, the values obtained for surface areas were comparable with the geometric surface areas.

Based on these studies, Gray and Guillet concluded that, under appropriate conditions, simple elution techniques of gas-adsorption chromatography can be applied to polymeric substrates to obtain sorption isotherms and hence to determine surface areas and enthalpies of adsorption. A primary requisite for this application is that bulk sorption into the polymer does not contribute to the retention of the adsorbate on the gas chromatographic column. Thus only compounds having little tendency to dissolve in or swell the polymer in the time scale of the gas chromatographic experiment are useful as adsorbates. Also, bulk sorption can be reduced by working at column temperatures well below the glass transition temperature of the polymer under study, where the rate of penetration is slower.

Columns of relatively low internal surface area are necessary to avoid kinetic effects on peak shape, which would invalidate the method used to calculate adsorption isotherms. In view of the low column area and the resultant use of large, relatively nonvolatile adsorbate compounds, the agreement between the geometric surface areas of the polymers and the areas measured by the BET treatment of experimental isotherms is satisfactory. The use of relative peak retention volumes, measured at effectively zero coverage, provides an alternative and more rapid method of surface area determination. In this case, no restriction on the surface area of the column is necessary; in fact as the specific surface area of the polymer is increased, the linear distribution region becomes more accessible and measurement of retention values becomes more accurate.

When a linear distribution region is accessible at low adsorbate concentrations, enthalpy data on adsorbate–polymer interactions can be derived readily from the variation in peak retention volumes with temperature. However, the results for n-hexanol on PMMA reveal that due allowance must be made for possible adsorbate–adsorbate interactions. Theoretically, when the distribution isotherms are nonlinear, the temperature dependence of the isotherm shape should make possible the calculation of adsorption enthalpies as a function of surface coverage. However, the accuracy of adsorption enthalpies calculated by this method from nonlinear isotherms appears to be unsatisfactory. Nevertheless, a significant amount of information on the properties of polymer surfaces can be obtained by simple and rapid gas chromatographic experiments.

VIII. DETERMINATION OF DIFFUSION CONSTANTS

If column parameters are selected carefully, it is possible to utilize a GLC method to estimate diffusion constants for probe molecules in polymeric substrates. The method utilizes the van Deemter (59) relation

$$H = A + \frac{B}{u} + Cu \qquad (34)$$

where the plate height H can be defined as $\sigma^2_{(x)}/x$, where $\sigma_{(x)}$ is the standard deviation of a peak at a point distance x from the start of the column, u is the linear carrier gas flow rate, and A is a constant independent of flow rate.

In the second term, which describes the diffusional spreading of the vapor as it is eluted in the gas phase, $B = 2\gamma D_g$, where D_g is the diffusion coefficient of the vapor in the gas phase and γ is a constant less than unity introduced to allow for the tortuous path followed by the gas flow through a packed column. In the simple van Deemter approach, we can write

$$C = \frac{8}{\pi^2} \frac{d_f^2}{D_2} \frac{k}{(1+k)^2} \tag{35}$$

where d_f is the thickness of the stationary phase; D_2 is the diffusion coefficient of the vapor in the stationary phase; the partition ratio $k = (t_r - t_m)/t_m$, where t_r is the retention time for injection for the vapor; and t_m is the time for the carrier gas to pass through the column.

Of the many factors that lead to peak broadening, only the kinetics of diffusion of the volatile probe in the polymer are of interest here. If the broadening factors are independent, then the overall variance, σ^2_{total} is the sum of the individual variances, $\sum_i \sigma_i^2$. Since the actual peak spreading is measured by the standard deviations σ_i, the total broadening is less than the sum of the individual standard deviations

$$\sum_i \left(\sigma_i^2 \right)^{1/2} \leqslant \sum_i \sigma_i \tag{36}$$

Thus the total broadening is mainly due to the factor producing the largest individual broadening, and by a suitable choice of conditions the factor of interest—namely, diffusion in the stationary phase—can be made to predominate. One important condition is that the amount of vapor injected be as small as possible; at infinite dilution, the partition isotherms may approach linearity, and also the simple model for diffusion assumed in the deviation of the van Deemter equation may be adequate.

Gray and Guillet (60) have recently reported data on the diffusion coefficients for n-tetradecane, n-decane, and benzene in low-density polyethylene and natural rubber by this GLC method. Their approach involved coating the polymer on carefully screened glass beads to give nearly uniform particle size and coating thickness. The value of H was then determined from the relation

$$H = \frac{l}{5.54} \left(\frac{d}{t_r} \right)^2 \tag{37}$$

where t_r is the indicated retention time from injection to peak maximum, d is the measured peak width at half-height, and l is the column length.

The experimental determination of d and t_r was performed in triplicate for each flowrate, and an average plate height was calculated. The linear portion of a graph of H against $\bar{u}$ was used to calculate C in the van Deemter equation.

Typical van Deemter curves are plotted in Figure 18, and diffusion constants determined for low-density polyethylene are summarized in Table XIII. The values for benzene agree quite well with those obtained by McCall and Slichter (61) by desorption measurements. The data for n-tetradecane above the melting point of polyethylene gave a linear Arrhenius plot indicating an activation energy for diffusion of 16.2 kcal/mole, which is similar to that for viscous flow.

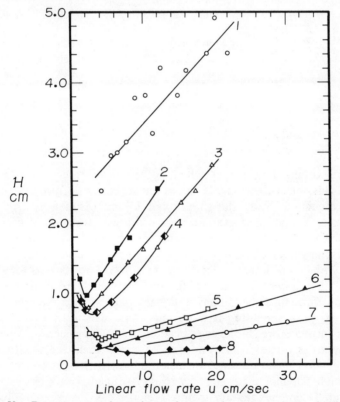

Fig. 18. Van Deemter curves for various polymer–probe systems: 1, T_{10}, benzene–PE at 27°C; 2, T_{35}, benzene–natural rubber at 25°C; 3, T_{40}, benzene–PE at 25°C; 4, T_{34}, benzene–natural rubber at 23°C; 5, T_{15}, n-nonane–PE at 50°C; 6, T_{44}, decane–natural rubber at 40.2°C; 7, T_{48}, dodecane–natural rubber at 685°C; 8, T_{13}, hexadecane–PE at 200° C.

TABLE XIII

Diffusion Coefficients from Gas Chromatographic
Measurements on Low-Density Polyethylene

Probe	Temperature, °C	van Deemter C term, sec × 10³	k	$k/(1+k)^2$	D, cm/sec × 10⁸
n-Tetradecane	125	20 ± 7	20.4	0.045	~0.85
	140	13.0 ± 0.06	11.3	0.075	2.2
	150	10.3 ± 0.4	7.83	0.100	3.7
	160	12.5 ± 0.6	5.47	0.131	4.1
	170	8.6 ± 0.5	3.88	0.163	7.4
n-Decane	30	30.0 ± 1.8	34.3	0.028	0.35
	50	24.6 ± 0.7	13.2	0.066	1.03
	60	31.8 ± 0.7	10.0	0.083	1.00
	65	29.9 ± 1.5	8.0	0.099	1.28
	80	38.2 ± 1.0	5.3	0.133	1.34
Benzene	25	114 ± 6	1.50	0.240	0.82

These results reveal that when polyethylene is coated evenly on spherical glass beads, reasonable values for penetrant diffusion coefficients can be calculated, using the simple van Deemter equation, from measurements of gas chromatographic peak width as a function of carrier gas flow rate. The success of this simple approach is perhaps surprising in view of the discrepancies often found with conventional gas chromatographic stationary phases of lower molecular weight (62). Two reasons for this are possible. First, rates of diffusion through polymers are generally much lower than through conventional stationary phases; thus the relative contribution to peak spreading is more important. Second, the very viscous polymers form a more stable, even bead coating. The gas chromatographic measurement of diffusion rates appears to complement the usual sorption and permeation methods, since diffusion at low penetrant concentrations and through polymer melts may be readily studied. However, use of the simple van Deemter equation to interpret the gas chromatographic data may only be valid for relatively nonpolar penetrants and polymers at temperatures well above the glass transition temperature of the polymer. Otherwise, strong polymer–penetrant interactions may result in thermodynamic peak spreading and a diffusion coefficient that is dependent on both time and penetrant concentration.

IX. OTHER APPLICATIONS

It is clear from the foregoing text that inverse gas chromatography provides a powerful tool for the study of polymer substrates. Since polymer substrates are usually stable in an inert atmosphere in a GLC column, many experiments can be carried out on the same sample. However, the method is not restricted to stable materials. For example, the curing of nonvolatile paint compositions can be followed by changes in the retention times for suitable probe molecules (63). Studies have also been reported by Perrault on the curing of propellant–binder systems in a GLC column, the progress of the cure being followed by changes in retention time of a hydrocarbon probe (64).

In a similar manner, the cross-linking of polyethylene by γ-radiation was observed by changes in the retention diagram as a function of radiation dose (65). Figure 19 gives typical retention diagrams for linear

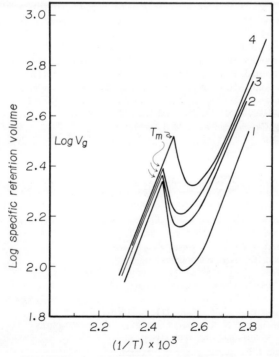

Fig. 19.   Retention diagrams for irradiated linear polyethylene: curve 1, unirradiated; curve 2, irradiated 2.5 μrad; curve 3, irradiated 5.0 μrad; curve 4, irradiated 20 μrad.

polyethylene after various doses of γ-radiation. It is clear that a substantial reduction in crystallinity occurs before there is much alteration of the melting point T_m of the polymer. It is evident that substantial changes occur in the retention data and that, in principle, the method could be used to follow the effects of a cross-linking reaction.

References

1. J. E. Guillet, W. C. Wooten, and R. L. Combs, *J. Appl. Polym. Sci.*, **3**, 61 (1960).
2. E. A. Radell and H. C. Strutz, *Anal. Chem.*, **31**, 1890 (1959).
3. R. S. Lehrle and J. C. Robb, *Nature*, **183**, 1671 (1959).
4. J. Strassburger, G. M. Brauer, M. Tryon, and A. F. Forziati, *Anal. Chem.*, **32**, 454 (1960).
5. J. E. Guillet, "Gas Chromatography of Polymers," *Soap Chem. Spec.*, December 1965.
6. J. Dhanraj and J. E. Guillet, *J. Polym. Sci., Part C*, **23**, 433 (1968).
7. L. Michajlov, P. Zugenmaier, and H. J. Cantow, *Polymer*, **9**, 325 (1968).
8. J. E. Guillet, *J. Macromol. Sci.—Chem.*, **A4**, 1669 (1970).
9. O. Smidsrød and J. E. Guillet, *Macromolecules*, **2**, 272 (1969).
10. A. Lavoie and J. E. Guillet, *Macromolecules*, **2**, 443 (1969).
11. A. N. Stein and J. E. Guillet, *Macromolecules*, **3**, 102 (1970).
12. J. R. Conder, D. C. Locke, and J. H. Purnell, *J. Phys. Chem.*, **73**, 700 (1969).
13. D. E. Martire and J. H. Purnell, *Trans. Faraday Soc.*, **62**, 710 (1966).
14. D. D. Patterson, Y. B. Tewari, H. P. Schreiber, and J. E. Guillet, *Macromolecules*, **4**, 356 (1971).
15. B. E. Eichinger and P. J. Flory, *Trans. Faraday Soc.*, **64**, 2035 (1968).
16. J. R. Conder and J. H. Purnell, *Trans. Faraday Soc.*, **64**, 1505 (1968).
17. J. R. Conder and J. H. Purnell, *Trans. Faraday Soc.*, **64**, 3100 (1968).
18. J. R. Conder and J. H. Purnell, *Trans. Faraday Soc.*, **65**, 824 (1969).
19. J. R. Conder and J. H. Purnell, *Trans. Faraday Soc.*, **65**, 839 (1969).
20. N. F. Brockmeier, R. W. McCoy, and J. A. Meyer, *Macromolecules*, **5**, 464 (1972).
21. C. L. Young, *Chromatogr. Rev.*, **10**, 129 (1968).
22. H. P. Schreiber, Y. B. Tewari, and D. Patterson, *J. Polym. Sci.,* Polym. Phys. ed., **11**, 15 (1973).
23. D. Patterson, *Macromolecules*, **2**, 672 (1969).
24. W. R. Summers, Y. B. Tewari, and H. P. Schreiber, *Macromolecules*, **5**, 12 (1972).
25. Y. B. Tewari and H. P. Schreiber, *Macromolecules*, **5**, 329 (1972).
26. G. M. Bristow and W. G. Watson, *Trans. Faraday Soc.*, **54**, 1567 (1958).
27. R. D. Newman and J. M. Prausnitz, *J. Phys. Chem.*, **76**, 1492 (1972).
28. W. E. Hammers and C. L. De Ligny, *Rec. Trav. Chim. Pays-Bas*, **90**, 912 (1971).
29. K. H. Reichert, *J. Oil Colour Chem. Ass.*, **54**, 887 (1971).
30. H. K. Frensdorff, private communication.
31. E. Kovats, *Helv. Chim. Acta*, **41**, 1915 (1958).
32. R. Kaiser, *Chromatographie in der Gasphase*, vol. III/2, Mannheim, 1969, Bibl. Inst., p. 182.
33. L. Rohrschneider, *Z. Anal. Chem.*, **211**, 18 (1965).
34. M. Schneider and K. H. Reichert, *Dtsch. Farb. Z.*, **24**, 492 (1972).
35. R. F. Boyer, *Rubber Rev.*, **36**, 1303 (1963).
36. A. Lavoie and J. E. Guillet, paper in preparation.
37. J. R. Wallace, P. Kozak, and F. Noel, *Soc. Plast. Eng. J.*, **26**, 43 (1970).

38. J. E. Guillet, unpublished work.
39. K. Ateya, B. Chabert, J. Chauchard, and G. Edel, *C.R. Acad. Sci. Paris*, **274**, 506 (1972).
40. M. Galin and J. E. Guillet, *J. Polym. Sci.,* in press.
41. R. Hayakawa, T. Nishi, K. Arisawa, and Y. Wada, *J. Polym. Sci., Part A-2*, **5**, 165 (1967).
42. J. E. Guillet and D. Pinchin, unpublished work.
43. S. Nakamura, S. Shindo, and K. Matsuzaki, *Polym. Lett.*, **9**, 591 (1971).
44. V. R. Alishoev, V. G. Berezkin, and Yu. V. Mel'nikova, *Zh. Fiz. Khim.*, **39**, 200 (1965).
45. A. G. Altenau, R. E. Kramer, D. J. McAdoo, and C. Merrit, *J. Gas Chromatog.*, **3**, 96 (1966).
46. D. G. Gray and J. E. Guillet, *Macromolecules*, **4**, 129 (1971).
47. A. N. Stein, D. G. Gray, and J. E. Guillet, *Brit. Polym. J.*, **3**, 175 (1971).
48. A. V. Kiselev and Ya. I. Yashin, *Gas-Adsorption Chromatography*, Plenum Press, New York, 1969.
49. A. B. Littlewood, *Gas Chromatography*, 2nd ed., Academic Press, New York, 1970.
50. D. H. James and C. G. S. Phillips, *J. Chem. Soc.*, 1066 (1954).
51. E. Gluekauf, *J. Chem. Soc.*, 1302 (1947).
52. D. De Vault, *J. Amer. Chem. Soc.*, **65**, 532 (1943).
53. D. G. Gray and J. E. Guillet, *Macromolecules*, **5**, 316 (1972).
54. A. V. Kiselev, Le Zung, and Yu. S. Nikitin, *Kolloid Zh.*, **33**, 224 (1971).
55. S. Brunauer, *The Adsorption of Gases and Vapors*, vol. 1, Princeton University Press, Princeton, N.J., 1945.
56. E. M. Nelsen and F. T. Eggertsen, *Anal. Chem.*, **30**, 1387 (1958).
57. G. N. Fadeyev and T. B. Gavrilova, *Zh. Fiz. Khim.*, **42**, 3075 (1968).
58. A. V. Kiselev, *Proceedings of the International Industrial Congress on Surface Activity*, Butterworths, London, 1957, p. 168.
59. J. J. van Deemter, F. J. Zuiderweg, and A. Klinkenberg, *Chem. Eng. Sci.*, **5**, 271 (1956).
60. D. G. Gray and J. E. Guillet, *Macromolecules*, in press.
61. D. W. McCall and W. P. Slichter, *J. Am. Chem. Soc.*, **80**, 1861 (1958).
62. J. C. Giddings, *Dynamics of Chromatography*, Part I, Dekker, New York, 1965.
63. J. E. Guillet and G. Davis, unpublished data.
64. G. Perrault, paper delivered at Chemical Institute of Canada Conference, Quebec, June 1972.
65. J. Slivinskas, B. McAneney, and J. E. Guillet, paper in preparation.

Gas Chromatographic Studies
of Complexing Reactions

C. A. WELLINGTON, *Department of Chemistry,*
University College of Swansea, Swansea, Wales

I. INTRODUCTION

Over the last 20 years, gas chromatography has played an extraordinarily large part in research work in chemistry and chemical engineering and in control of processes in the chemical industry. In gas-liquid chromatography it has usually been sufficient to use a pure liquid absorbed on a solid support, since the partition coefficients of the components in mixed solutes have been sufficiently different to allow a reasonable separation.

However, when separation is required between two solutes of very similar structure, a single pure liquid is sometimes not sufficient. The analytical possibilities of introducing to the liquid an additive that can form complexes with the solutes was recognized very early by Bradford, Harvey, and Chalkley (1), who used silver nitrate columns for olefin analysis. Although these columns have been widely used, the exploitation of the idea of specific cationic interaction with solutes has not really commenced, in spite of the work of Phillips and his co-workers (2, 3).

Special complexation effects for nonionic species were noted by Langer (4, 5), Norman (6), and Langer and Purnell (7), but the phenomenon observed was a direct complexation between the solute and the pure liquid phase, not selective complexation due to the inclusion of a special additive to the liquid phase. It is under the latter conditions that physicochemical measurements of complexation can be carried out, because (a) the concentration of additive can be varied and (b) the formation constant for the complex can be obtained from the subsequent change in retention volume of the solute. It is an attractive proposition that the selectivity of a column can be changed in a quantitative manner by measurements of the formation constants. Since this would improve the predictability of GLC, it would seem to be a good investment of time both for chemical research and industry to obtain a comprehensive list of the more important formation constants.

Technically the GLC method is in itself an attractive means of obtaining formation constants, the only limitation being that the complexation must be between two species with significant differences in volatility. It has been common practice to run GLC experiments in thermostated ovens at room temperature and above; but there is no reason for not having columns that consist of relatively volatile components and conducting the experiments in a good refrigerated compartment. Thus, technically, all that is required is a good thermostated compartment, an efficient detector, and satisfactory control of the pressure and flow of carrier gas. A number of columns are then required with differing concentrations of additive B in an inert solvent S with a total liquid volume V_l. These are used in turn, and the retention volumes V_R (corrected for the retention volume of a noncomplexing material, e.g., air) of the appropriate solute samples A are measured under the same conditions on each column. In the simplest procedure, the apparent partition coefficients K_R are calculated from the equation

$$V_R = K_R V_l$$

where V_l is the total volume of liquid on the column.

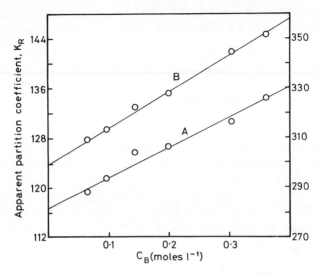

Fig. 1. Plots of apparent infinite-dilution partition coefficient K_R against concentration C_B of di-*n*-propyl tetrachlorophthalate in squalane: curve A, for benzene (left-hand scale); curve B, for toluene (right-hand scale), at 70°C (25).
By courtesy of the *Journal of the Chemical Society.*

Neglecting activity coefficient corrections, the variation of K_R with C_B is given by

$$K_R = K_R{}^0(1 + K_1'C_B)$$

where $K_1' = C_{AB}/(C_A C_B)$ and gives the molar formation constant of the complex AB.

A typical plot of K_R against C_B, shown in Figure 1, is taken from the work of Cadogan and Purnell (25) on the complexation between di-*n*-propyl tetrachlorophthalate in squalane and benzene and toluene as the volatile solutes. The K_1' values (M^{-1}) were found to be 0.416 ± 0.027 and 0.491 ± 0.013, respectively. The formation constant is given here in molar concentration units, and several papers have recently been published on the correct formulation of formation constants. This is discussed in detail later, along with the limitation of the GLC and other methods for obtaining formation constants. Moreover the precautions necessary in obtaining formation constants by GLC are also outlined before the results in the literature are given.

The further advantages of the GLC method can be summarized by quoting Purnell (8).

"How then can the GC method be regarded as offering advantage over the usual methods? The answers might be the following:

1. Since GC is a sensitive separation technique we can use trivial amounts of relatively rare and/or impure compounds.
2. In extension of 1, we can, by using mixtures, make many measurements in a single experiment.
3. Temperature can be varied at will, which is not often the case with conventional methods.
4. Nonaqueous systems are ideal for GC, whereas in other methods they offer considerable difficulty. This point is evidenced by the paucity of data for stability constants in nonaqueous media in the literature.
5. From 4 it follows that hydrolyzable or water-insoluble complexes may be studied, and so previously unexplored regions are opened up.
6. The method is technically simple."

II. DEFINITIONS OF FORMATION CONSTANT OF COMPLEXATION

It is perhaps unfortunate that this chapter is written at a time when the whole subject of the proper definition of formation constants is being called into question. As one set of authors stated, "There is a great deal of confusion in the literature on the proper measure of the formation constant of complexation." It is hoped that the following is an unbiased guide to the situation as it stands at the moment.

The problem is a general one, and it is not restricted to complexation studies by GLC. For simplicity, it is expedient to consider in the first instance the simplest case of a $1:1$ complex formed in solution:

$$A + B \rightleftarrows AB \tag{1}$$

Homer et al. (9) stated that, in all circumstances, the chemical potential of any component i of a liquid mixture may be represented in terms of its mole fraction by the following equation:

$$\mu_i = \mu_i^0 + RT \ln \gamma_i^R x_i \tag{2}$$

where μ_i^0 is the chemical potential of i as a pure liquid at the same temperature and pressure and γ_i^R is the (Raoult) activity coefficient. Thus $\gamma_i^R \to 1$ as $x_i \to 1$ (pure liquid i). Therefore the equilibrium condition for Eq. 1 is

$$\Delta\mu = \mu_{AB} - \mu_A - \mu_B = 0 \tag{3}$$

and so we have

$$\mu_{AB}{}^0 - \mu_A{}^0 - \mu_B{}^0 = \Delta\mu^0 = -RT\ln\left[\frac{x_{AB}}{x_A x_B}\frac{\gamma_{AB}{}^R}{\gamma_A{}^R\gamma_B{}^R}\right] \tag{4}$$

In Eq. 4, three different standard states are involved, namely, pure AB, pure A, and pure B. Clearly, in any solution containing AB, A, and B, it is theoretically impossible for the respective activity coefficients to approach unity simultaneously. This is unfortunate since, as Eon and Karger (10) pointed out, the logarithm term in Eq. 4 [which they term K_{eq}] is a thermodynamic constant which is independent, by definition, of the nature of the solvent.

Similarly (9, 10), the chemical potential of species i is given by

$$\mu_i = \mu_i{}^\theta + RT\ln x_i\gamma_i{}^H \tag{5}$$

where $\mu_i{}^\theta$ is given by

$$\mu_i{}^\theta = \lim_{x_i\to 0}(\mu_i - RT\ln x_i) \tag{6}$$

Thus the components in Eqs. 2 and 5 obey Henry's law, $p_i = K_i x_i\gamma_i{}^H$ ($\gamma_i{}^H\to 1$ as $x_i\to 0$) or Raoult's law.

Thus $\mu_i{}^\theta$ is independent of x_i *but dependent on the nature of the environment* (9).

The value of a formation constant, in terms of mole fractions, for the complexation reaction of Eq. 1 is nearly independent of x_A and x_B only under certain conditions. These are: (1) the equilibrium values of x_B, x_B, and x_{AB} are extremely small ($\mu_i\to\mu_i{}^\theta$)—that is, when the thermodynamic equilibrium constant K_x^*, given by

$$RT\ln K_x^* = RT\ln\left(\frac{x_{AB}}{x_A x_B}\frac{\gamma_{AB}{}^H}{\gamma_A{}^H\gamma_B{}^H}\right) = \mu_A{}^\theta + \mu_B{}^\theta - \mu_{AB}{}^\theta \tag{7}$$

is dependent on the nature of the environment, or (2) the mole fraction of one component (say B) is close to unity and that of A (and AB) is extremely small, when,

$$RT\ln K_x = RT\ln\left(\frac{x_{AB}}{x_A x_B}\frac{\gamma_{AB}{}^H}{\gamma_A{}^H\gamma_B{}^R}\right) = \mu_A{}^\theta + \mu_B{}^0 - \mu_{AB}{}^\theta \tag{8}$$

If the activity coefficients can be accurately assessed, it becomes purely a matter of convenience whether Raoult's law or Henry's law is taken as the basis for the definition of formation constants. Eon and Karger (10), in selecting the Raoult's law basis, have not argued this point correctly. However, under different experimental conditions when activity coefficients are ignored or only estimated, it is expedient to choose one or other of these laws as the preferred basis.

Clearly under condition (1) either molar or molal concentration scales could be used and the formation constant will have correspondingly different values but will depend on the environment (solvent) used. Perhaps a more important matter to explore is the possible use of volume fractions (ϕ_i). These are used in dealing with nonelectrolyte solutions because they allow for the effect of size differences on the energy of mixing in a more nearly adequate way than do mole fractions (11). Thus disparity in molar (or molal) volumes (usually between A or B and the solvent S) is associated with asymmetry of all the excess functions of mixing; in such circumstances, solution behavior is best treated by using volume fractions.

$$\mu_i = \mu_i^0 + RT \ln \gamma_i^V \phi_i \tag{9}$$

where ϕ_i is the volume fraction of i in the environment and γ_i^V the corresponding activity coefficient.

As previously, two standard states can be used:

$$\mu_i = \mu_i^0 + RT \ln \gamma_i^{(V,R)} \phi_i \tag{10}$$

or

$$\mu = \mu_i^\theta + RT \ln \gamma_i^{(V,H)} \phi_i \tag{11}$$

where Eqs. 10 and 11 are the analogs of Eqs. 2 and 5, respectively.

Thus we have

$$K_V = \frac{\phi_{AB}}{\phi_A \phi_B} \frac{\gamma_{AB}^{(V,H)}}{\gamma_A^{(V,H)} \gamma_B^{(V,R)}}$$

and

$$K_V^* = \frac{\phi_{AB}}{\phi_A \phi_B} \frac{\gamma_{AB}^{(V,H)}}{\gamma_A^{(V,H)} \gamma_B^{(V,H)}}$$

Such formation constants may be particularly important in dealing with complexation in GLC experiments since, by the nature of the experiment, the reactant (commonly referred to as the liquid component) is a high-boiling compound, often with a high molecular weight and a large molar volume.

It is interesting to compare this formulation with that using molar concentrations, stated by Kuntz, Gasparro, Johnston, and Taylor (12) to be the most satisfactory in dealing with equilibria in liquid systems.

If in any system the total volume of the liquid phase at equilibrium is denoted by V_{tot} and $\overline{V}_i$ represents the partial molar volume of species i, the volume fraction of species i is given by

$$\phi_i = \frac{n_i \overline{V}_i}{V_{tot}} = C_i \overline{V}_i \tag{12}$$

where C_i is the molar concentration of i in the equilibrium mixture.

Now the chemical potential of species i can also be written

$$\mu_i = \mu_i{}^C + RT \ln C_i \gamma_i{}^{(c)} \tag{13}$$

where $\mu_i{}^C$ is the chemical potential the solute i would have in a $1M$ solution if that solution behaved according to the ideal dilute rule. This is satisfactory for dilute solutions in i; that is, C_i is small.

It is thus more satisfactory to compare the formation constant in terms of molar concentrations K_C^* under solution conditions 1 (given in connection with Eq. 7), which is equivalent to the concentrations of A, B, and AB being simultaneously as small as possible. For this condition we can write

$$K_V^* = \frac{\phi_{AB}}{\phi_A \phi_B} \frac{\gamma_{AB}^{(V,H)}}{\gamma_A^{(V,H)} \gamma_B^{(V,H)}} = \frac{C_{AB}}{C_A C_B} \frac{\overline{V}_{AB}}{\overline{V}_A \overline{V}_B} \frac{\gamma_{AB}^{(V,H)}}{\gamma_A^{(V,H)} \gamma_B^{(V,H)}} \tag{14}$$

and since

$$K_C^* = \frac{C_{AB}}{C_A C_B} \frac{\gamma_{AB}^{(C)}}{\gamma_A^{(C)} \gamma_B^{(C)}} \tag{15}$$

we have

$$K_V^* = K_C^* \frac{\overline{V}_{AB}}{\overline{V}_A \overline{V}_B} \frac{\gamma_{AB}^{(V,H)}}{\gamma_A^{(V,H)} \gamma_B^{(V,H)}} \frac{\gamma_A^{(C)} \gamma_B^{(C)}}{\gamma_{AB}^{(C)}} \tag{16}$$

If the solutions are sufficiently dilute that the activity coefficients are constant or negligible, we can write

$$K_V^* = K_C^* \frac{\overline{V}_{AB}}{\overline{V}_A \overline{V}_B} \tag{17}$$

Under these conditions, since the partial molar volumes are reasonably constant at constant temperature and pressure, the molar concentration scale can be satisfactorily used for dilute solutions. The advantage of its use as a scale is that, despite differing molar volumes of the constituents of a liquid mixture, the formation constant is reasonably independent of concentration in dilute solutions.

The foregoing details of theory can be applied to any method of measuring stability constants, which means that direct comparisons can be made between values obtained by different methods. However, before considering the GLC method in more detail, it may be helpful to summarize the considerations outlined previously and to make short comments on other methods currently employed to measure formation constants. In summary:

1. It does not seem to be possible to measure stability constants that are independent of the environment (particularly the solvent used),

2. The conditions for study of the solutions can conveniently be chosen as one of the following:

(a) A and B (and therefore AB) in small amounts, solvent S in excess,

(b) One component B in large excess (mole fraction $0.9 \rightarrow 1.0$), A (and therefore AB) in small constant amounts.

If the molar volumes of A and B and S or B and S (at constant A) differ significantly, it would seem to be important to use volume fractions in determining the formation constant; otherwise, mole fractions can be used. Solution conditions (a) or (b) can be used; if condition (a) is adhered to, molar concentrations or volume fractions are appropriate in all circumstances.

III. TECHNIQUES FOR DETERMINING FORMATION CONSTANTS

A. Nuclear Magnetic Resonance Spectroscopy

NMR spectroscopy has been used to measure formation constants, particularly by Foster and his co-workers (13, 14). Anomalies do exist (see, e.g., Ref. 14), as suggested in the results obtained by Homer et al. (9) for the benzene–chloroform reaction in a series of hydrocarbon solvents. These authors paid careful attention to the solution conditions used for the study, and their thermodynamic treatment was outlined previously. They employed the Benesi–Hildebrand (15) equation instead of the Scott equation (16) (see later discussion) but under solution conditions when the mole fraction of reactant being varied was close to unity. Indeed, if the results (e.g., with cyclohexane as solvent) presented in their paper are treated by the Scott method, essentially similar values are obtained for the formation constants in mole fraction units. The authors, while using an internal reference for their NMR measurements, made extensive corrections for screening effects of the nuclei present in the mixture. Although these corrections could make a difference of 7%, they were less important than refinements made on the basis of the nature of the solvent. In terms of mole fractions, K_x for the chloroform-benzene complexation at $306.6°K$ was found to be 1.06 with cyclohexane as solvent, but -0.14 with hexadecane. The corresponding K_C values were 0.137×10^{-3} and 0.163×10^{-3} $m^3/mole$. Thus the results appear to be anomalous; the authors examined the assumption that $\mu_A{}^\theta$, $\mu_{AB}{}^\theta$ and $\gamma_B{}^R$ are constant in the range $x_B = 0.9$–1.0 and discussed the effect of reducing x_B and thus of introducing more molecules of solvent S. To a reasonable approximation, one molecule of S would behave as V_S/V_B molecules of inert B. On a macroscopic basis, the amounts of S (n_S moles) in the equilibrium expression [e.g., $x_B = n_B/(n_B + n_S + n_A + n_{AB})$] must be quantified in terms of the effective number of moles relative to B so that they replace n_S by $V_S n_S/V_B$.

The Benesi–Hildebrand expression for an NMR experiment in which the observed shift (Δ_{obs}) of a nucleus on molecules B is observed as a function of the mole fraction of B (x_B) is given by

$$\frac{1}{\Delta_{obs}} = \frac{1}{K_x x_B \Delta_C} + \frac{1}{\Delta_C} \qquad (18)$$

where Δ_C is the shift of the nuclei in B in pure complex. Substituting the number of moles of each species present, and ignoring the small amounts of A and AB present, we have

$$\frac{1}{\Delta_{obs}} = \frac{n_B + n_S}{K_x n_B \Delta_C} + \frac{1}{\Delta_C} \qquad (19)$$

and it was the plot of $1/\Delta_{obs}$ against $(n_B + n_S)/n_B$ which gave the values of K_x quoted earlier.

Introducing the refinement for the size of the solvent molecules, we have

$$\frac{1}{\Delta_{obs}} = \frac{n_B + V_S n_S/V_B}{n_B K_x \Delta_C} + \frac{1}{\Delta_C} \qquad (20)$$

Plotting $1/\Delta_{obs}$ against $[n_B + (V_S n_S/V_B)]/n_B$ gave "K_x" = 1.40 and 1.69, instead of 1.06 and -0.14 for cyclohexane and hexadecane, respectively. Since $[n_B + (V_S n_S / V_B)] / n_B = (n_B V_B + n_S V_S) / n_B V_B = 1 / \phi_B$, this procedure is simply the use of volume fraction in place of mole fraction, and the refined "K_x" values are simply a form of the volume fraction formation constants K_V.

The model adopted by Homer et al. is a simple one, but it appears satisfactorily to refine results which, on a mole fraction basis, seemed to be discrepant. This being so, it is perhaps unnecessary to assume solvation of the interacting species as has been proposed by Carter, Murrell, and Rosch (17). These workers have defined the solvation reaction as

$$AS_n + BS_m \rightleftharpoons ABS_p + qS$$

and the ratio of the equilibrium constant for the solvation reaction to that in the absence of solvation is $(x_S)^q$, where x_S is the mole fraction of solvent in the system. One evident difficulty is that of estimating the value of q; but a procedure to evaluate it was adopted by Carter, Murrell, and Rosch.

B.   Spectrophotometry

Because of the general availability of ultraviolet and visible spectrophotometers, the bulk of the tabulated values of formation constants have been measured by this technique. The procedure and many of the results can be found in Foster's book (13).

It is assumed in the more straightforward studies that the Beer–Lambert law is obeyed and that the concentration units used are commonly moles per liter, moles per 10^3 g of solvent, or moles per 10^3 g of solution. The values of the formation constant are thus obtained in similar units and, as has already been discussed, the use of molar units is perfectly satisfactory for dilute solutions. Kuntz, Gasparro, Johnston, and Taylor (12) justified the use of molarity as the correct concentration scale both theoretically and experimentally. They considered the interrelationship between the molar and mole fraction scales, and they showed that only when the molar volume of the solvent and that of the interacting species in greater concentration are equal will *both* the molar and the mole fraction formation constants be constants over any range of concentration.

For the bulk of the results obtained by spectrophotometry, the Benesi–Hildebrand equation (15) has been used to evaluate the data. Since this involves extrapolation to infinite concentration in one of the species, it is clear that the values are likely to be less satisfactory than those obtained using the Scott equation (16).

General criteria for the determination of formation constants by spectrophotometry have been discussed by Person (18). He stated that reliable equilibrium constants can only be obtained when the equilibrium concentration of the complex is approximately equal to the equilibrium constant of the most dilute component (usually the acceptor A). If the complex is weak, very high concentrations of donor B must be used to reach this state. If this condition is used then, as Homer et al. (9) have shown, the Benesi–Hildebrand treatment may be more satisfactory.

Person considered a 1 : 1 complex with $C_B \gg C_A$ and thus $C_B \cong C_B{}^0$ and $C_A = C_A{}^0 - C_{AB}$. If the complex is weak, and C_B is not high enough, $C_A \approx C_A{}^0$. Thus the absorbance $D \cong \epsilon l K C_B C_A$, and D varies *linearly* with increasing initial donor concentration, since C_A is effectively constant. In this region, a straight line of zero slope and intercept $1/K\epsilon$ will be obtained from the following Scott equation plot of $C_A C_B / D$ against C_B:

$$\frac{C_B C_A l}{D} = \frac{1}{K\epsilon} + \frac{C_B}{\epsilon}$$

When C_B is large, all A will be complexed; therefore, further addition of B does not change D. Thus it is only in the intermediate range of $C_D{}^0$ from $0.1/K$ to $9.0/K$ that the conditions are suitable for the determination of K.

The difficulties particularly affecting the spectrophotometric determination of formation constants are those of overlapping peaks in the spectrum, contact change transfer, and deviations from the Beer–Lambert law. A discussion on the limitations imposed by these factors on the validity of the formation constants in the literature is outside the scope of this review. However, these limitations do exist, and they could well be responsible for the lack of agreement between spectrophotometric and GLC formation constants.

Some work has been done on these factors. For example, a method for determining the formation constants whose charge-transfer band is masked by the absorption of another species has been devised by Foster (19), and the validity of the method has been demonstrated by Corkill, Foster, and Hammick (20). Moreover, Emslie, Foster, Fyfe, and Horman (21), in considering the possibility that the Beer–Lambert law may not be obeyed by the complex, have discussed the case of the extinction coefficient that varies linearly with the concentration of the component in excess, usually the donor. They showed that a Benesi–Hildebrand plot would still be linear, the gradient being unaffected but the intercept being reduced. Thus a value of K lower than the true value would be obtained.

Contact charge transfer was postulated by Mulliken and Orgel (22) in an attempt to explain the zero formation constants found by the simple Benesi–Hildebrand analysis of the new absorption bands produced when *n*-heptane is added to solutions of iodine in perfluoroheptane (23). The effect arises from a mere collisional encounter between free A and B molecules without any specific orientation or interaction. Prue (24) has theoretically determined a value for the formation constant due to contact charge transfer and finds it could be as high as 0.2 mole/l. Thus a contribution of up to this magnitude could be present in formation constants obtained from charge transfer spectra.

IV.   CHROMATOGRAPHIC FORMATION CONSTANTS

Initially, it is convenient to consider the simplest system where a 1:1 complex is formed between GLC additive B—dissolved in solvent S and absorbed on to solid support Z—and solute A, which is partitioned between the gas and liquid phases. The following procedure has been used (25) to evaluate the formation constant. We write the equilibria

$$A(g) \overset{K_R^0}{\rightleftharpoons} A(l) \tag{21}$$

$$A(l) + B(l) \overset{K_1}{\rightleftharpoons} AB(l) \tag{22}$$

where K_R^0 is the partition coefficient of uncomplexed A between S and the gas phase and K_1 is the stoichiometric formation constant of AB in S. The apparent gas chromatographic partition coefficient K_R, *assuming no volume change on dissolution of A*, is given by (25)

$$K_R = K_R^0 (1 + K_1 C_B) \tag{23}$$

where C_B is the additive concentration. We evaluate K_R from the fully corrected net peak maximum retention volume, through the familiar equation for GLC at infinite dilution

$$V_R = K_R^\infty V_l \tag{24}$$

where V_l is the total volume of solvent in the column. The several K's of Eq. 23 can be evaluated meaningfully by this procedure only if it is established that V_R is independent of sample size, since Eq. 23 is not otherwise applicable.

This idealized treatment has been used extensively, and the results obtained are discussed later. It is important to realize the approximations that have been made. Let us outline these, some of which have been mentioned by Purnell (8), before considering the advantages and the measurements that have been performed.

A. Precautions and Limitation of the Method

1. Equilibration

Gas chromatography is a flow technique, and the theory given previously is developed on the assumption that equilibrium is established. Thus, in general, the theoretical treatment requires that complexing be far faster than the chromatographic process. To date, there does not seem to be any report of a study of flow-rate dependence of the formation constants obtained. However, before thermodynamic constants are derived for any system, it is important to establish the independence of carrier gas flow rate of the measurements. One particular type of system involving slow equilibration has been treated theoretically by Klinkenberg (26), and the complexity of the mathematics provides an incentive to ensure that proper equilibration is established under the chosen experimental conditions.

If equilibration were not properly established, presumably the apparent gas chromatographic partition coefficient K_R would be low, since the retention volume would be low. If the variation of K_R with C_B were then affected in the same sense, it would mean that the value of the formation constant was too low. However, from the comparisons made with other methods to date (see later), the evidence is that the formation constant values measured by GLC are higher rather than lower. For example, Eon, Pommier, and Guiochon (27) have determined the formation constant for the complex between 2-ethyl thiophene and di-n-butyl tetrachlorophthalate with squalane as a solvent. By GLC, the value obtained was 0.37 l/mole, and that by ultraviolet spectroscopy was 0.23 l/mole. For the complex between trinitrofluorenone and ortho-xylene in di-n-butyl phthalate and in di-iso-ortho-phthalate as solvents, Meen (28) found values of 0.71 and 0.77, respectively, by GLC, whereas by ultraviolet spectroscopy the respective values are 0.07 and 0.11. Meen (28) commented that since such differences are outside experimental error, a more fundamental explanation for them must be found. If the GLC procedure suffered from lack of equilibration, we would expect the discrepancy for any corrected values to be worse.

2. Solute Volume

In the usual calculation procedure for GLC, the volume of solute A is ignored; in other methods, however, it can be included in the total volume of the solution when molar concentrations are calculated. Advantages of the GLC method are that sensitive detectors can be used and the amount of A required is very small. Cadogan and Purnell (25) made a special mention of this in their experimental section, and they used samples of less than 0.1 μl. Such a small volume will cause a negligible difference if ignored in the total volume of the solution. However, it is important to keep the volume of solute injected small to ensure that any nonideality due to the solute concentration in the column is minimized. Cadogan and Purnell established that the retention volume is independent of the amount of solute injected. This presumably means that at less than 0.1 μl the system is close to ideal behavior. They used columns 60 cm long and 1/4 in. o.d.; assuming 15% of the total liquid phase on the solid support, then the total liquid volume on the column would be 2 cm^3. Since the solutes used were benzene, toluene, and xylenes, 0.1 μl would be equivalent to 10^{-6} mole. Thus the concentration of solute over the whole column would be $5 \times 10^{-4} M$. However, the solute is not spread over the whole column. The equilibria refer only to that portion of the column containing solute and, immediately after injection, this is a portion of carrier gas that is virtually pure solute of total volume of 0.02 cm^3. Since the partition coefficients between the gas and liquid (squalane) phases are often 250 (25, 30), this results in an extremely high concentration of solute in the liquid phase.

Very few authors have been concerned about these factors. In *Gas Chromatography*, Purnell (29) discussed the effect of sample size on column performance. He calculated the concentration of solute in solvent (assuming a very high partition coefficient) for the maximum permissible feed-band width (Gaussian) if this is to have a negligible effect on the emergent solute band. Thus, this calculation gives a lower limit for the concentration of solute in the liquid phase for the injection of a 1-mg sample of solute of molecular weight of 100 for a 1-m-long column containing 1 g of solvent, the column efficiency being equivalent to 1000 theoretical plates. The calculated concentration is 12 mole% for an injection bandwidth at half-height of 1.5 cm, that for the eluted peak being 6.5 cm. This concentration is clearly much above the limit for which Henry's law might apply. With more efficient sample injection, the feed band is likely to be much narrower, and even at 0.1-mg samples (similar to 0.1-μl injections) it is very doubtful if behavior at the first part of the column is likely to be satisfactory in any sense, except in eluted peak shape. It is necessary therefore that a systematic study be made on the effect of size of solute injection on the formation constants and also on the width of the feed band by, for example, varying the temperature of the injection port in the case of injection of a liquid. Except for statements such as that made by Cadogan and Purnell (25), little attention so far has been paid to the possible effect of the solute sample size on the measured formation constants.

3.   Interfacial Phenomena

The difficulties inherent in the NMR method involve making appropriate corrections for shielding effects of the nuclei present and providing a suitable reference. Ultraviolet spectroscopy work is accompanied by problems of proper interpretation of the changes in the absorption spectra, whereas GLC measurements can become highly complex because of the interfaces that are present. Almost no fundamental work seems to have been done on the effects on the formation constants of (a) the interface between the solid support and the liquid phase, and (b) the interface between the liquid phase and the carrier gas. If the solid support had a surface area of 2 to 4 m^2/g, the average film thickness for a column containing 12 g of solid support with 2 cm^3 of liquid phase would be 700 to 350 Å. Such thin films are very unlikely to behave like bulk liquids of the same composition; hence comparison with methods, such as spectroscopic procedures, where formation constants are studied in the bulk solutions are likely to show considerable differences. Liao and Martire (31), Cadogan and Purnell (32), and Pecsok and Gump (33) have noted liquid surface effects for alcohol solutes on nonpolar liquid phase. In such studies the corrected retention volume (V_g^0) values vary with the liquid phase loadings. If the V_g^0 values varied with the composition of the additive—solvent phase at constant loading, the value obtained for the formation constant would be abnormally high. It is difficult to check this effect directly, but Cadogan and Purnell (25) have tested their columns with methyl cyclohexane as a noncomplexing solute in place of their aromatic solutes. In their tests, the apparent partition coefficient did not vary with the composition of di-n-propyl tetrachlorophthalate in squalane.

If the liquid film at the gas-liquid interface is thin, it is important to determine how the surface tension of the liquid phase is affected by the change in the concentration C_B of additive B. For dilute solution, the surface excess of B, Γ_B, is given by the Gibbs adsorption isotherm

$$\Gamma_B = -\frac{1}{RT}\left(\frac{\partial \gamma}{\partial \ln C_B}\right)_{T,p} \tag{25}$$

where γ is the surface tension of the solution. If the surface tension of the solution decreases with an increase in concentration of B, then $(\partial \gamma / \partial \ln C_B)$ is negative and Γ_2 is positive; that is, there is an excess of additive B at the surface. This situation could again lead to spuriously high values for the formation constant, if the solid support was coated with a thin film. Meen, Morris, and Purnell (34) found that on columns of β,β'-thiodipropionitrile in squalane, adsorption took place at the gas-liquid interface. Like Cadogan and Purnell (32), they did not correct for the effect of additive B on the formation constant K_1, but corrected for the effect of solute sample size on retention volumes, as outlined below.

A number of systems display peak asymmetry, as well as a marked dependence of retention volume on solute sample size (31, 32, 34). In such systems, elaborate procedures (35–37) are required to obtain meaningful values of the retention volume V_N. A succinct account of the procedure, given by Cadogan and Purnell (32), is based on the treatment of Conder (37). Cadogan and Purnell pointed out that, in the absence of surface effects, the fully corrected net retention volume V_R is given by the product of the apparent infinite-dilution partition coefficient K_L and the column liquid volume V_L:

$$V_R = K_L V_L \qquad (26)$$

Eq. 26 being identical to Eq. 24, since $K_R \equiv K_L$. In the presence of surface effects, the extended form of Eq. 24 is (35, 36)

$$V_N = K_L V_L + K_I A_I + K_S A_S \qquad (27)$$

where K_I and K_S are the infinite-dilution partition coefficients relevant to liquid-gas and solid-gas interface adsorption, respectively, and A_I and A_S are the corresponding surface areas. Since K_L is the same as the K_R of Eq. 23, we can write

$$K_L = K_L^0 (1 + K_1' C_B) \qquad (28)$$

from which the required formation constant can be evaluated if K_L is known as a function of liquid additive B.

Equation 27 can be rewritten as

$$V_N / V_L = K_L + (K_I A_I + K_S A_S) / V_L \qquad (29)$$

and thus, if values of V_N corresponding to infinite dilution can be obtained, a plot of V_N / V_L against V_L^{-1} should give a curve extrapolating to K_L at $V_L^{-1} = 0$. It is thus necessary to use extra columns of varying solvent–support ratio (i.e., V_L) for each value of C_B employed.

Furthermore, more experiments are needed where marked peak asymmetry persists at the smallest possible solute sample size. There is no simple route to the determination of the infinite-dilution value of V_N; but also because peak maximum retention is not necessarily thermodynamically significant in these circumstances, methods such as those discussed by Liao and Martire (31) must be employed to obtain V_N from asymmetric peaks. Conder's method (37) has proved to be satisfactory and has received general acceptance (31, 32).

In Conder's method it is necessary to obtain a large eluted peak from each of a series of columns (arbitrarily numbered $j = 0, 1, 2$) identical in all respects (e.g., C_B) but for the total volume of the liquid phase V_L. For one column $j = 0$, a point is chosen on the diffuse side of the peak h_0 mm above the baseline, and the retention volume (V_{N_0}) is measured from that point to the maximum of the air peak. For the other peaks the corresponding heights $h_1, h_2, \ldots, h_j$ and retention volumes $V_1, V_2, \ldots, V_j$ are found such that $h_0/V_{N_0} = h_1/V_{N_1} = \cdots h_j/V_{N_j}$. To obtain h_1 and V_{N_1} it is necessary to guess a point, calculate the ratio, and adjust h_1 (and thus V_{N_1}) until the ratio corresponds to h_0/V_{N_0}. Only two successive approximations are necessary. Conder has shown that these V_{N_j} values are the retention volumes corresponding to elution at a fixed gas phase concentration. Choosing now a different initial value of h_0 (and thus V_{N_0}), the procedure is repeated, using the same eluted peaks, and for each column a new V_{N_j} value is obtained, corresponding to a different gas phase concentration of solute. In this way for each column V_L a series of V_N values is obtained for different gas phase concentrations corresponding to the h values. Plotting V_N against h, and extrapolating to $h=0$, gives the infinite-dilution value of V_N corresponding to a particular V_L. Repeating this for all the columns gives the V_N and V_L values needed for evaluation of K_L by way of Eq. 29 by plotting V_N/V_L against V_L^{-1} and extrapolating to $V_L^{-1}=0$.

Cadogan and Purnell (32) used an alternative procedure in which they plotted a series of h and V_N values for each column in turn. This can be done on a single graph giving a different curve for each column (value of V_L). Straight lines through the origin, corresponding to fixed values of h_j/V_{N_j} can be drawn to intersect the curves. The foregoing procedure can now be followed again or, alternatively, V_{N_j}/V_L can be plotted against V_L^{-1}, a curve being obtained for each value of h_0. This gives a series of curves that extrapolates to the same value of K_L at $V_L^{-1}=0$ since, then, all practical sample sizes correspond to infinite dilution. Thus the difficulties due to adsorption of the solute at the interface (assumed to be independent of C_B) can be overcome, and plots of K_L against C_B values give straight lines for $1:1$ complexes, whose slopes give formation constants for each solute sample used.

4. The Partition Coefficient

The apparent partition coefficient is usually defined in publications as the ratio of the total concentration of solute in the liquid phase to the concentration of solute in the gas phase. That is

$$K_R' = (C_A^{\text{total}})_l/(C_A)_{\text{gas}} \qquad (30)$$

Since the total concentration of A in the liquid phase is the sum of the concentration of complexed (AB) and uncomplexed A, we have

$$K_R' = (C_{AB} + C_A)_l/(C_A)_g \qquad (31)$$

Since, in GLC experiments, only the solute A is volatile, it is more convenient to write Eq. 31 as

$$K_R' = (C_{AB} + C_A)/C_g \tag{32}$$

Thermodynamically, it is necessary to introduce activity coefficients; thus we have

$$K_R = (\gamma_{AB}C_{AB} + \gamma_A C_A)/\gamma_g C_g \tag{33}$$

assuming that, for the liquid phase,

$$a_A^{total} = a_A + a_{AB}$$

and

$$K_R = \frac{(1 + \gamma_{AB}C_{AB}/\gamma_A C_A)\gamma_A C_A}{\gamma_g C_g} \tag{34}$$

Using Eq. 15 as

$$K_c^* = C_{AB}\gamma_{AB}/(C_A C_B \gamma_A \gamma_B)$$

we have

$$K_R = (1 + K_c^* \gamma_B C_B)K_R^0 \tag{35}$$

where

$$K_R^0 = \gamma_A C_A/\gamma_g C_g \tag{36}$$

The commonly used (28, 30) expression for K_R^0 is that derived by Purnell (Ref. 29, pp. 206–207):

$$K_R^0 = C_A/C_g = RT/\left(p_A^0 \overline{V}_s \gamma_A\right) \tag{37}$$

where p_A^0 is the vapor pressure of pure solute A at temperature T, $\overline{V}_s$ is the molar volume of the solvent S, and γ_A is the Raoult law activity coefficient for A in S. It is important to recognize the approximations used in deriving this expression. The most obvious is that the solute concentration is low enough so that the mole fraction of solute A can be approximated thus:

$$x_A = n_A(n_A + n_S) \approx n_A/n_S = C_A \overline{V}_s \tag{38}$$

As Purnell (29) pointed out, this approximation is good when x_A is low (perhaps 0.01). However, as has been discussed earlier, unless extremely small solute samples are used, x_A can be 0.12 or even very much higher, especially just after injection. It is perhaps less important in the correctness of the formulation of K_R^0 but, if x_A is at all significant, this must lead to errors in the absolute values of the activity coefficients calculated (30) from measured K_R^0 values using the rewritten form of Eq. 37,

$$\gamma_A = RT / \left(K_R^0 p_A^0 \overline{V}_s \right) \qquad (39)$$

This correction should certainly be studied along with the activity coefficient corrections used by Cadogan and Purnell (25) and Meen, Morris, and Purnell (34), and the more elaborate treatment of Eon and Karger (10).

In the derivation of Eq. 38 it is not essential to consider that the vapor is ideal and fugacity corrections can be applied (p_A^0 being replaced by f^0). To derive Eq. 38, however, Raoult's law,

$$p_A = p_A^0 \gamma_A x_A, \qquad (40)$$

is assumed in a region where Henry's law is applicable, since x_A has to be small enough to make the approximation of Eq. 38 tenable. Thus the γ_A's are essentially Henry's law activity coefficients.

Since in GLC experiments the solute A must be volatile and the solvent involatile, the latter is usually a large molecule, the solute being relatively small. It is very likely that the solute has a small molar volume ($\overline{V}_A$) and the solvent a large one ($\overline{V}_S$). Under these conditions (see section on formation constants), in order that Eq. 40 is able to give an adequate account of the variation of p_A with change of amount of A in the liquid phase, γ_A is likely to be an important factor. Under such conditions, and introducing fugacity of A (f_A) for completeness, a better equation to account for the behavior would be

$$f_A = f_A^0 \gamma_A^{(V)} \phi_A \qquad (41)$$

where ϕ_A is the volume fraction of A in S, the liquid phase, and $\gamma_A^{(V)}$ is the volume fraction activity coefficient.

Now if n_A and n_S represent the number of moles of solute A and solvent S, of partial molar volumes $\overline{V}_A$ and $\overline{V}_S$, respectively, present in the liquid phase, we can write

$$\phi_A = n_A \overline{V}_A / \left(n_A \overline{V}_A + n_S \overline{V}_S \right) = \overline{V}_A n_A / V_{liquid} = \overline{V}_A C_A \qquad (42)$$

The concentration of A in the gas phase is given by

$$C_g f_A / RT = f_A{}^0 \gamma_A^{(V)} \phi_A / RT = f_A{}^0 \gamma_A^{(V)} \overline{V}_A C_A / RT \tag{43}$$

Since $K_R{}^0 = C_A / C_g$, we have

$$K_R{}^0 = RT / \left(f_A{}^0 \gamma_A^{(V)} \overline{V}_A \right) \tag{44}$$

Interestingly, Eq. 44, derived without assumptions similar to that in Eq. 38, shows that Eq. 37 is probably better than it appears because, if Raoult's law in terms of mole fraction of A is not a satisfactory description of systems having components with large differences in molar volumes, the assumption that n_A is much smaller than n_S improves the inadequacy of the mole fraction version of Raoult's law!

Let us return to the apparent partition coefficient given in Eq. 32 (idealized) and Eq. 33. If the formation constant is required in molar concentration units (which was shown to be adequate from the discussion on formation constants), the procedure is straightforward, and the thermodynamic formation constant can be obtained from Eq. 35. Eon and Karger (10) preferred to take the formation constant in terms of mole fraction units, while using the partition coefficient in concentration units. Thus we can write

$$C_A^{\text{total}} = C_A + C_{AB} = (x_A + x_{AB}) / \left(x_A \overline{V}_A + x_B \overline{V}_B + x_S \overline{V}_S + x_{AB} \overline{V}_{AB} \right)$$

$$C_g = f_A{}^0 \gamma_A x_A / RT$$

Thus

$$K_R' = \frac{RT(1 + K_x^* x_B \gamma_A \gamma_B / \gamma_{AB})}{f_A{}^0 \left(x_A \overline{V}_A + x_B \overline{V}_B + x_S \overline{V}_S + x_{AB} \overline{V}_{AB} \right)}$$

Unfortunately, the partial molar volumes of the constituents are not equal, and the term in parentheses in the denominator of the equation for K_R' is not a constant; therefore, it is necessary to plot the product of this term (or an approximation to it) and K_R' against x_B in order to obtain K_x^*. Eon and Karger recognized that, in correcting the formation constant K_x^* by evaluating activity coefficients, allowance must be made in the athermal terms for the different molar volumes of the species present. They applied the simplified Flory–Huggins equation as given by Guggenheim (38). This gives a difference

$$K_1 = K_x^* V_S - V_B \frac{1 - \exp\left(\overline{V}_A / \overline{V}_B\right)}{\exp\left(\overline{V}_A / \overline{V}_B\right)} \tag{45}$$

between $K_1 (= C_{AB} / C_A C_B)$ and K_x^*.

Most workers, however, have evaluated K_1; thus useful comparisons can be made between the results of different investigators.

B. Complex Interactions: Systematic Classification

So far, only the simplest 1:1 complex formation has been considered. However, the ideas that have been presented can be applied to more complex systems, and it is only necessary to further express the behavior of such systems by idealized equations in order to give a satisfactory classification. Purnell (8) proposed a classification and deduced the idealized equations for each class.

Class A

Solute A reacts with additive B to give complexes of the following types:

$$\text{(1)} \quad BA_n \qquad \text{(2)} \quad AB_m \qquad \text{and} \qquad \text{(3)} \quad B_m A_n$$

where both n and m are not smaller than 1. In each type of system, several complexes may coexist (e.g., BA, BA_n, BA_{n+1}).

Purnell gave the equilibria as

$$nA(l) + B(l) \overset{K_1}{\rightleftharpoons} BA_n(l)$$

followed by a series of further additions of A

$$BA_{n+i-1}(l) + A(l) \overset{K_1}{\rightleftharpoons} BA_{n+i}(l)$$

$$C_A^{\text{total}}(l) = C_A + nC_{BA_n} + \cdots + (n+i)C_{BA_{n+i}} + \cdots$$

$$K_R = C_{A(l)}^{\text{total}} / C_g$$

$$= K_R^0 \left[1 + nK_1 C_B C_A^{n-1} + \cdots + (n+i)K_1 K_2 \cdots K_i C_B C_A^{(n+i-1)} + \cdots \right] \tag{46}$$

Thus for a 1:1 complex we have

$$K_R = K_R^0(1 + K_1 C_B) \tag{47}$$

Of course the behavior of Eq. 47 is by far the most common; but Eon, Pommier, and Guiochon (30) have found that two successive associations occur in squalane with pyrrole derivatives as the solute A and dibutyl-tetrachlorophthalate as the additive B. They have estimated K_1 and K_2 for these equations.

For class A(2):

$$K_R = K_R^0(1 + K_1 C_B^m + K_1 K_2 C_B^{m+1} + \cdots) \tag{48}$$

For class A(3):

$$K_R = K_R(\text{Eq. 46}) + K_R(\text{Eq. 48}) - K_R^0$$

Class B

Solute A reacts with solvent S to give complexes of the following types:

(1) SA_n (2) AS_p and   (3) $S_p A_n$

with n and p not smaller than 1. Several complexes may coexist.

Here the system is similar to class A(1) with S replacing B, so that we write

$$K_R = K_R^0[1 + nK_1 C_S C_A^{n-1} + \cdots + (n+i)K_1 K_2 \cdots K_i C_S C_A^{n+i-1} + \cdots] \tag{49}$$

As Purnell pointed out, K_R^0 is unmeasurable because it represents the partition coefficient of uncomplexed solute in the complexing solvent. The occurrence of such behavior can only be inferred from exceptionally high K_R values or from unexpected solvent selectivities. Barker, Phillips, Tusa, and Verdin (39) have obtained a good estimate of relative stability constants by assuming that, for a range of related solutes, relative values of K_R^0 for uncomplexed solute in the complexing liquid parallel those in "normal" solvents. Thus these investigators obtained a graph of K_R against K_R^0 for the solutes which reflected the influence of the bracketed term in Eq. 49. The expressions for class B(2) and B(3) follow directly from those of classes A(2) and A(3), respectively, by replacing C_B by C_S.

Class C

The solute A reacts with itself as follows:

(1)   A polymerizes in solution   or   (2)   A depolymerizes in solution

In the simplest situation, no interaction takes place between A and any other species in solution.

For class C(1):

If K_i represents the equilibrium constant for the process

$$(A)_i(l) + A(l) \rightleftarrows (A)_{i+1}(l)$$

then

$$K_R = K_R^0 [1 + 2K_1 C_A + \cdots + (i+1)(K_1 K_2 \cdots K_i) C_A^i + \cdots] \quad (50)$$

For C(2), where $n = 2$:

$$K_R = K_R^0 \left(1 + \tfrac{1}{2} K_1^{1/2} C_{A_2}^{-1/2}\right) \quad (51)$$

where C_{A_2} is the concentration of dimer in the liquid phase. Purnell recasted the expression in terms of the concentration of dimers in the gas phase and remarked that study of such systems is likely to offer considerable technical difficulty.

Class D

Additive B reacts with solvent S to give complexes of the following types:

(1) $SB_{m,m+1}$, etc.   and   (2) $BS_{p,p+1}$, etc.

By the very nature of solubility of B in S, such complexes are likely to occur often in practice. As Purnell pointed out, if there is only an interaction of B with S, systems of this type are very difficult to treat theoretically unless drastic assumptions are made. The only simple case is that in which the complex forms "precipitate," thus allowing the retention volume of "inert" solute to measure the residual solvent volume. Thus we can write

$$V_R^0 / V_R = V_l^0 / V_l = C_S^0 / C_S = R_V \quad (52)$$

and so we have

$$C_S / C_S^0 = 1 - (K_1 C_A / C_S^0)(C_S^p + K_2 C_S^{p+1} + K_2 K_3 C_S^{p+2} + \cdots) \quad (53)$$

In the case of $p = 1$, and no other complexes formed, we have

$$K_1 = (1 - R_V) / (R_V C_B)$$

and since $(C_A^0 - C_A) = (C_S^0 - C_S)/p$, we also have

$$K_1 = (1 - R_V) / \{R_V[C_B^0 + C_S^0(R_V - 1)]\} \quad (54)$$

For reactions of class D(2):

$$C_S/C_S^0 = 1 - K_1(C_S/C_S^0)(C_B^m + K_2C_B^{m+1} + \cdots) \qquad (55)$$

and for the simple case of $m = 1$, and no higher complex formed, we have $K_1 = (1 - R_V)/(R_V C_B)$, which is of course the same as for D(1).

An extension of Purnell's classification D can be considered and is as follows.

Class E

Additive B and solute A react with solvent S, and the AB complexes formed also react with solvent S to form "solvated" species BS_m, AS_n, and ABS_p. This system has been discussed previously, and the analysis has been done by Carter, Murrell, and Rosch (17). The GLC equation for this solvated complexing system was considered by Meen (28).

For the process

$$AS_n + BS_m \rightleftarrows ABS_p + qS$$

$$K_1^{\text{solv}} = (C_{ABS_p} x_s^q)/(C_{AS_n} C_{BS_m})$$

This form of definition is adopted so that the equilibrium constant has the same dimensions that would describe the constant in the corresponding unsolvated equilibrium.

The complexing reaction is the simple A(1) type and so, from Eq. 47 we have

$$K_R = K_R^0 \left[1 + (K_1^{\text{solv}} C_{BS_m})/x_s^q \right]$$

where x_s^q can be estimated by the procedure of Carter, Murrell, and Rosch (17). In the GLC case, it is usually assumed that the concentration of A is lower than that of B (if the sample injected is kept to a minimum); thus we write

$$x_s^q = 1 + C_B^0 q(n+1)/C_S^0$$

Therefore

$$K_R = K_R^0 \left[1 + K_1^{\text{solv}} C_B^0 (1 - C_B^0 q(n+1)/C_S^0) \right] \qquad (56)$$

However, if C_B^0 is much smaller than C_S^0, Eq. 56 reduces to Eq. 47, and the role of solvation can only be detected as a nonlinear variation (negative deviation) of K_R with C_B in the more concentrated solutions, K_R increasing less rapidly with C_B. In class A(1) systems, with n greater than 1, the deviations are to increased K_R values (positive deviation). There does not seem to be any straightforward report yet of systems behaving in a class E manner. The deviations from linearity for olefin complexes with silver nitrate in ethylene glycol in the study of Muhs and Weiss (40), which they attributed to "salting out" as the concentration of silver nitrate increases, may in fact be due to class E behavior. However, Muhs and Weiss did not maintain constant ionic strength, and this may account for part of the deviation.

C. Results

1. Early Work

The first quantitative study of equilibria by GLC was that of du Plessis and Spong (41) in which ammonia was passed into columns containing silver nitrate in benzyl cyanide. The results were interpreted in terms of the formation of a lower stable ammine $AgZ(NH_3)_x$ and a higher unstable one of the silver–benzyl cyanide moiety AgZ

$$AgZ(NH_3)_x + 0.5NH_3 \rightleftharpoons AgZ(NH_3)_{x+0.5}$$

The dissociation pressure of the higher ammine was estimated to be 5.7×10^{-3} atm.

2. Silver Nitrate–Glycol Systems

Thereafter, the quantitative data published concerned the formation constants between unsaturated organic solutes and the silver ion in ethylene glycol. The first is due to Gil-Av and Herling (42). They did not adopt the procedure outlined in Section I but used a column containing $1.77M$ $AgNO_3$ in ethylene glycol to determine K_R and a second column with $1.77M$ $NaNO_3$ to determine $K_R{}^0$. Muhs and Weiss (40) did adopt the procedure given in Section I, and it is interesting that they found significant negative deviations from linearity in their plots of K_R against C_B (above $C_B = 1.0M$). The same behavior was reported by Cvetanovic et al. (43). As we noted earlier, this effect might be due to class E behavior or to "salting out" of the positive silver ions; or, possibly, it occurs because ionic strengths were not kept constant in these studies. The solute sample sizes were reasonable at $0.5 \ \mu l$ or less (Muhs and Weiss found no change of retention time with variation in sample size and all peaks were symmetrical), but the concentrations of silver ions were unnecessarily high—0.8 to $4M$. Surprisingly, none of the authors mentioned correcting the retention volume for the gaseous volume of the column. The results are given in Table I.

TABLE I

Molar Formation Constants K_1 and Partition Coefficients
at Infinite Dilution K_R^0
for Silver Nitrate–Ethylene Glycol

	Ref. 40		Ref. 43				
	40°C		40°C	25°C	0°C		
Solute	K_1 M^{-1}	$(K_R^0)^a$	K_1 M^{-1}	K_1 M^{-1}	K_1 M^{-1}	$-\Delta H$, kJ/mole	$-\Delta S$ J/(mole)(°K)
Ethylene	22.3	(0.1)	13.5	17.5	30.6	14.6	25.1
Propene	9.1	(0.4)	5.9	7.5	13.3	14.6	31.4
But-1-ene	7.7	(0.9)	6.8	8.8	16.1	15.5	33.5
Pent-1-ene	4.9	(1.9)	6.8	6.7	12.1	15.1	33.9
Hex-1-ene	4.3	(3.5)					
Oct-1-ene	2.6	(13.1)	3.2[b]				
3-Methyl-but-1-ene	5.1	(1.5)	6.1	8.0	15.2	16.3	36.8
3,3-Dimethyl-but-1-ene	3.6	(2.2)					
3-Methyl-pent-1-ene	3.4	(2.8)					
3-Ethyl-pent-1-ene	1.4	(5.2)					
4-Methyl-pent-1-ene	2.8	(2.7)					
3-Methyl-hex-1-ene	2.7	(5.3)					
4-Methyl-hex-1-ene	2.3	(6.0)					
5-Methyl-hex-1-ene	3.1	(5.9)					
cis-But-2-ene	5.4	(1.1)	3.9	4.9	8.7	14.2	34.3
cis-Pent-2-ene	4.3	(2.3)	4.4	5.6	10.1	15.1	35.6
cis-Hex-2-ene	3.1	(4.3)					
cis-Hept-2-ene	2.6	(8.1)					
cis-Oct-2-ene	2.2	(14.6)					
cis-Hex-3-ene	3.9	(4.1)					
cis-Hept-3-ene	2.7	(7.6)					
cis-4-Methyl-pent-2-ene	3.1	(2.8)					
cis-4,4-Dimethyl-pent-2-ene	2.7	(5.3)					
cis-2,2-Dimethyl-hex-3-ene	2.5	(8.9)					
trans-But-2-ene	1.4	(1.0)	1.4	1.6	2.6	10.9	32.2
trans-Pent-2-ene	1.1	(2.3)	1.4	1.8	2.8	12.1	36.4
trans-Hex-2-ene	0.8	(4.0)					
trans-Hept-2-ene	0.6	(7.6)					
trans-Oct-2-ene	0.4	(14.6)					
trans-Hex-3-ene	1.0	(4.1)					
trans-Hept-3-ene	0.6	(7.6)					
trans-Oct-4-ene	0.5	(13.4)					
trans-4-Methyl-pent-2-ene	0.7	(3.4)					
trans-2-Methyl-hex-3-ene	0.6	(5.5)					
trans-4,4-Dimethyl-pent-2-ene	0.4	(4.4)					
trans-2,2-Dimethyl-hex-3-ene	0.3	(7.9)					
trans-2,5-Dimethyl-hex-3-ene	0.2	(6.5)					
2-Methyl-prop-1-ene	3.9	(0.9)					
2-Methyl-but-1-ene	3.0	(2.3)	3.5	4.4	8.0	14.6	36.0

TABLE I (continued)

| | Ref. 40 | | Ref. 43 | | | | |
| | 40°C | | 40°C | 25°C | 0°C | | |
Solute	K_1 M^{-1}	$(K_R^0)^a$	K_1 M^{-1}	K_1 M^{-1}	K_1 M^{-1}	$-\Delta H$, kJ/mole	$-\Delta S$, J/(mole)(°K)
2-Methyl-pent-1-ene	2.1	(3.7)					
2,3-Dimethyl-but-1-ene	2.4	(3.2)					
2,3,3-Trimethyl-but-1-ene	1.8	(5.4)					
2-Ethyl-but-1-ene	3.5	(4.2)					
2-Ethyl-hex-1-ene	2.1	(12.7)					
2-Methyl-but-2-ene	0.8	(2.2)	0.87	1.01	1.52	10.0	32.6
2-Methyl-pent-2-ene	0.6	(4.2)					
cis-3-Methyl-pent-2-ene	0.7	(4.2)					
trans-3-Methyl-pent-2-ene	0.7	(4.4)					
2,3-Dimethyl-but-2-ene	0.1	(4.5)					
Allene	0.8	(1.6)					
Methylallene	0.8	(3.5)					
1,3-Butadiene	4.2	(2.2)	3.6	4.5	7.6	13.4	32.2
1,4-Pentadiene	10.2	(2.9)					
1,5-Hexadiene	28.8	(5.1)*					
1,6-Heptadiene	14.7	(9.9)*					
1,7-Octadiene	11.3	(19.4)*					
1,8-Nonadiene	10.4	(36.5)*					
1,9-Decadiene	7.8	(76.3)*					
trans-1,3-Pentadiene	3.5	(5.1)					
cis-1,3-Pentadiene	4.4	(6.1)					

TABLE I (continued)

Solute	Ref. 40 40° $K_1(M^{-1})$	(K_R^0)	Ref. 42 30° K_1	(K_R^0)	γ_A	$10^4 K_a$
Isoprene	3.1	(4.4)				
2,3-Dimethyl-1,3-butadiene	1.9	(8.8)*				
2,4-Dimethyl-1,3-pentadiene	1.6	(12.8)*				
2,5-Dimethyl-2,4-hexadiene	0.8	(47.3)*				
2-Methyl-1,5-hexadiene	22.1	(10.2)*				
2,5-Dimethyl-1,5-hexadiene	13.3	(19.4)*				
trans-1,3,5-Hexatriene	5.1	(19.4)*				
cis-1,3,5-Hexatriene	4.7	(22.7)*				
Methylenecyclobutane	5.8	(5.3)	8.1	(7.2)*	290	43
1-Methyl cyclobutane			0.54	(6.3)*	140	38
Cyclopentene	7.3	(5.8)				
1-Methylcyclopentene	1.9	(8.9)	2.9	(13.0)*	165	175
3-Methylcyclopentene	2.1	(9.0)*	12.0	(10.0)*	145	810
4-Methylcyclopentene			5.5	(10.6)*	265	210
Methylenecyclopentane	4.0	(10.2)	6.0	(13.4)*	160	370
1-Ethylcyclopentene	2.3	(16.0)*	3.6	(24.3)*	285	125
3-Ethylcyclopentene			11.8	(192)*	285	415
Ethylidenecyclopentane	0.7	(21.8)	3.0	(55.8)	450	67
Allylcyclopentane	3.9	(24.9)*	7.1	(19.2)*		
Cyclopentadiene	4.6	(10.7)*				
Cyclohexene	3.6	(14.7)	7.7	(—)		
1-Methylcyclohexene	0.5	(21.3)	1.25	(27.0)	290	43
3-Methylcyclohexene	3.5	(16.5)	5.5	(23.0)*	290	210
4-Methylcyclohexene	3.8	(14.2)	5.1	(23.0)*	265	190
Methylenecyclohexane	6.0	(18.7)*	9.6	(24.3)*	270	350
1-Ethylcylohexene	0.5	(35.9)	1.3	(55.8)*	405	32
Ethyldienecyclohexene	1.6	(39.6)	3.0	(55.8)*	450	67
1-Isopropyl cyclohexene			1.05	(69.5)*	845	12.5
Vinylcyclohexane	5.9	(27.1)				
4,4-Dimethyl-1-cyclohexene	1.4	(21.2)*				
Allylcyclohexane	32.	(46.7)*				
1,3-Cyclohexadiene	8.9	(22.9)*				
1,4-Cyclohexadiene	4.9	(33.2)*				
1-Methyl-1,4-cyclohexadiene	3.3	(52.4)*				
4-Vinyl-1-cyclohexene	11.2	(45.5)*				
Dipentene	5.9	(126.0)*				
α-Phellandrene	5.1	(108.0)*				
Cycloheptene	12.8	(27.0)				
Cycloheptene	12.7	(27.0)*				
Cyclophetatriene	7.6	(89.2)*				
cis-Cyclöctene	14.4	(56.1)*				
trans-Cyclöctene	>1000	(56.1)*				
1,3-Cyclöctadiene	3.2	(78.7)*				
1,4-Cyclöctadiene	14.4	(90.7)*				
1,5-Cyclöctadiene	75.	(144)*				
Cyclöctatetraene	91.	(211)*				

TABLE I (continued)

| | Ref. 40 | | Ref. 45 | | | | |
Solute	40° $K_1(M^{-1})$	(K_R^0)	40° $K_1(K_R^0)$	25° $K_1(K_R^0)$	0° $K_1(K_R^0)$	40°C γ_A	40°C $10^4 K_a$
2-Norbornene	62.	(17.7)*					
2,5-Norbornadiene	33.7	(27.8)*					
2-Methylenenorbornane	4.3	(35.1)*					
Camphene	3.1	(61.2)*					
α-Pinene	1.1	(47.4)*					
β-Pinene	3.7	(84.5)*					
Benzene	0.1	(47.8)					
Toluene	0.1	(78.9)					
o-Xylene	0.1	(183)					
m-Xylene	0.1	(128)					
p-Xylene	0.1	(217)					
Ethylbenzene	0.1	(123)					
Pent-2-yne	—	—	1.8 (13.8)	2.5 (21.6)	3.5 (54.9)	54.3	331
* "	—	—	2.1 (32.1)	3.2 (17.5)	4.1 (48.7)	61.3	342*
Hex-2-yne	—	—	1.6 (21.7)	2.1 (36)	3.0 (105)	97.5	164
* "	2.0	(18.3)	2.1 (46.9)	2.9 (28.5)	3.7 (92.4)	118	178*
Hex-3-yne	—	—	2.2 (17.2)	3.1 (28)	4.8 (79)	110.7	199
* "	2.6	(15.2)	2.95 (45.5)	4.15 (22.4)	5.8 (67.8)	137	216*
Hept-2-yne	—	—	1.4 (37.7)	2.0 (67)	2.9 (224)	148	95
* "	1.6	(34.0)	2.05 (77.1)	2.9 (52.2)	3.85 (182.5)	187	110*
Hept-3-yne			2.0 (26.7)	2.8 (46.6)	4.0 (153)	183	109
* "	2.1	(25.1)	2.8 (66.5)	3.7 (38.2)	5.2 (125.5)	230	122*
Oct-2-yne	—	—	1.3 (63.7)	1.9 (123)	2.7 (480)	241	54
* "	—	—	2.0 (124)	2.8 (93)	3.6 (384.3)	314	63*
Oct-3-yne			1.9 (44.7)	2.6 (85)	3.9 (318)	309	61
* "			2.7 (108.5)	3.8 (64)	5.2 (261)	392	69*
Oct-4-yne			1.6 (42.6)	2.3 (80)	3.2 (331)	309	53
* "	1.5	(43.8)	2.45 (94.8)	3.2 (62.3)	4.35 (248.7)	398	62*

[a] The values marked with an asterisk (*) have been calculated from a single measurement and are thus less reliable.

[b] At 30°C; Ref. 43.

Schnecko (44), in using silver nitrate–ethylene glycol to study complexation of unsaturated nitrile solutes, kept the ionic strength constant by adding lithium nitrate. Although the concentration of silver nitrate was varied from 0 to $4M$ and 10-μl injections of solutes were used, the plots obtained were good straight lines and gave the results listed in Table II. Although the formation constants were much lower for the unsaturated nitriles compared with the olefins, it seems possible that the changing ionic strength resulted in deviations from linearity. Schnecko interpreted these lower values as being due to the electron-donating power of the olefinic double bond (to the silver ion) being reduced by CN substitution.

The results of Queignec and Wojtkowiak (45) are also given in Table I. They also did not keep the ionic strength constant, but they obtained good straight-line plots for the much more satisfactory silver nitrate–ethylene glycol concentration range, from 0 to $0.76M$. They also calculated the formation constants using the procedure of Gil-Av and Herling (42) from the results of the $0.76M$ column and one containing sodium nitrate of the same concentration. These appear for comparison in Table I, indicated by an asterisk.

Perhaps the most unusual numbers given in Table I are the extremely large solute activity coefficients. These have been calculated from the relationship (29)

$$\gamma_A = RT/\overline{V}_l\,p_A{}^0 K_R{}^0$$

based on Raoult's law, $p_A = p_A{}^0\gamma_A x_A$. These large values mean that extremely large corrections have to be made when applying Raoult's law.

Henry's law should be applied to these solutions, and the large values mean that the Henry's law constant is very different from the saturated vapor pressure of the solute $p_A{}^0$. Thus it does not seem correct to apply such activity coefficients (γ_A) to derive a "thermodynamic" formation constant (K_a) for the complexation.

Gil-Av and Herling have reported good agreement between the GLC formation constants for oct-l-ene (3.3) and cyclohexane (7.7) with those obtained by partition between $1.77M$ silver nitrate–glycol and cyclohexane of 3.2 and 7.2, respectively. Since the additive here is an inorganic ion (rather than an organic liquid with a surface tension differing substantially from that of the solvent) interfacial adsorption effects are unlikely to be important in this GLC work.

Wasik and Tsang (46) have measured formation constants for silver ion–olefin complexes with water as the solvent. At the temperatures used, water bled from the column, but by presaturating the carrier gas, Wasik and Tsang were able to prevent a concentrating of the silver ion in

TABLE II

Silver Nitrate–Ethylene Glycol System at 61°C (44)

Solute	K_1, M^{-1}	K_R^0
Acrylonitrile	0.703	85
α-Chloroacrylonitrile	0.283	51
Methylacrylonitrile	0.774	65
Crotonitrile	0.940	165
Methyl methyacrylate	0.0	51

solution. This technique allows investigators to use a wide range of solvents. Unlike in ethylene glycol, the silver ion in water becomes completely dissociated in these experiments. Thus it is to be expected that the formation constants will be greater in this system, as indeed Wasik and Tsang found. The results are given in Table III, along with literature values obtained by static partition methods and the silver nitrate–ethylene glycol results of Cvetanovic et al. (43) at 25°C (see Table I).

Wasik and Tsang considered the agreement with the static method to be good, except for the aromatics, where the solubility of these compounds in the water solvent presented a problem. Except for ethylene, the values show the same trend as do those for the glycol solvent. By using very heavily loaded columns (up to 52% by weight), they minimized surface effects. In fact, their peaks show only a small amount of tailing.

Wasik and Tsang also measured the retention volumes of some deuterated species on a column 3.5M in silver nitrate. They calculated the isotope effect as the ratio of the retention times, which are given in Table IV, along with values in ethylene glycol solvent from the work of Cvetanovic et al. (43).

3. Charge-Transfer Complexes

a. Alkyl Tetrachlorophthalates. Langer et al. (4, 5) showed the possibility of using tetrachlorophthalate esters as complexing agents in GLC systems. Cadogan and Purnell (25) proceeded to study quantitatively the complexation between di-n-propyl tetrachlorophthalate (DNPTCP), dissolved in squalane and aromatic hydrocarbons solutes. They kept the DNPTCP concentration as low as possible (0–0.36M) and demonstrated that the retention volumes of the solutes were independent of solute sample volumes, if these were less than 0.1 μl. The straightforward proce-

TABLE III

Molar Formation Constants $K_1(M^{-1})$ and Partial Coefficients K_R^0 for Silver Nitrate–Water System

Solute	0°C $K_1(K_R^0)$	13.2°C $K_1(K_R^0)$	23.2°C $K_1(K_R^0)$	25°C K_1	Static method, 25°C	Glycol system,[a] 25°C	ΔH, kJ/mole	ΔS, J/(mole)(°K)
Ethylene	203 (0.20)	127 (0.14)	92 (0.11)	85	85[b]	17.5	−23	−38
Propylene			85	79	87[b]	7.5		
2-Me-prop-l-ene			55	51	71[c]	—		
But-l-ene			120	110	118[c]	8.8		
cis-But-2-ene			90	83	62[c]	4.9		
trans-But-2-ene			29	27	25[c]	1.6		
Benzene	1.86	1.45	1.58		2.41[d]	0.1[e]		
Toluene	0.43	0.52	1.19		2.95[d]	0.1[e]		

[a] Ref. 43.
[b] Ref. 47.
[c] Ref. 48.
[d] Ref. 49.
[e] Values of Muhs and Weiss (40) at 40°C.

TABLE IV

Isotope Effects on Ratio of Retention Times
for 3.5M Silver Nitrate–Water System

Solutes	Temperature, °C	Ratio	Ratio (43)
$C_2D_4–C_2H_4$	0	1.168	1.14
	23.2	1.142	1.113
$CHDCHD–C_2H_4$	0	1.086	1.07
cis-C_4D_8-2–cis-C_4H_8-2	0	1.119	1.08
$trans$-C_4D_8-2–$trans$-C_4H_8-2	0	1.144	1.09
$C_6D_6–C_6H_6$	0	1.086	
	13.2	1.079	
	23.2	1.080	

TABLE V

Di n-Propyl Tetrachlorophthalate in Squalane at 70°C (25)

Solute	K_1, M^{-1}	$-\Delta H^0,$ kJ/mole	$-\Delta S,$ J/(mole)(°K)	$K_R{}^0$	γ_f^∞
Benzene	0.416 ± 0.027	9.6	35.6	116.8	0.629
Toluene	0.491 ± 0.013	9.6	33.4	299	0.657
o-Xylene	0.542 ± 0.023	9.6	32.6	906	0.690
m-Xylene	0.468 ± 0.016	11.3	39.3	764	0.683
p-Xylene	0.492 ± 0.018	11.7	39.7	753	0.667

dure gave good straight-line plots and the slope of each gave the value of K_1 appearing in Table V, and the intercept yielded the corrected partition coefficients between solute and solvent. Measurements of K_1 at 60, 70, and 80°C gave the thermodynamic values quoted.

The activity coefficients of the aromatic hydrocarbons in squalane (γ_f^∞) were calculated using the following relationship, based on Raoult's law:

$$\gamma_p^\infty = RT / \left(K_R{}^0 \overline{V}_S\, p_A{}^0 \right)$$

where $\overline{V}_S$ is the molar volume of the solvent; fugacity corrections were then made. The authors discussed these values at length and proved that they are consistent not only with the work of Locke (50) but also with a combination of Hildebrand–Scatchard and Flory–Huggins theories of solutions.

Eon, Pommier, and Guiochon (30) have studied complexation of a variety of solutes with dibutyl tetrachlorophthalate in squalane. They presented their results in terms of the formation constant in mole fraction units. The problems of this approach were discussed previously in this chapter. The only further comment needed is, perhaps, a mention of how their results were obtained. The equation they derived relating K_R and x_B, which is more complex than that relating K_R and C_B, is

$$v_{BS}{}^0 K_R = v_S{}^0 K_R{}^0 [1 + (\psi + K_{eq}^*) x_B]$$

where $v_{AS}{}^0$ is the molar volume of the mixture of B and S of mole fraction x_B and ψ is an activity coefficient term $[= (\gamma_{A(S)}^\infty / \gamma_{A(B)}^\infty) - 1]$, where the activity coefficients are for A in pure solvent and A in pure additive B at infinite dilution of A. Using the Flory–Huggins approach, they calculated ψ from the relationship

$$\psi = v_A{}^0 \frac{\exp v_B{}^0 / v_A{}^0}{(v_S{}^0 \exp v_B{}^0 / v_S{}^0)} - 1$$

where the v^0 are the respective molar volumes.

The plot of $v_{BS}{}^0 K_R$ against x_B gives excellent straight lines from $x_B = 0$ to $x_B = 1$ for furan, 2-methylthiophene, and 2,5-dimethyl thiophene; but, for the pyrroles, the plots were distinctly curved. From the latter plots, Eon et al. derived formation constants for 1:1 and 1:2 complexes. Their paper (30) gives all the apparent partition coefficients obtained; thus $K_1(M^{-1})$ were calculated. The summary of their results is given in Table VI wherein only $K_R{}^0$ values are quoted.

Langer, Johnson, and Conder (51) have done extensive work on several alkyl tetrachlorophthalates with a variety of solutes. Their paper contains a large number of measurements of specific retention volumes of the solutes on columns coated with the pure alkyl tetrachlorophthalates. The activity coefficients are given for each solute in all the alkyl tetrachlorophthalates. It is not necessary to quote the values here. However, it is from these data that the authors calculated formation constants. They considered that, for the complexation process, an alkyl tetrachlorophthalate consists of an active tetrachlorophthalate moiety set in a matrix of alkyl chains which are inert in the charge-transfer and aromatic nucleus interactions. The concentration of the active part of the molecule per unit volume is varied by changing the size of the alkyl chains. The situation is analogous to a class A interaction, since the alkyl matrix plays the role of the solvent and C_B the concentration of the active group becomes a variable equal to C_S, the concentration of the solvent in moles per unit volume.

TABLE VI

Equilibrium Constants K^*_{eq} and Thermodynamic Functions of Complexation Reactions with Di-Butyl tetrachlorophthalate (30)

Solute	K^*_{eq} 60.3°	70.3°	80.3°	90.3°	100.3°	ΔG^* at 80.3°C	ΔH^*, kcal/mole	ΔS^*, cal/(mole)(°K)	$K_R{}^0$
Cyclopentene	7×10^{-3}	6×10^{-3}	4.8×10^{-3}	4.2×10^{-3}	3.5×10^{-3}	0.11	−8.20	−26.3	39.3
Cyclopentadiene	0.42	0.29	0.21	0.15	0.11	3.65	−28.4	−90.8	30.9
Diethyl ether	6×10^{-2}	1.7×10^{-2}	5×10^{-3}	2×10^{-3}	6×10^{-4}	0.55	−1.00	−4.39	23.0
Divinyl ether	0.50	0.47	0.46	0.44	0.42	0.49	−1.03	−4.30	14.3
2,5-Dihydrofuran	0.545	0.521	0.499	0.479	0.461	0.26	−1.39	−4.65	56.5
Furan	0.776	0.730	0.690	0.653	0.620	0.39	−1.95	−6.62	17.2
2-Methylfuran	0.680	0.624	0.575	0.533	0.495	0.50	−0.069	−1.62	45.7
Tetrahydrothiophene	0.491	0.489	0.488	0.487	0.485	0.016	−1.79	−5.52	270
Thiophene	0.928	0.858	0.796	0.742	0.694	0.024	−2.29	−7.14	88
2-Methylthiophene	0.866	0.783	0.712	0.651	0.597	0.25	−2.20	−6.93	212
3-Methylthiophene	0.843	0.765	0.698	0.640	0.590	0.023	−2.42	−7.40	228
2,5-Dimethylthiophene	0.881	0.792	0.716	0.651	0.595	0.38	−1.55	−5.47	482
2-Ethylthiophene	0.662	0.618	0.580	0.545	0.515	0.012	−1.63	−5.02	446
2-Chlorothiophene	0.946	0.884	0.838	0.770	0.726	0.32	−2.00	−6.45	337
2,5-Dichlorothiophene	0.748	0.685	0.630	0.583	0.541	0.039	−2.29	−6.59	1080
2-Bromothiophene	1.15	1.04	0.946	0.864	0.794	−0.052	−2.40	−6.66	672
3-Bromothiophene	1.33	1.19	1.07	0.980	0.896	−0.33	−3.86	−10.0	772
3,4-Dibromothiophene	2.16	1.84	1.60	1.35	1.16	−0.12	−3.24	−8.84	6190
2-Iodothiophene	1.51	1.33	1.18	1.02	0.90	−0.46	+0.78	+3.52	1610
Pyrrole K_1	1.81	1.87	1.93	2.00	2.06	0.045	−6.09	−17.4	103
Pyrrole K_2	1.57	1.20	0.94	0.74	0.59	−0.12	+1.56	+4.75	
1-Methylpyrrole K_1	1.04	1.11	1.19	1.26		1.18	−15.87	−48.3	} 126
1-Methylpyrrole K_2	0.715	0.355	0.183	0.099					
Benzene	0.882	0.747	0.683	0.627	0.580	0.27	−2.16	−6.88	87.2

The results obtained at 100°C are plotted in Figure 2; the numbers correspond to the solute numbers appearing in Table VII.

Although the results, especially for the di-octyl-tetrachlorophthalate, were less complete at 110°C, Langer et al. did obtain values at this temperature; the results are given in Table VIII.

If the assumptions regarding the alkyl chains are sound, there should be a correlation between the results of Tables V, VII, and VIII. At the least, however, we must admit that no such correlation has been conclusively proved. As Figure 2 clearly reveals, the concentrations are very high, as they would have to be, unless extremely bulky alkyl groups were used. It is improbable that simple behavior is likely at such high concentrations. Moreover, the assumption has been made that the type of alkyl groups present have no effect on, for example, the charge-transfer characteristics of the tetrachlorophthalate moiety.

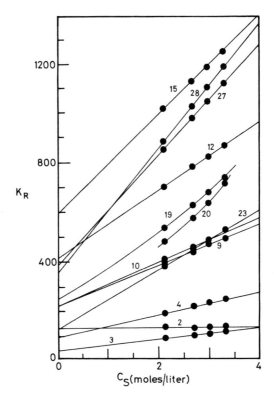

Fig. 2.   Plots (51) of K_R versus C_s (concentration of tetrachloroterephthalate) for a variety of solutes (numbers identified in Table VII).

TABLE VII

Formation Constants in Tetrahaloterephthalates at 100°C (51)

Compound	K_R^0	K_1, l/mole
1. Methylcyclohexane	96	0
2. Heptane	63	0
2a. Octane	130	0
2b. Decane	570	0
3. Benzene	35	0.675
4. Toluene	90	0.562
5. Ethylbenzene	178	0.458
6. n-Propylbenzene	354	0.405
7. n-Butylbenzene	752	0.386
8. o-Xylene	247	0.488
9. m-Xylene	221	0.378
10. p-Xylene	214	0.427
11. 1-Methyl-2-ethylbenzene	457	0.415
12. 1-Methyl-3-ethylbenzene	416	0.327
13. 1-Methyl-4-ethybenzene	418	0.360
14. 1,2,3-Trimethylbenzene	698	0.416
15. 1,2,4-Trimethylbenzene	593	0.339
16. 1,3,5-Trimethylbenzene	551	0.210
17. Isopropylbenzene	262	0.462
18. t-Butylbenzene	427	0.456
19. Styrene K_1	239	0.491
19a. Styrene K_2	—	0.091
20. Phenylacetylene	—	—
21. Anisole	—	—
22. Fluorobenzene	27	1.15
23. Chlorobenzene	131	0.908
24. Bromobenzene	276	0.832
25. Iodobenzene	710	0.727
26. o-Chlorotoluene	361	0.592
27. m-Chlorotoluene	385	0.578
28. p-Chlorotoluene	360	0.698
29. o-Dichlorobenzene	798	0.42
	($K_2 = 0.086$)	
30. m-Dichlorobenzene	671	0.44
	($K_2 = 0.028$)	
31. p-Dichlorobenzene	632	0.50
	($K_2 = 0.08$)	
32. o-Chlorobromobenzene	1150	0.83
33. m-Chlorobromobenzene	1314	0.48
34. p-Chlorobromobenzene	865	1.05

TABLE VIII

Formation Constants in Tetrahaloterephthalates at 110°C (51)

Compound	$K_R{}^0$	K_1, M^{-1}
4. Toluene	75	0.435
5. Ethylbenzene	134	0.433
7. n-Butylbenzene	507	0.398
8. o-Xylene	192	0.418
9. m-Xylene	169	0.335
10. p-Xylene	165	0.368
13. 1-Methyl-4-ethylbenzene	299	0.346
14. 1,2,3-Trimethylbenzene	522	0.347
15. 1,2,4-Trimethylbenzene	435	0.296
16. 1,3,5-Trimethylbenzene	391	0.199
17. Isopropylbenzene	188	0.455
19. Styrene	166	0.63(?)
21. Anisole	123	1.24(?)
25. Iodobenzene	537	0.613
26. o-Chlorotoluene	286	0.48(?)
27. m-Chlorotoluene	294	0.49(?)
28. p-Chlorotoluene	282	0.578(?)
31. p-Dichlorobenzene	398	0.696

(b)   Trinitro   Fluorenone.   Meen, Morris, and Purnell (34) have studied the complexation of 2,4,7-trinitro-9-fluorenone (TNF) with aromatic hydrocarbons in a series of solvents, one being β,β'-thiodipropionitrile and the rest being alkyl phthalate esters. The partition coefficients and the molar formation constants were obtained in the usual manner. Care was taken to keep the concentration of additive as low as possible ($C_B = 0$–$0.2M$), and solute samples were kept to a minimum. Good straight lines were obtained for plots of K_R against C_B and from the $K_R{}^0$ values; activity coefficients for the aromatic hydrocarbons in the solvents used were calculated, and fugacity corrections were applied. We have

$$\ln\gamma_f^\infty = \ln\gamma_p^\infty - p_A{}^0\left(B_{22} - V_A{}^0\right) + \bar{p}\left(2B_{12} - V\right)/RT$$

where $V_A{}^0$ is the solute molar volume, V is its partial molar volume, $\bar{p}$ is the average column total pressure, B_{22} is the second virial coefficient for gaseous solute, and B_{12} is the corresponding cross-coefficient for solute–carrier gas.

TABLE IX

Formation Constants, $K_1(M^{-1})$ Corresponding with $K_R{}^0$ Values and γ_f^∞ Data for Trinitro Fluorenone in the Named Solvents at 60°C (34)

	Solutes									
	Toluene		Ethyl benzene		o-Xylene		x-Xylene		p-Xylene	
Solvents[a]	$K_1(K_R{}^0)$	γ_f^∞	$K_1(K_R{}^0)$	γ_f^∞	$K_1(K_R{}^0)$	γ_f^∞	$K_1(K_R{}^0)$	γ_f^∞	$K_1(K_R{}^0)$	γ_f^∞
DNPP	0.237 (532.5)	1.182	0.161 (1158.5)	1.352	0.319 (1603.6)	1.326	0.331 (1210.2)	1.397	0.281 (1252.8)	1.404
DIPP	0.193 (520.6)	1.192	0.128 (1141.5)	1.353	0.308 (1558.8)	1.345	0.291 (1177.4)	1.415	0.262 (1217.7)	1.429
DNAP	0.430 (518.7)	0.944	0.348 (1144.1)	1.066	0.493 (1616.0)	1.024	0.476 (1243.1)	1.060	0.466 (1277.3)	1.072
DIOP	0.240 (473.9)	0.779	0.184 (1051.4)	0.874	0.331 (1492.1)	0.836	0.318 (1161.4)	0.854	0.293 (1188.7)	0.868
DNNP	0.289 (439.1)	0.773	0.226 (976.8)	0.865	0.421 (1393.1)	0.824	0.416 (1080.2)	0.844	0.378 (1119.0)	0.848
DNBM	0.339 (559.2)	1.067	0.295 (1288.7)	1.227	0.458 (1795.5)	1.195	0.453 (1392.6)	1.225	0.415 (1430.6)	1.241
TDPN	0.509 (187.7)	6.247	0.376 (304.2)	9.518	—	—	0.670 (341.2)	9.473	0.675 (343.4)	9.164

[a]Key to abbreviations: Phthalate esters: DNPP, di-n-propyl; DIPP, di-iso-propyl; DNAP, di-n-amyl; DIOP, di-isoctyl; DNNP, di-n-nonyl. DNBM, di-n-butyl maleate, TDPN, $\beta\beta'$-thiodipropiomitrile.

TABLE X

Temperature Dependence of K_1 and Thermodynamic Data
for Alkyl Benzene–TNF Complexes in DIOP (34)

	K_1, M^{-1}			$-\Delta H,$	$-\Delta S,$
Solute	44	50	60°C	kJ/mole	J/(mole)(°K)
Toluene	0.338	0.271	0.240	13.8	53.1
Ethylbenzene	0.216	0.212	0.184	9.2	41.8
o-Xylene	0.468	0.409	0.331	18.8	66.1
p-Xylene	0.438	0.400	0.318	18.0	63.6
m-Xylene	0.399	0.361	0.293	17.2	61.1

Only in the case of β,β'-thiodipropionitrile (TDPN) as solvent were the peaks asymmetric and solvent loading dependent; thus the special techniques discussed earlier (35–37) for obtaining K_R values were employed. The results obtained are listed in Tables IX and X.

Meen et al. examined the correlation between the K_1 and the $\gamma_f^\infty(=\gamma_A)$ values, initially on the basis that γ_{AB}/γ_B may be independent of solvent. Since $K_C = K_1\gamma_{AB}/(\gamma_A\gamma_B)$, a plot of log K_1 against log γ_A (which is assumed to be = log γ_f^∞) should give a straight line. Yet the plot is very far from being linear. However, Meen et al. assumed that K_1 might be unaffected by entropy changes attending the solution of the alkyl benzenes. Moreover,

$$\gamma_f^\infty = \gamma_a^\infty \gamma_t^\infty$$

where γ_a^∞ is the athermal contribution to the activity coefficient, which can be calculated from the Flory–Huggins equation $\ln\gamma_a^\infty = [1-(1/r)] - \ln r$ (r being the molar volume ratio of solvent to alkyl benzene).

Then, using the previous assumption regarding failure of entropy changes to affect K_1, along with the equation just given for γ_f^∞, Meen and his colleagues plotted log K_1 against log γ_t^∞ and obtained several linear relationships for the ester solutions. For di-iso-octyl phthalate (DIOP), all the data for the xylenes fell on a common straight line, whereas those for toluene and ethyl benzene fell on another. Thus Meen concluded that, for closely related solvents, γ_{AB}/γ_B may be either solvent independent or solvent dependent in a systematic way.

Eon and Karger (10) recalculated these results to obtain formation constants in mole fraction units (see Section IV.C.3.a), since they believed that in terms of activities, such formation constants are, by definition, independent of the nature of the solvent. As pointed out in Section II this is only true if the system could have a standard state in which each component was close to being pure. The plots of log K_x against log (activity coefficient) do not appear in Ref. 10, but the scatter would be no less than that in the plots of Meen et al.

One important point has not yet been properly explored, but it must be covered, especially in view of the results for the Ag^+–olefin complexes —namely, the nature of the activity coefficients derived from K_R^0 values. It must be realized that the derivation depends on Raoult's law, but Henry's law is a better approximation in the GLC experiments. Thus the Henry's law constant is not necessarily equal to the saturated vapor pressure of pure solute; moreover, it is important to know whether the behavior of the solutions should be related to mole fractions, concentrations, or volume fractions of solute.

4. Hydrogen-Bonding Studies

By the nature of hydrogen-bonding studies, interfacial interaction is often present, resulting in asymmetry in the eluted peaks, which have retention volumes that are also dependent on sample size. As has been outlined earlier, there is a sufficient understanding of the problems to allow formation constants to be measured in such systems. Cadogan and Purnell (32) used such methods to obtain the formation constants listed in Table XI for the interactions between aliphatic alcohols and didecyl sebacate dissolved in squalanes.

In their investigation of the interaction of monofunctional organic solutes with additives dodecan-1-ol and lauronitrile in squalane as solvent, Littlewood and Willmott (52) found that, in columns containing pure squalane or squalane with a low concentration of additive, there was evidence that retention volumes were influenced by adsorption of the solute on the Celite support. This, however, was avoided by a silanizing treatment of the support. In their theory, Littlewood and Willmott carefully considered the possible interactions that might occur. The alcohol additives are complexed in solution, but when the solute is also an alcohol, interaction need not take place only at end groups. They argued that the nonpolar part of the additive does not grossly affect the retention of the solutes. Thus the concentration unit used is the weight proportion of the polar group in the solvent, and the results obtained are given in these units in Table XII.

TABLE XI

Formation Constants and Thermodynamic Data
for the Interaction of Aliphatic Alcohols and Didecyl Sebacate in Squalane

Solute	K_1, M^{-1}			$-\Delta H°$, kJ/mole	$-\Delta G°$, kJ/mole	$-\Delta S°$, J/(mole)(°K)
	50°C	60°C	70°C			
Propan-2-ol	1.80	(1.63)	1.48	8.8	1.4	22
Butane-1-ol	1.89	1.67	1.51	10.0	1.4	26
Butane-2-ol	1.48	1.31	1.18	10.5	0.75	29
2-Methyl propan-1-ol	1.77	1.57	1.37	11.7	1.3	31
2-Methyl propan-2-ol	1.48	1.32	1.18	10.5	0.75	29
Pentan-2-ol	1.51	1.37	1.21	10.0	0.88	27
3-Methyl butan-1-ol	1.83	1.65	1.51	8.8	1.4	22
1,1-Dimethyl propan-1-ol	1.23	1.10	1.00	9.6	0.25	28

TABLE XII

Formation Constants K_1 (inverse weight fraction of hydroxyl group)
for Dodecan-1-ol in Squalane with Alcohols (52)

Solute	K_1 (inverse weight fraction of hydroxyl group)		
	40°C	56°C	80°C
Methanol	—	—	64
Ethanol	137	84	64
n-Propanol	148	95	65
Isopropanol	160	110	65
n-Butanol	170	84	63
Isobutanol	153	117	65
sec-Butanol	152	110	66
tert-Butanol	169	98	62
1-Pentanol	—	123	68
Isopentanol	—	117	65
2-Pentanol	—	97	60
3-Pentanol	125	85	49
2-Methyl-1-butanol	—	107	64
3-Methyl-2-butanol	134	84	55
tert-Pentanol	162	86	57
Neopentanol	—	101	—

In the case of polar but nonalcoholic solutes, simple theory cannot be used because of the association of dodecan-1-ol in the liquid phase. Here interactions take place mainly with unbonded end groups, the concentration of which is given by $C^* = \left[(1 + 4K_H C)^{1/2} - 1\right]/2K_H$, where K_H is the equilibrium constant for self-hydrogen-bonding in dodecan-1-ol of concentration C. The plots of relative retention volume against C^* gave good linear sections for all the solutes, except methyl ethyl ketone, and thus the formation constants K_1 given in Table XIII were obtained.

Solutions of lauronitrile in squalane behaved toward both organic nitriles and nitroalkanes in a fashion similar to that exhibited by dodecanol and alcohol solutes. Thus the formation constants were obtained from plots of retention volumes against concentration of nitrile groups; the results appear in Table XIV.

Again for weak solutes in lauronitrile, the self-association of the latter was important, and it appeared that chainlike polymers were present and that C^* is as given earlier. The formation constants obtained are presented in Table XV.

Martire and Riedl (53), investigating alcohol solutes in solutions of di-n-octyl ether and di-n-octyl ketone in n-heptadecane, also found that retention volumes were dependent on solute sample size. They used an empirical procedure to obtain meaningful retention volumes. The initial and peak retention times were measured and plotted against sample size. It was assumed that the peak maximum retention time was independent of support effects as long as the initial retention time remained constant; therefore, the peak maximum retention times of those peaks which had constant initial retention time were extrapolated to zero sample size. The formation constants which obtained are given in Tables XVI and XVII; the thermodynamic data appear in Table XVIII.

The activity coefficient of the electron donor γ_B was calculated from Eq. 57, based on the van Laar and Gibbs–Duhem equations

$$\gamma_B = \frac{\overline{V}_g^{OB} M^B}{\overline{V}_g^{OA} M^A} \tag{57}$$

where $\overline{V}_g^{OB}$ and $\overline{V}_g^{OA}$ are the specific retention volumes for a nonpolar solute in the electron-donor liquid phase and in the reference nonpolar solvent, respectively, and M^A and M^B are the molecular weights of the reference solvent and the electron donor, respectively. This equation is only valid if the components in the solution have the same molar volumes, which they do in these systems. The results obtained (K_c) when the K_1 values were corrected for this activity coefficient are given in Tables XIX to XXI.

TABLE XIII

Values of K_1 for Weakly Polar Solutes with Dodecan-1-ol in Squalane (52)

Solute	K_1 (inverse weight fraction of hydroxyl group)		
	40°C	56°C	80°C
Diethyl ether	50	35	33
Di-*n*-propyl ether	32	24	22
Ethyl acetate	78	41	25
n-Propyl acetate	64	41	25
n-Butyl acetate	60	41	25
n-Propyl chloride	25	18	—
Ethyl bromide	27	18	—
n-Propyl bromide	25	12	16
Ethyl iodide	25	18	16

TABLE XIV

Values of K_1 and ΔH for Interaction of Nitriles and Nitroalkanes with Lauronitrile (52)

Solute	K_1 (inverse weight fraction of cyanide group)			ΔH, kcal/mole
	40°C	56°C	80°C	
Ethyl cyanide	67	59	46	2.1
n-Propyl cyanide	57	52	40	2.0
Nitromethane	99	71	48	3.9
Nitroethane	69	59	46	2.2
Nitropropane	58	46	39	2.1

TABLE XV

Values of K_1 for Weakly Polar Solutes
with Lauronitrile (52)

| Solute | K_1, (inverse weight fraction of cyanide group) | | |
	40°C	56°C	80°C
1-Hexene	8	6	5
1-Heptene	8	6	5
Diethyl ether	22	18	13
Di-n-propyl ether	16	13	9
n-Propyl bromide	28	—	—
n-Propyl chloride	29	18	—
Ethyl acetate	55	36	—
n-Propyl acetate	47	36	—
n-Butyl acetate	43	36	—

TABLE XVI

Equilibrium Constants K_1 with Di-n-Octyl Ether
in n-Heptadecane (53)

| Solute | K_1 | | | |
	22.5°C	30°C	40°C	50°C
Methanol	6.41	4.99	4.17	2.96
Ethanol	4.03	3.36	2.75	2.17
1-Propanol	3.77	3.21	2.61	2.12
2-Propanol	2.99	2.55	2.00	1.64
1-Butanol	3.74	3.27	2.48	1.93
2-Butanol	3.79	3.27	2.46	2.01
sec-Butanol	2.59	2.26	1.69	1.40
t-Butanol	2.05	1.76	1.36	1.10

TABLE XVII

Equilibrium Constants K_1 with Di-n-Octyl Ketone in n-Heptadecane (53)

Solute	50°C	60°C	70°C	80°C
			K_1	
Methanol	7.58			
Ethanol	5.77	4.91	4.41	
1-Propanol	5.28	4.33	3.74	3.28
2-Propanol	4.53	3.65	3.25	2.81
1-Butanol	5.10	4.18	3.49	3.03
2-Butanol	5.09	4.17	3.46	3.00
sec-Butanol	3.80	3.19	2.73	2.41
t-Butanol	3.49	2.98	2.60	2.20

TABLE XVIII

Enthalpy Changes ($\Delta H'$) of Hydrogen-Bond Formation (53)

Solute	$\Delta H'$, kcal/mole	
	Di-n-octyl ether	Di-n-octyl ketone
Methanol	-5.12 ± 0.48	
Ethanol	-4.22 ± 0.14	
1-Propanol	-3.97 ± 0.06	-3.79 ± 0.19
2-Propanol	-4.21 ± 0.09	-3.73 ± 0.29
1-Butanol	-4.66 ± 0.30	-4.20 ± 0.14
2-Butanol	-4.51 ± 0.23	-4.27 ± 0.14
sec-Butanol	-4.40 ± 0.28	-3.67 ± 0.13
t-Butanol	-4.40 ± 0.28	-3.65 ± 0.13

TABLE XIX

Equilibrium Constants K_C with Di-n-Octyl Ether
in n-Heptadecane (53)

Solute	K_C, 1/mole,			
	22.5°C	30°C	40°C	50°C
Methanol	1.91	1.52	1.30	0.938
Ethanol	1.20	1.02	0.857	0.688
1-Propanol	1.12	0.976	0.814	0.672
2-Propanol	0.890	0.776	0.623	0.520
1-Butanol	1.11	0.994	0.773	0.612
2-Butanol	1.13	0.994	0.767	0.637
sec-Butanol	0.771	0.687	0.527	0.444
t-Butanol	0.610	0.535	0.424	0.349

TABLE XX

Equilibrium Constants K_C with Di-n-Octyl Ketone
in n-Heptadecane (53)

Solute	K_C, 1/mole,			
	50°C	60°C	70°C	80°C
Methanol	2.84			
Ethanol	2.16	1.85	1.65	
1-Propanol	1.98	1.63	1.41	1.23
2-Propanol	1.70	1.37	1.22	1.05
1-Butanol	1.91	1.57	1.31	1.13
2-Butanol	1.91	1.57	1.30	1.12
sec-Butanol	1.42	1.20	1.03	0.900
t-Butanol	1.31	1.12	0.978	0.822

TABLE XXI

Enthalpy and Entropy Changes for Hydrogen-Bond
Formation with Di-n-Octyl Ether
in n-Heptadecane (53)

Solute	ΔH, kcal/mole	ΔS, eu
Methanol	-4.69 ± 0.50	-14.6 ± 1.7
Ethanol	-3.79 ± 0.16	-12.5 ± 0.5
1-Propanol	-3.52 ± 0.10	-11.7 ± 0.3
2-Propanol	-3.77 ± 0.10	-13.0 ± 0.3
1-Butanol	-4.21 ± 0.32	-14.0 ± 1.0
2-Butanol	-4.08 ± 0.22	-13.6 ± 0.7
sec-Butanol	-3.96 ± 0.27	-13.9 ± 0.9
t-Butanol	-3.93 ± 0.15	-14.3 ± 0.5

Apelblat (54) has studied the self-association of di-n-butyl phosphoric acid (DBP) and mono-n-butyl phosphoric acid (MBP) by measuring the specific retention volumes of a variety of diluents. The liquid phases (DBP and MBP) denoted by B exist as a mixture of B and B$_r$ which, with the solute A, represents the model of the solution considered by Apelblat. From the specific retention volumes of A, we can determine γ_A^∞, which is defined by

$$\gamma_A^\infty = 273 R / \left(V_g p_A{}^0 M_S \right)$$

Unfortunately, the peaks were asymmetric with acetone, methanol, and water; the asymmetry increased with sample size, and no corrections were made. However, the other diluents gave symmetrical peaks.

Assuming ideal associated solutions, we can write

$$\lim_{x_B \to 1} \gamma_B = 1$$

$$\lim_{x_B \to 1} \gamma_A = 1 + (r-1) x_r = \gamma_A^\infty$$

where x_r is the mole fraction of the r-mer at equilibrium; namely,

$$r B \underset{}{\overset{K_r}{\rightleftharpoons}} B_r \quad \text{and} \quad K_r = x_r / \left(1 - x_r \right)^r$$

Hence the association constant can be obtained. The interpretation of the results depends on the value of r deduced. Partial results appear in Tables XXII and XXIII. Since there are many factors in the interpretation of the results, these tables are merely a guide; for a complete discussion of the results, reference should be made to the original work.

5. Other Systems

a. Fused Salts. A notable new application of the GLC technique has been made by Juvet, Shaw, and Khan (55) in the study of fused salts. They used antimony trichloride, niobium pentachloride, and tantalum pentachloride as volatile solutes in a series of alkali metal chlorides with the appropriate alkali metal tetrachloroaluminate and tetrachloroferrate as "solvents"; 1 mg of solute samples was used, and the eluted peaks were quite symmetrical. The apparent partition coefficients K_R were measured, and the results are given in Table XXIV.

The processes involving the solute that occurred in the liquid phase were as follows:

$$SbCl_3 + Cl^- \overset{K_1}{\rightleftharpoons} SbCl_4^-$$

$$MSbCl_4 + Cl^- \overset{K_2}{\rightleftharpoons} SbCl_5^{2-}$$

Thus for K_1 only we have

$$(K_R - K_R^0)/K_R^0 = K_1 C_{MCl}$$

and for K_1 and K_2 we can write

$$(K_R - K_R^0)/K_R^0 C_{MCl} = K_1 + K_1 K_2 C_{MCl}$$

Since the plots of K_R against C_{MCl} were linear at higher values of C_{MCl}, a 1:1 complex was assumed. The curvature at low C_{MCl} was attributed to the known formation of the $Al_2Cl_7^-$ ion, which is suppressed when C_{Cl^-} and therefore C_{MCl} is increased by way of

$$2AlCl_4^- \overset{K_d}{\rightleftharpoons} Al_2Cl_7^- + Cl^-$$

$$2FeCl_4^- \rightleftharpoons Fe_2Cl_7^- + Cl^-$$

The reaction $KAlCl_4 \rightleftharpoons KCl + AlCl_3$ does not occur. Thus K_d and $K_1(M^{-1})$ were evaluated by successive approximations; the results are given in Tables XXV and XXVI.

TABLE XXII

Activity Coefficients at Infinite Dilution γ_A^∞
in the diluent–DBP System and Values
of Self-Association Constants K_r (54)

| | γ_A^∞ | | | | | K_r, (mole |
| | Temperature, °C | | | | | fraction)$^{1-r}$ at |
Diluent (Solute)	25	45	60	80	r	25°C
Hexane	6.61	5.92	5.66	5.42	8	3.1×10^5
Benzene	2.14	2.13	2.20	2.27	3	7.2
Toluene	2.18	2.30	2.64	2.84	3	8.6
Cyclohexane	5.29	4.74	4.62	4.36		
Carbon tetrachloride	2.81	2.75	2.69	2.59		
Chloroform	1.28	1.38	1.47	1.59	3	0.22
Carbon disulphide	2.98	3.17	3.48	3.68		
Acetone	0.675	0.734	0.808	0.856		
Methanol	0.328	0.400	0.447	0.517	1	2.05
Water	1.31	1.32	1.38	1.42	2	0.65
					3	0.26

TABLE XXIII

Activity Coefficients at Infinite Dilution γ_A^∞
in the diluent–MBP System and Values
of the Self-Association Constant (54)

| | γ_A^∞ | | | | | K_r, (mole |
| | Temperature, °C | | | | | |
Diluent	25	45	60	80	r	fraction)$^{1-r}$
Hexane	33.4	26.2	22.2		48	1.5×10^{24}
Benzene	12.5	11.0	10.5		18	4.4×10^8
Cyclohexane	34.3	28.1	24.7			
Carbon tetrachloride	17.8	15.1	13.8			
Chloroform	8.89	8.69	8.79		8	4.5×10^4
Acetone	0.027	0.040	0.058	0.079	1	37.5 (1440)
Methanol	0.026	0.037	0.046	0.062		
Water	0.160	0.200	0.224	0.259		

286

TABLE XXIV

Thermodynamic and Retention Data of $SbCl_3$ in Various Melt Mixtures (55)

Liquid phase	Molar concentration of MCl added	V_g, ml/g	ρ_L, g/cm^3	K_R	$-\Delta H_S$	$-\Delta H_m$
KFeCl$_4$	0.000 KCl	19.8	1.890	77.2	11.0	−0.6
(at 290°C)	0.161	26.9	1.890	104.8	13.0	−2.6
	0.235	31.8	1.889	123.9	13.4	−3.0
	0.482	60.3	1.888	234.8	15.5	−5.1
	0.665	74.8	1.888	291.2	16.4	−6.0
KAlCl$_4$	0.00 KCl	13.8	1.667	47.3	11.4	−1.0
(at 289°C)	0.0912	17.4	1.667	57.7	12.1	−1.7
	0.182	20.6	1.668	70.7	13.6	−3.2
	0.237	25.2	1.668	86.5	14.0	−3.6
	0.286	28.4	1.669	97.6	14.8	−4.4
	0.391	37.0	1.669	127.1	14.8	−4.4
TlAlCl$_4$	0.00 TlCl	6.5	2.890	40.5	11.7	−1.3
(at 315°C)	1.029	8.1	3.010	52.5	12.9	−2.5
	1.517	9.3	3.063	61.4	13.1	−2.7
	2.639	12.5	3.182	85.7	13.9	−3.5
	3.295	13.9	3.252	97.4	14.7	−4.3

TABLE XXV

Disproportionation Constants for Potassium and Thallium
Tetrachloroaluminates and Potassium Tetrachloroferrate (55)

Melt	Temperature, °C	Equilibrium	K_d	$K_R{}^0$
KFeCl$_4$	290	$(Fe_2Cl_7{}^-)(Cl^-)/(FeCl_4{}^-)^2$	4×10^{-4}	10 ± 3
KAlCl$_4$	289	$(Al_2Cl_7{}^-)(Cl^-)/(AlCl_4{}^-)^2$	3×10^{-4}	10 ± 3
TlAlCl$_4$	315	$(Al_2Cl_7{}^-)(Cl^-)/(AlCl_4{}^-)^2$	6.5×10^{-3}	27 ± 5

TABLE XXVI

Formation Constants for Tetrachloroantimonates in Various Melts (55)

Melt	Temperature, °C	Equilibrium	K_1, l/mole
$KFeCl_4$	290	$(KSbCl_4)/(SbCl_3)(KCl)$	40 ± 10
$KAlCl_4$	289	$(KSbCl_4)/(SbCl_3)(KCl)$	40 ± 10
$TlAlCl_4$	315	$(TlSbCl_4)/(SbCl_3)(TlCl)$	0.8 ± 0.2

TABLE XXVII

Stability Constants and Heats of Formation of Aromatic–$LaCl_3$ Complexes (56)

	Stability constant K_1, at 200°C, m^2/mole	Enthalpy of complex formation, $-\Delta H_f$, at 200°C, kcal/mole
Benzene	2.09×10^5	3.6
Toluene	4.37×10^5	4.8
Ethylbenzene	5.06×10^5	4.8
Isopropyl benzene	4.43×10^5	4.5
Fluorobenzene	3.00×10^5	4.6
Chlorobenzene	2.17×10^5	2.8
Bromobenzene	2.84×10^5	3.1
Iodobenzene	2.75×10^5	6.6
1-Hexene	1.62×10^5	—

b. Surface Complexes.   In gas-solid chromatography it is of interest to modify the salt adsorbent in order to obtain selective separations. Cadogan and Sawyer (56) have quantified the effect of the addition of lanthanum chloride ($LaCl_3$) to modify the performance of sodium chloride on silica gel and graphon (a graphitized carbon) for the separation of aromatic and unsaturated hydrocarbons.

In gas-solid chromatography, the apparent partition coefficient of an adsorbate is related to the corrected retention volume by the expression

$$K_R = V_R / A$$

where A is the surface area of the adsorbent on the column. Hence a plot of V_R/A against surface concentration C of $LaCl_3$ illustrates the effect of this additive (to NaCl). The slopes of these plots, which were linear, gives an equilibrium constant K_1, since

$$K_R = K_R^0 (1 + K_1 C)$$

The adsorbate–$LaCl_3$ formation constant must be K_1. Since C is the surface concentration, the units of K_1 in Table XXVII are square meters per mole.

D. Summary

This review is based on the formation constants that can be obtained from GLC measurements. The rationalization of the different values has occupied very many pages of discussion in the articles quoted. Although these arguments are interesting, they are semiempirical. It was felt that the inclusion of such discussions would have made this chapter much too long. Again, many more pieces of thermodynamic data, especially those related to excess functions of solution, could have been included; but here, too, the data given were restricted to values directly related to formation constants.

Much work needs to be done in this field. A few workers having set up the necessary apparatus have produced a considerable amount of data and have established that GLC is an efficient method of measuring formation constants. The precautions needed in designing experiments have been mentioned throughout, and although data from different methods of measurement must be correlated if we are to establish the validity of each of these methods, GLC ranks as a very competitive technique for such measurements.

References

1. B. W. Bradford, D. Harvey, and D. E. Chalkley, *J. Inst. Petrol.*, **41**, 80 (1955).
2. D. W. Barber, C. S. G. Phillips, G. F. Tusa, and A. Verdin, *J. Chem. Soc.*, 18 (1959).
3. G. P. Cartoni, R. S. Lowrie, C. S. G. Phillips, and L. M. Venanzi, in *Gas Chromatography 1960* R. P. W. Scott, ed., Butterworths, London, 1960, p. 273.
4. S. H. Langer and P. Pantages, *Anal. Chem.*, **30**, 1889 (1958).
5. S. H. Langer, C. Zahn, and G. Pantazopolos, *J. Chromatogr.*, 3, 154, (1960).
6. R. O. C. Norman, *Proc. Chem. Soc.*, 151 (1958).
7. S. H. Langer and J. H. Purnell, *J. Phys. Chem.*, **67**, 263 (1963).
8. J. H. Purnell, in *Gas Chromatography 1966*, A. B. Littlewood, ed., Institute of Petroleum, London, 1967, p. 3.
9. J. Homer, M. H. Everdell, C. J. Jackson, and P. M. Whitney, *J. Chem. Soc. Faraday. Trans. II*, 874 (1972).
10. C. Eon and B. L. Karger, *J. Chromatogr. Sci.*, **10**, 140 (1972).
11. See, for example, J. H. Hildebrand and R. L. Scott, *Regular Solutions*, Prentice-Hall, Englewood Cliffs, N.J., 1962, p. 11, pp. 137.
12. I. D. Kuntz, Jr., F. P. Gasparro, M. D. Johnston, Jr., and R. P. Taylor, *J. Am. Chem. Soc.*, **90**, 4778 (1968).
13. R. Foster, *Organic Charge Transfer Complexes*, Academic Press, New York, 1969.
14. R. Foster and C. A. Fyfe, *Progress in N.M.R. Spectroscopy*, Vol. 4, Pergamon Press, Oxford.
15. H. A. Benesi and J. H. Hildebrand, *J. Am. Chem. Soc.*, **71**, 2703 (1948).
16. R. L. Scott, *Rec. Trav. Chim.*, **75**, 787 (1956).
17. S. Carter, J. N. Murrell, and E. J. Rosch, *J. Chem. Soc.*, 2048 (1965).
18. W. B. Person, *J. Am. Chem. Soc.*, **87**, 167 (1965).
19. R. Foster, *Nature*, **175**, 222 (1954).
20. J. M. Corkill, R. Foster, and D. Ll. Hammick, *J. Chem. Soc.*, 1202 (1955).
21. P. H. Emslie, R. Foster, C. A. Fyfe, and I. Horman, *Tetrahedron*, **21**, 2843 (1965).
22. L. E. Orgel and R. S. Mulliken, *J. Am. Chem. Soc.*, **79**, 4839 (1957).
23. D. F. Evans, *J. Chem. Phys.*, **23**, 1426, 1429 (1954).
24. J. E. Prue, *J. Chem. Soc.*, 7534 (1965).
25. D. F. Cadogan and J. H. Purnell, *J. Chem. Soc. A*, 2133 (1968).
26. A. Klinkenberg, *Chem. Eng. Sci.*, **15**, 255 (1961).
27. C. Eon, C. Pommier, and G. Guiochon, *C. R. Acad. Sci. Paris*, **270**C, 1436 (1970).
28. D. L. Meen, Ph.D. thesis, University of Wales, 1971.
29. J. H. Purnell, *Gas Chromatography*, Wiley, London, 1962, p. 167.
30. C. Eon, C. Pommier, and G. Guiochon, *J. Phys. Chem.*, **75**, 2632 (1971).
31. H.-L. Liao and D. E. Martire, *Anal. Chem.*, **44**, 498 (1972).
32. D. F. Cadogan and J. H. Purnell, *J. Phys. Chem.*, **73**, 3489 (1969).
33. R. L. Pecsok and B. H. Gump, *J. Phys. Chem.*, **71**, 2202 (1967).
34. D. L. Meen, F. Morris, and J. H. Purnell, *J. Chromatogr. Sci.*, **9**, 281 (1971).
35. J. R. Conder, D. C. Locke, and J. H. Purnell, *J. Phys. Chem.*, **73**, 700 (1969).
36. D. F. Cadogan, J. R. Conder, D. C. Locke, and J. H. Purnell, *J. Phys. Chem.*, **73**, 708 (1969).
37. J. R. Conder, *J. Chromatogr.*, **39**, 273 (1969).
38. E. A. Guggenheim, *Mixtures*, Oxford University Press, Oxford, 1954, Chapters X and XI.
39. D. W. Barber, C. S. G. Phillips, G. F. Tusa, and A. Verdin, *J. Chem. Soc.*, 18 (1959).
40. M. A. Muhs and F. T. Weiss, *J. Am. Chem. Soc.*, **84**, 4697 (1962).
41. L. A. Du Plessis and A. H. Spong, *J. Chem. Soc.*, 2027 (1959).

42. E. Gil-Av and J. Herling, *J. Phys. Chem.*, **66**, 1208 (1962).
43. R. J. Cvetanovic, F. J. Duncan, W. E. Falconer, and R. S. Irwin, *J. Am. Chem. Soc.*, **87**, 1827 (1965).
44. H. Schnecko, *Anal. Chem.*, **40**, 1391 (1968).
45. R. Queignec and B. Wojtkowiak, *Bull. Soc. Chim. France*, 3829 (1970).
46. S. P. Wasik and W. Tsang, *J. Phys. Chem.*, **74**, 2970 (1970).
47. K. N. Trueblood and H. J. Lucas, *J. Am. Chem. Soc.*, **74**, 1338 (1952).
48. F. R. Hepner, K. N. Trueblood, and H. J. Lucas, *J. Am. Chem. Soc.*, **74**, 1333 (1952).
49. L. J. Andrews and R. M. Keefer, *J. Am. Chem. Soc.*, **71**, 3644 (1949).
50. D. C. Locke, personal communication.
51. S. H. Langer, B. M. Johnson, and J. R. Conder, *J. Phys. Chem.*, **72**, 4020 (1968).
52. A. B. Littlewood and F. W. Willmott, *Anal. Chem.*, **38**, 1031 (1966).
53. D. E. Martire and P. Riedl, *J. Phys. Chem.*, **72**, 3478 (1968).
54. A. Apelblat, *J. Inorg. Nucl. Chem.*, **31**, 483 (1969).
55. R. S. Juvet, Jr., V. R. Shaw, and M. A. Khan, *J. Am. Chem. Soc.*, **91**, 3788 (1969).
56. D. F. Cadogan and D. T. Sawyer, *Anal. Chem.*, **43**, 941 (1971).

Chemical Reactor Applications
of the Gas Chromatographic Column

STANLEY H. LANGER AND JAMES E. PATTON,
Department of Chemical Engineering
University of Wisconsin, Madison, Wisconsin, and
Research Laboratories, Eastman Kodak, Rochester, New York

I.   INTRODUCTION

During the passage of chemical compounds through gas chromatographic columns, a number of chemical reactions can take place. The exploitation of this capability to allow the use of the chromatographic column as a chemical reactor with unique and valuable characteristics is the central theme of this discussion. As we shall see, the chromatographic column already has served as a chemical reactor for a variety of purposes. In several areas, the present discussion is limited to a few examples and experimental observations because some applications have developed slowly. However, these applications can be expected to expand as important experimental factors are identified and as recognition of the reactor potential of the column becomes more widespread.

Four obvious factors contribute to the special character of the chromatographic reactor: (a) the presence of concerted separation and reaction processes throughout the column, (b) the ease of measuring the amount of material introduced to the column and eluted from it, (c) the capacity for quantitatively handling volatile products and reactants readily, and (d) the relatively simple manipulation and control of gas chromatographic equipment. Under proper conditions, a kinetic rate constant and reactant concentration dependence can be determined. The nature of the chromatographic process with constant dilution of substances eluted, together with the inherent presence of longitudinal diffusion and a pressure gradient, favors kinetic studies of first-order and pseudo-first-order reactions with respect to the volatile reactant, although special complex systems have been considered. For preparative and identification purposes, requirements are less stringent.

To concentrate discussion on chromatographic column reactor application and potential, coverage of related areas is limited here. For instance, the use of auxiliary reactor equipment before and after the chromatographic column is certainly well known and such techniques have been discussed on a number of occasions (e.g., Refs. 1–3). Pyrolyses, as well as related chemical and physical treatments before and after the columns, have been described (1–3). Other systems involving reaction in a flow or batch reactor followed by chromatographic analysis are also widely used, and most require little additional discussion here.

Where a knowledge of common gas chromatographic principles is assumed in our discussion, we have frequently referenced the gas chromatographic texts in English most available to us (3–6). However, any good general treatment of chromatographic principles should include similar material.

The chromatographic reactor approach involves the introduction of a reactant or reaction mixture into the column as a pulse. Introduced materials are distributed between the mobile (gas or liquid) and stationary phases, as determined by characteristic partition coefficients. The mobile phase, which may also be a reactant, continuously sweeps any reactant or product through the column at velocities determined by relative distribution in the mobile phase. The stationary phase may be a liquid coated on a solid support or an adsorbent solid. When partition coefficients in the reactor column differ for reactants and products, separation tends to occur. We would expect it to be possible to adapt many different types of chromatographic columns for use as reactors with either gas or liquid mobile phases.

In considering chromatographic equipment for reactor applications, careful account must be taken of apparatus specifications. For reaction kinetic data, as with other physicochemical measurements, temperature control is critical. Much commercial chromatographic equipment cannot be controlled with limited temperature gradients, to within 0.1 to 0.3°C (the range that allows significant data to be obtained). Such equipment should be checked, depending on the intended application. Conder (7) and others (4, 6) have reviewed apparatus requirements for physical measurements.

II. THE GAS CHROMATOGRAPHIC REACTOR

A. Description and Characteristics

Specifically, we can define a gas chromatographic reactor (GCR) as a chromatographic column in which a solute or several solutes are intentionally converted, partially or totally, to products during their residence in the column. The solute reactant or reaction mixture is injected into the GCR as a pulse. Both conversion to product and separation take place in the course of passage through the column; the device is truly both reactor and chromatograph.

A hypothetical reactor chromatogram for a volatile reactant, injected

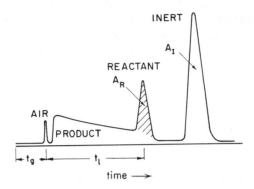

Fig. 1. Hypothetical reactor chromatogram for formation of more volatile product from volatile reactant. Areas represented as A.

as a pulse, forming a more volatile product, appears in Figure 1. The major distinction between a reactor chromatogram and a conventional chromatogram is the tailing of the product peak into the reactant peak. Obviously the relative positions of the reactant and product peaks can be reversed if the product is retained longer than is the reactant; the reactant–product portion of the chromatogram then would be the mirror image of Figure 1. If the injected material reacts with the column packing to give nonvolatile product, the reactor chromatogram is distinguished only by variation with residence time of the amount of reactant eluted (8, 9).

After the mixture of reactants and inerts (if present) is injected into the GCR, individual components are partitioned independently between the gas and stationary (liquid or adsorbent solid) phases. The carrier gas sweeps reactants, products, and inert substances through the column at velocities proportional to the fraction of time each spends in the gas phase. Diffusion may broaden individual peaks. Reactant is consumed continuously by reaction in one or both phases. Assuming the column to be isothermal and homogeneous, a complete material balance on a thin section of the gas chromatographic column at position x could be represented by

$$f_g\left(\frac{\partial C_g}{\partial t}\right)+f_s\left(\frac{\partial C_s}{\partial t}\right)=f_g D_g(x)\left(\frac{\partial^2 C_g}{\partial x^2}\right)+f_s D_s\left(\frac{\partial^2 C_s}{\partial x^2}\right)$$

$$-f_g\left\{\frac{\partial}{\partial x}\left[u(x)C_g\right]\right\}-f_g r_g-f_s r_s \qquad (1)$$

The material balance on the reactant in the stationary phase would be

$$\frac{\partial C_s}{\partial t} = k^{\bullet}(KC_g - C_s) - r_s \tag{2}$$

where C_g, C_s = concentration of reactant in the gas and stationary phases, respectively

$D_g(x)$ = gas phase interdiffusivity of reactant-carrier gas at x

D_s = liquid (stationary) phase diffusivity of reactant-stationary phase

f_g, f_s = volume fraction of the gas and stationary phases, respectively

$k^{\bullet}$ = mass transfer coefficient between gas and stationary phase

K = partition coefficient for the reactant

r_g, r_s = rate of reactant depletion in the gas and stationary phases, respectively

$u(x)$ = linear velocity of the carrier gas at x

t = time measured from injection

x = position in the column measured from inlet

Unless otherwise noted, concentrations with only a single subscript (to indicate the phase) and unsubscripted partition coefficients are understood to refer to a *reactant* of transient concentration.

The description of the real GCR with Eqs. 1 and 2 and other relations, together with boundary and initial conditions, can be made complete. There are two principal disadvantages to pursuing this tack. One is that, before the reactor can be used to investigate the rate expressions r_g and r_s, at least $k^{\bullet}$ and diffusion coefficients must be determined by way of other experiments. The second, and more serious, is that no analytic solution is available for the equations as presented.

Rigorous material balance equations are complex and difficult to solve. All solutions have assumed that the pressure and the velocity of the gas phase are uniform throughout the column. This presents problems for non-first-order reactions in which the conversion is dependent on partial pressure of reactant. Since percentage of conversion is independent of pressure for first-order reactions, these are the most amenable to the chromatographic approach. Most treatments also assume linear distribution isotherms with the rates of adsorption, desorption, and mass transfer much greater than rates of reaction. Although this assumption is justified in many cases, it does impose limitations on the treatment and application of the chromatographic reactor. The treatments that do include commensurate rates of adsorption (10–12) are complicated, and practical application is formidable except under special conditions. The inclusion of diffusion results in a second-order partial differential equation whose solution is complex, even if an average diffusivity is used (11) or if the reaction rate is zero (13, 14).

Mathematical treatments of nonequilibrium chromatography with diffusion and constant rate of flow have been described by Lapidus and Amundson (13), Yamazaki (14), Giddings (15–17), and Grubner (18). These authors do not include chemical reactions in their analyses, although their treatments can be extended in that direction.

Kocirik, using statistical moments, extended the Lapidus and Amundson treatment of chromatographic conditions, including constant diffusivity, nonequilibrium mass transfer, and constant pressure, to incorporate first-order reactions in both the mobile and the stationary phases (11). This approach is discussed in a separate section.

B. The Ideal Chromatograhic Reactor

The gas chromatographic reactor perhaps might best be appreciated by defining and characterizing an ideal chromatographic reactor (ICR). This permits comparison with conventional ideal reactors—namely, batch, plug flow, and continuous stirred tank. The ICR concept is not limited to gas chromatographic reactors, although this language is used here. It should be applicable with other types of chromatography as well. In fact, the assumption of incompressible flow is most applicable with a liquid mobile phase.

In the ideal chromatographic reactor, the following conditions prevail:

1. A reacting pulse is swept through the column and resulting products are instantaneously separated from the pulse, implying an infinite difference in retention volumes. The instant separation feature contrasts with the concept of instant, uniform dispersion in the continuous stirred tank reactor.

2. The column is isothermal. Solution and reaction heat effects are negligible.

3. Reaction rates are chemically controlled. Thus mass transfer, adsorption, and desorption cannot be rate limiting to any extent. When these processes are relatively fast, the partition between gas and stationary phases is constant and can be treated simply.

4. The height of a theoretical plate approaches zero. As a result, peak spreading and axial diffusion are not important. (The residence time of any portion of the pulse varies little.)

5. The column is homogeneous in composition throughout. The

stationary phase composition does not change except during the passage of small concentrations of eluted material. However, this does not always preclude compressibility of the gas phase due to the pressure drop across the column. In other cases, the mobile phase should be incompressible so that its linear velocity throughout the column is constant.

6.   The distribution isotherm for the reactant between the gas and liquid phases is linear (i.e., the partition coefficient is independent of concentration).

The ICR concept presents a model for operation of real chromatographic reactors. Fortunately, with some attention to basic principles and the use of small, narrow pulses, real reactors often can be operated with the features just enumerated. In some circumstances, broad pulses are used for special reasons and certain advantages of the reactor are lost.

C.   The Appeal of the Gas Chromatographic Reactor

Given the present characterization of the GCR, it seems reasonable to note the features that make it especially appealing to chemical kineticists and reactor specialists:

The GCR system is well suited for the study of reactions involving volatile products and reactants. Such reactions are especially difficult to study in batch reactors, where the material balance may be complicated by loss of either product or reactant. The GCR, by its nature, is suitable also for small samples where reactant material may be available to only a limited extent. Small sample use minimizes or eliminates solution or reaction heat effects that might complicate ordinary studies. Furthermore, reactants may not have to be completely pure because of the separation capability of the chromatographic column.

Reversible reactions can often be studied. Instantaneous, or even rapid, separation removes product from the reactant zone so that the rate of the forward reaction alone can be measured. Reverse reaction to form reactant further down the column will not affect the original reactant peak if a faster- or slower-moving intermediate material is formed. The monotonic formation of such intermediate throughout the column also leads to a dilution effect, which may inhibit reverse reaction. If two different products are formed, they may be separated, thus preventing the reverse reaction from occurring. Constant separation of products from reactants makes it possible to study reactions that might involve product inhibition or autocatalysis.

Residence or reaction time and temperature are readily controlled and varied in well-designed apparatus. Flow-rate adjustment will determine residence time and, therefore, reaction time.

Reactor systems have the added attraction of potential use of the reactor column for physicochemical determinations that may complement and aid in the interpretation of kinetic data (4, 7). The GCR may contact a catalytic solid surface, a liquid supported on a solid surface, or a combination of both, as the medium for reaction in the stationary phase. In the case of a linear adsorption isotherm on a solid, the adsorption equilibrium constant K_A is related under certain conditions to the chromatographic retention volume per gram of catalyst for a pulse V_R^T by (19–22)

$$K_A = \frac{V_R^T}{RT} \tag{3}$$

where V_R^T is corrected to column temperature as well as for column dead space and pressure drop. Thus the adsorption characteristics for reactants on a solid catalyst, as well as a rate constant for conversion in a chromatographic column, can be conveniently measured and correlated. Such adsorption data frequently are needed in catalytic chemistry, and elaborate equipment and effort may be required to obtain them in other ways. There are many chromatographic methods for measuring adsorption characteristics and a distribution coefficient K, where the simple pulse method is not applicable (6, 22). A nonlinear adsorption isotherm, of course, would not give a constant retention volume for varying sample sizes (19, 21).

Gas chromatographic retention volumes for one gram of liquid phase are similarly related to the partition coefficient for the liquid-gas combination in chromatographic columns by (3–7, 23)

$$V_R^T = KV_L \tag{4}$$

where V_L is the volume of one gram of liquid, K is the partition coefficient, and V_R^T is the corrected retention volume per gram (24) at column temperature. Such data allow the further calculation of activity coefficients for both reactants and products (23, 25, 26). Thus the gas chromatographic column employed as a reactor can be used simultaneously for the determination of activity coefficients (4–6, 23, 24) and solution thermodynamic data for reactant as well as for kinetic parameters. Since solution data for reactants and products are rarely available for comparison with reaction kinetic data, the GCR has special promise for the eventual conciliation of thermodynamic and kinetic data.

D.   Early Work; Irreversible Reactions

Bassett and Habgood (20) first reported a reactor chromatogram for the catalytic isomerization of cyclopropane to propylene as shown in

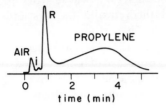

Fig. 2. Reactor chromatogram (20) for first-order isomerization, cyclopropane (R) →propylene, on 4.2 g of nickel zeolite (Linde Molecular Sieve 13× exchanged with Ni^{2+}) at 225°C; flow, 91 cm^3 He/min; i represents propane impurity.

Figure 2 (cf. Figure 1). However, for their actual kinetic work the effluent from another chromatographic catalytic reactor was trapped and analyzed with a separate analytical column. These investigators recognized the possibility of obtaining an adsorption equilibrium constant as well as a surface kinetic rate, provided other processes were not rate limiting. With the assumption of a linear adsorption isotherm, they derived an equation relating the first-order heterogeneous reaction rate constant to fractional conversion, flow rate, and the chromatographic retention volume on the catalyst.

Gil-Av and Herzberg-Minzly (8, 9) and later Berezkin et al. (28) expanded the scope of the GCR to certain bimolecular reactions, or actually pseudo-first-order reactions, by using one reactant in large excess in the liquid phase. Gil-Av studied the Diels–Alder reaction between chloromaleic anhydride and several aliphatic dienes, using the nonvolatile liquid anhydride as the stationary phase in the column. Berezkin extended the work by employing a liquid phase of maleic anhydride dissolved in tricresyl phosphate. Second-order rate constants were calculated from the concentration of maleic anhydride in the liquid phase.

In our own laboratories, Yurchak and Langer refined the chromatographic reactor technique with the study of three first-order reactions. These included the depolymerization of trioxane to formaldehyde in a polyphenyl ether column, the depolymerization of paraldehyde to acetaldehyde in a tricresyl phosphate column, and the decomposition of β-pinene to several products, also on a tricresyl phosphate column. When the percentage of liquid phase was increased, the calculated trioxane depolymerization rate constant was decreased, despite the absence of gas phase reaction. This suggested that catalytic impurities on the firebrick support influenced reaction rates, which were affected by increased mass transfer resistance to the surface. For the poorly defined β-pinene reaction with a number of products, chromatograhic reactor application was complex, as expected, although an order of magnitude estimation of initial rate constant could be made (29, 31).

Kallen and Heilbronner (32), whose work paralleled that of Bassett and Habgood, treated theoretically the first-order reaction A→B in a chromatographic reactor by adding a first-order reaction to the plate theory of Martin and Synge (24, 33) and others (34, 35). Significantly, this takes into account the peak broadening caused by the chromatographic process itself; therefore, molecules of a given substance in an initially narrow pulse have varying residence times in a column. Kallen and Heilbronner simulated chromatograms for a hypothetical reaction on a thousand-plate column by assuming different combinations of rate constants. They ignored variation in detector sensitivity to different materials and assumed that the amounts of reactant and product, respectively, could be determined independent of each other.

Nakagaki and Nishino (36) extended the Kallen and Heilbronner discontinuous plate treatment by integrating the distribution functions. The result is an equation for the first-order rate constant in terms of the number of theoretical plates, the residence time in the column, and the conversion of reactant.

Roginskii, Yanovskii, and Gaziev (37) were led to the study of the chromatographic reactor by an extension of the work of Bassett and Habgood. They noted that reactions such as A⇌P+Q become viable in the chromatographic column with separation of P and Q, thus hindering reverse reaction. They specifically investigated the catalytic dehydrogenation of cyclohexane to benzene and the tendency to drive the reaction to completion by separation of hydrogen from cyclohexane. In the course of this work, they anticipated, and realized experimentally, many of the advantages of running reactions under "chromatographic conditions" for catalytic reactions. Gaziev et al. (38) then considered conversion and determination of rate constants for zero, first-order, and second-order reactions. They also attempted to determine the effect of order on the shape of the exit peak for rectangular and triangular initial pulses. For chromatographic conditions, constant partial pressure of reactant was assumed for zero and second-order reactions. This is a difficult assumption to justify in real columns if a narrow pulse of reactant is introduced to the column.

III. STUDY OF FIRST-ORDER REACTIONS

A. Introduction

In order to use the GCR for the determination of chemical kinetic rate constants, it is necessary to model the column reactor so that conversion, as indicated by the ratio of output to input of reactant, can be related quantitatively to residence time in the column. The standard models for the chromatographic column are the "continuous flow" and "plate" models. (References to both treatments are given in Section II.) The continuous flow model recognizes the nature of the flow in the column and considers a number of the processes taking place over an infinitesimally thin section of the column before integrating over the complete column length. The "plate" model assumes that all processes take place repeatedly and discontinuously in a regular manner in separate uniform sections or plates in a multicompartmented column.

The subsections that immediately follow deal mainly with a liquid stationary phase, subscripted by l, but treatments can be extended readily to consider reaction on a solid surface phase. First-order reactions are considered initially because they are least sensitive to the dilution and spreading of the reactant over the column occurring during the chromatographic process. Their nature makes them most suited for kinetic studies in the column.

As considered here, both models lead to the same equation for relating conversion to a rate constant. Before proceeding to Section III.C, readers not concerned with details of the mathematical treatment may wish to note Eq. 15, which is a result expected intuitively by experienced workers in kinetics.

B. Mathematical Models for the Chromatographic Reactor

1. Continuous Flow Model

The characteristics of the ICR, including those disregarding spreading processes, make it possible to present a differential equation for the GCR. An analytical solution is available for first-order reactions. The solution is validated to a great extent because first-order reactions are not sensitive to column-broadening processes for a reactant pulse if there is not a great spread in residence time for the reactant.

The material balances of Eqs. 1 and 2 are simplified by the characteristics of the ICR described earlier. Specifically;

1. Kinetic factors and temperature are constant throughout the column and are position invariant (characteristics 2 and 5, Section II.B).
2. Diffusion fluxes may be ignored (characteristic 4).
3. Gas and liquid phase concentrations of reactant are related by a linear constant (characteristics 3 and 6)

$$C_l/C_g = K \tag{5}$$

throughout the column, where C_l is concentration in the liquid phase.

4.   The rate of reaction is controlled by a first-order rate law in both phases (characteristic 3) expressed by

$$r_l = k_l C_l \tag{6a}$$

$$r_g = k_g C_g \tag{6b}$$

The simplified equation for a section of the reactor of unit volume then becomes

$$f_g\left(\frac{\partial C_g}{\partial t}\right) + f_l\left(\frac{\partial C_l}{\partial t}\right) = -f_g\left\{\frac{\partial}{\partial x}[u(x)C_g]\right\} - f_g r_g - f_l r_l \tag{7}$$

| accumulation of reactant | = | net change of reactant due to flow | + | change in reactant due to chemical reaction in gas and liquid phases |

If Eqs. 5 and 6 are substituted in Eq. 7, we have

$$(f_g + f_l K)\left(\frac{\partial C_g}{\partial t}\right) = -f_g\left\{\frac{\partial}{\partial x}[u(x)C_g]\right\} - (f_g k_g + f_l k_l K)C_g \tag{8}$$

Reactant may be introduced into the column as a pulse of arbitrary shape. Hence the boundary condition at the inlet of the column $(x=0)$ is

$$C_g(0, t) = \phi(t)$$

If no reactant remains in the column from previous injections, the initial condition at zero time and at distance x from the inlet, is

$$C_g(x, 0) = 0$$

Equation 8 can be solved with Laplace transforms and these boundary conditions (31, 38–40) to give the following expression for C_g at point x and time t:

$$C_g(x, t) = \frac{u(0)}{u(x)} \phi[t - \alpha\tau(x)] \exp[-\beta\tau(x)] \tag{9}$$

where residence time up to x is

$$\tau(x) = \int_0^x \frac{dx'}{[u(x')]}$$

and

$$\alpha = \frac{f_g + f_l K}{f_g}$$

$$\beta = \frac{f_g k_g + f_l k_l K}{f_g}$$

(10)

The concentrations of reactant in the gas phase at the column inlet [where $\tau(0) = 0$] and at the outlet [where $x = L$ and $\tau(L) = t_g$)] are then, respectively,

$$C_g(0,t) = \phi(t)$$

(11a)

$$C_g(L,t) = \frac{u(0)}{u(L)} \phi(t - \alpha t_g) \exp(-\beta t_g)$$

(11b)

The total weight of reactant, of molecular weight M, entering a column having cross-sectional area a_g (corrected for porosity), is W_{in} or

$$W_{in} = a_g u(0) M \int_0^\infty C_g(0,t)\, dt$$

$$= a_g u(0) M \int_0^\infty \phi(t)\, dt$$

(12)

and the total amount leaving the column, W_{out}, is

$$W_{out} = a_g u(L) M \int_0^\infty C_g(L,t)\, dt$$

With Eq. 11b, we can write

$$W_{out} = a_g u(0) M \exp(-\beta t_g) \int_0^\infty \phi(t - \alpha t_g)\, dt$$

(13)

Since

$$\phi(t) = 0 \quad \text{for} \quad t < 0,$$

$\int_0^\infty \phi(t)\,dt$ must be $\int_0^\infty \phi(t - \alpha t_g)\,dt$ if the input is bounded. Thus Eqs. 12 and 13 can be combined to eliminate the integral, as follows:

$$W_{in}/W_{out} = \exp(\beta t_g)$$

Since Kf_g/f_g is equal to the ratio of the reactant residence time in the liquid phase to that in the gas phase, we have

$$Kf_l/f_g = t_l/t_g \tag{14}$$

when

$$\beta = \frac{k_g t_g + k_l t_l}{t_g}$$

Thus we use the expression

$$W_{in}/W_{out} = \exp(k_g t_g + k_l t_l) \tag{15}$$

The use of total weight ratios in Eq. 15 is most compatible with the types of signals obtained from the uncalibrated differential detectors commonly used in gas chromatographic systems. Detection in terms of absolute concentrations is unusual; with most apparatus, the integrated areas of chromatographic curves are proportional to the amounts present of specified materials.

Equation 15 presents a result that might be anticipated from the properties of first-order reactions. That is, this also would be the result for a slug of reacting material passing through a two-phase system without appreciable spreading.

There is an interesting analogy between the performance of the ICR and that of a plug flow reactor or tubular reactor. A pulse traveling through a chromatographic column resembles a slug of material passing through the plug flow reactor. For both, the velocity of pulse or slug can be governed by the flow rate of some other medium, such as carrier. Carrier material can be compressible. For first-order reactions, the two phases of the column can be treated as two reactors in series, the first a liquid-filled plug flow reactor and the second a gas-filled plug flow reactor, with volumes such that the reactant spends time t_l in the first and time t_g in the second. The plug flow reactor does not have the separation characteristics of the chromatographic reactor, but the flow and reaction schemes are analogous. If the first-order rate constants are k_l and k_g in the two phases, the fraction of reactant remaining (W_{out}/W_{in}) after passing through the first plug flow reactor is $\exp(-k_l t_l)$ (41). After the second column, only $\exp(-k_l t_l)\exp(-k_g t_g) = \exp(-k_l t_l - k_g t_g)$ remains. This corresponds to the conversion of the chromatographic reactor (Eq. 15). It is reassuring to find the agreement of Eq. 15 with the results for one of the "classical" ideal reactors. A similar treatment is not applicable with reactions other than first-order, however, since conversion becomes concentration dependent. Residence times, then, cannot be separated with validity into liquid and gas phase plug flow reactor components.

2. Plate Model

The plate model concept arises from treating a chromatographic column and distillation column similarly. In earlier distillation treatments, equilibrium between liquid and vapor phases was considered to exist on every plate throughout the column. Analogously, the chromatographic column is considered to be composed of sections or theoretical plates, the solute existing in each, in equilibrium between gas and liquid phases (3–6, 24, 33–35). In the discontinuous plate model, the carrier gas flows in increments of ΔV_g, the volume of gas associated with a single theoretical plate. In any plate, the distribution of solute between phases is determined by its partition coefficient; thus the fraction of reactant in the liquid phase F_l is expressed by

$$F_l = f_l K / (f_g + f_l K) \tag{16}$$

and the fraction in the gas phase is

$$F_g = 1 - F_l = f_g / (f_g + f_l K) \tag{17}$$

To the column on the first plate we add W_{in} g of reactant, partitioning this amount so that $F_l W_{in}$ g is in the liquid phase and $F_g W_{in}$ g is in the gas phase. For a column with many plates, the residence time in any particular plate is small. Assuming the residence time in each plate is Δt, the reactant loss from a first-order reaction in the liquid phase of the nth plate (32, 36), when $W_{l,n}$ is introduced to the plate, is

$$\Delta W_{l,n} / W_{l,n} = \exp(- k_l \Delta t) - 1$$

approximated by

$$\Delta W_{l,n} / W_{l,n} = - k_l \Delta t \tag{18}$$

Similarly, the loss in the gas phase is

$$\Delta W_{g,n} / W_{g,n} = - k_g \Delta t \tag{19}$$

The fractional conversion of reactant is the same on all plates and can be defined $x_l = k_l \Delta t$ in the liquid phases and $x_g = k_g \Delta t$ in the gas phases. After reaction, ΔV_g of carrier gas enters the column and sweeps the $F_g(1-x_g)W_{in}$ grams of reactant in the gas phase to the second plate where redistribution between phases takes place. The remaining $F_l(1-x_l)W_{in}$ grams in the liquid phase of the first plate then is repartitioned. The entire process of reaction, flow, and redistribution is repeated with each increment of flow. A myriad of material balances is saved by the fact that the total amount of reactant on the nth plate of a column (32, 33, 34, 36) after passage of v volumes of carrier gas is given by the nth term of the binomial distribution $[(1-x_l)F_l+(1-x_g)F_g]^v W_{in}$. The amount emerging from an N-plate column after passage of v volumes is that in the gas phase of the Nth plate after the passage of $v-1$ volumes.

$$W_{out}(v) = F_g(1-x_g)W(N, v-1)$$

$$= \frac{(v-1)!F_l^{v-N}(1-x_l)^{v-N}F_g^N(1-x_g)^N}{(N-1)!(v-N)!} W_{in} \tag{20}$$

For $(F_g + x_l F_l)$ small relative to unity and v large relative to N, the following approximations are valid (4, 5, 32–34, 36):

$$F_l^{v-N}(1-x_l)^{v-N} = (1-F_g-x_lF_l)^{v-N}$$

$$\approx e^{-(v-N)(F_g+x_lF_l)} \tag{21}$$

and

$$\frac{(v-1)!}{(v-N)!} \approx (v-N)^{N-1} \tag{22}$$

Equation 20 then becomes

$$\frac{W_{out}(v)}{W_{in}} = \frac{(v-N)^{N-1}F_g^N(1-x_g)^N e^{-(v-N)(F_g+x_lF_l)}}{(N-1)!} \tag{23}$$

Multiplying numerator and denominator of Eq. 23 by $(1+x_lF_l/F_g)^{N-1}$ gives a function similar to the Poisson distribution multiplied by $F_g^N(1-x_g)^N/(F_g+x_lF_l)^{N-1}$.

$$\frac{W_{out}(v)}{W_{in}} = \frac{F_g^N(1-x_g)^N}{(F_g+x_lF_l)^{N-1}} \frac{[(v-N)(F_g+x_lF_l)]^{N-1}e^{-[(v-N)(F_g+x_lF_l)]}}{(N-1)!}$$

$$\tag{24}$$

The total conversion is the summation of all the incremental $W_{out}(v)$ / W_{in} contributions beginning with $v = N$, the number of volumes needed to elute a nonadsorbed material.

$$\frac{W_{out}}{W_{in}} = \sum_{v=N}^{\infty} \frac{W_{out}(v)}{W_{in}} \qquad (25)$$

$$= \frac{F_g^N (1-x_g)^N}{(F_g + x_l F_l)^{N-1}} \sum_{v=N}^{\infty} \frac{[(v-N)(F_g + x_l F_l)]^{N-1} e^{-[(v-N)(F_g + x_l F_l)]}}{(N-1)!}$$

Since the summation is the sum of the Poisson distribution (unity) divided by $F_g + x_l F_l$,

$$W_{out}/W_{in} = \frac{F_g^N (1-x_g)^N}{(F_g + x_l F_l)^N} \qquad (26)$$

With some algebraic manipulation (36) this can be rewritten

$$1 - (W_{out}/W_{in})^{1/N} = \frac{x_l F_l + x_g F_g}{x_l F_l + F_g} \qquad (27)$$

The number of volumes necessary to elute the reactant maximum is found by setting the first derivative of Eq. 23 equal to zero and solving for v_{max}, as follows:

$$v_{max} = \frac{N}{F_g + x_l F_l} \qquad (28)$$

The average residence time of the reactant peak t_R measured from injection is

$$t_R = v_{max} \Delta t$$

$$= \frac{N \Delta t}{F_g + x_l F_l} \qquad (29)$$

Combining Eqs. 27 and 29 with $x_l = k_l \Delta t$ and $x_g = k_g \Delta t$ gives

$$t_R (k_l F_l + k_g F_g) = N [1 - (W_{out}/W_{in})^{1/N}] \qquad (30)$$

but $F_l t_R = t_l$ and $F_g t_R = t_g$ (36); thus we have

$$k_l t_l + k_g t_g = N [1 - (W_{out}/W_{in})^{1/N}] \qquad (31)$$

Equation 31 provides an alternate to Eq. 15 for calculation of first-order rate constants. For reaction in the liquid phase only, where k_g is zero, a plot of $N(1 - W_{out}/W_{in})^{1/N}$, versus residence time, should give k_l.

The number of plates in a column can be calculated from

$$N = 5.545 (D/d_{1/2})^2 \qquad (32)$$

where D is the distance on the chromatogram from injection to the maximum of the reactant peak, and $d_{1/2}$ is the reactant peak width at one-half its maximum height (3–6).

The limit of the plate theory result with increase in the number of plates can be found with the aid of l'Hôpital's rule, as follows:

$$\lim_{N \to \infty} (k_l t_l + k_g t_g) = \lim_{N \to \infty} \frac{\left[1 - (W_{out}/W_{in})^{1/N} \right]}{N^{-1}}$$

$$= \lim_{N \to \infty} \frac{(W_{out}/W_{in})^{1/N} \ln (W_{out}/W_{in}) N^{-2}}{-N^{-2}}$$

$$= -\ln (W_{out}/W_{in}) \lim_{N \to \infty} (W_{out}/W_{in})^{1/N}$$

$$= \ln (W_{in}/W_{out}) \qquad (33)$$

This limit corresponds to the expression obtained for the ICR model (Eq. 15). Where the number of column plates in a column exceeds 1000, the difference between the limiting form and the exact expression is less than 0.1%.

Although plate theory has been used to study reactions on a solid surface (32, 36) and in a liquid phase (29, 30), the possible reduction of the plate theory equation to either the continuous flow model solution or Bassett and Habgood's solution (20) was not recognized. A comparison of results from Eqs. 15 and 31 was made for the paraldehyde decomposition reaction (forming acetaldehyde) in tricresyl phosphate at 100°C (29). Under operating conditions, the $1/4 \times 48$ in. column used gave 1000 to 1500 plates. The rate constant measured from the slope of Eq. 31 was 1.48×10^{-3} sec^{-1}, in good agreement with the value 1.52×10^{-3} sec^{-1} obtained by the inert standard method with the same chromatograms. The difference probably reflects errors of measurement on the chromatograms.

The importance in the similarity of the continuous flow model and plate model equations lies in the difference in their development. The continuous flow model does not consider peak spreading, perhaps an overidealized assumption. The discontinuous plate model includes the inherent spreading of a chromatographic peak, assuming a constant residence time in each plate. The similarity of the results helps justify each treatment, provided the chromatographic column is operated efficiently. This requires a narrow injection pulse as well as many plates.

C. Methods for Determining Rate Constants for First-Order Irreversible Reactions

With the foregoing equations and several other approaches, the GCR can be used to study first-order, irreversible reactions in both liquid and gas phases. The principal techniques for determining chemical rate constants are the inert standard, product curve, reactant–product ratio, and stopped-flow methods. Each of these employs different data from the reactor chromatogram, as well as different assumptions concerning their significance. Consequently, each method has advantages and disadvantages, and these methods are described and illustrated with examples.

1. Inert Standard Method

Often in quantitative gas chromatography, an inert substance is added quantitatively to a mixture to provide an "internal standard" for analyses (3–6). This compensates for the inaccuracies in measurement of the very small injected pulse ordinarily used. Since differential detectors (e.g., thermal conductivity, flame ionization, and electron capture) produce signals proportional to solute gas phase concentration, the area of a recorded peak on a chromatogram is proportional to the weight of solute passing the detector. If W_i and A_i are the weight and peak area of solute i, the sensitivity is

$$S_i = W_i / A_i \qquad (34)$$

The ratio of solute weight to inert weight is directly proportional to the ratio of their respective areas on a chromatogram. When using a reaction mixture containing an inert standard, the inert area from the resulting chromatogram is proportional to the initial weight of reactant, and the reactant area is proportional to the weight of unconverted reactant. Substituting Eq. 34 into Eq. 15 yields

$$\ln(A_R^* S_R^* / A_R S_R) = k_l t_l + k_g t_g \qquad (35)$$

where the subscript R denotes reactant. The quantities in the denominator refer to a detector located at the column outlet; an asterisk indicates that the quantity refers to an imaginary detector at the column inlet. Addition of an inert substance with sensitivity S_I to the reaction mixture will give another peak of area A_i. Adding $\ln(S_I A_I)$ to both sides of Eq. 35, rearranging, and using Eq. 34 gives

$$\ln(A_I/A_R) = k_l t_l + k_g t_g + \ln(W_I/W_{R,\text{in}}) + \ln(S_R/S_I) \tag{36}$$

The initial weight ratio of inert to reactant $W_I/W_{R,\text{in}}$ remains constant with a standard reaction mixture, regardless of the sample volume introduced to the column. Sensitivities S_R and S_I depend on the bridge current of the thermal conductivity cell, the temperature, the carrier gas, and to a lesser extent, the flow rate (3–6). Fortunately, the ratio of sensitivities does not change significantly, and for given conditions, $\ln(S_R/S_I)$ is considered invariant.

An apparent rate constant k_{app}, can be defined by

$$k_{\text{app}} \equiv k_l + (t_g/t_l)k_g \tag{37}$$

For a given column and temperature, t_g/t_l is also constant. Equation 36 then can be rewritten (42)

$$\ln(A_I/A_R) = k_{\text{app}} t_l + \ln\frac{W_I/S_I}{W_{R,\text{in}}/S_R} \tag{38}$$

With $\ln(A_I/A_R)$ and t_l (the liquid phase residence time) measured experimentally, the invariance of the other logarithmic term k_{app} can be evaluated from a series of reactor chromatograms derived at different flow rates. A value for $W_{R,\text{in}}$ is not needed.

Both liquid and gas phase rate constants can be determined if rates in the two phases are commensurate and the reactions are homogeneous. Since the ratio t_g/t_l decreases as the percentage of liquid phase increases, kinetic runs on columns with different percentages of stationary phase produce apparent rate constants which are linear combinations of k_l and k_g. Gas and liquid phase constants are evaluated by simultaneously solving the equations for k_{app}. Since t_g/t_l is frequently less than 0.2, k_{app} is approximately k_l if the liquid phase rate constant is an order of magnitude greater than the gas phase constant. When the rate in one phase predominates, the rate constant in the other phase cannot be determined accurately. Owing to mass transfer complications, constants for liquid phase reactions catalyzed by the surface of the solid support usually depend on the liquid loading; therefore, the heterogeneous rate constant cannot be found by simply varying the column composition. Such constants can be identified however, as noted previously.

A study of the dissociation of *endo*-dicyclopentadiene (DCPD) to give cyclopentadiene (CPD) (40, 43, 44), with hexatriacontane liquid phase, is illustrative of the inert standard approach.

$$\rightleftharpoons 2 \qquad\qquad (39)$$

endo-DCPD   CPD

The dilution and product removal effects described earlier allow the reverse reaction to be neglected with a sample of one microliter or less.

A set of reactor chromatograms obtained at 180°C but varying flow rates is presented in Figure 3. The reactant residence time in the liquid phase t_l is measured from the air peak A to the maximum in R. In theory, the interference of the product wave P in the reactant peak causes the maximum in R to be eluted before the mean residence time by an amount of time that increases with conversion (29). Here, any error amounts to a few tenths of one percent, and no correction is required (36, 39). Where retention volumes of product and reactant are close, it may be advisable to take this into account by reconstructing and subtracting the product trace.

The gas phase residence time for all components t_g is measured from the time of injection to A. The inert peak area A_I is measured by any integrating method. The reactant peak area A_R must be corrected for product interference in the peak, approximated as area under the dotted lines in Figure 3 or between AB' and the base line in Figure 4. A geometric correction has been recommended by Yurchak (29). For typical eluted reactant peaks in Figure 3, a reasonable estimate of the product contribution to the reactant peak is found by taking the area between AF and the baseline as illustrated in Figure 4. The line AF bisects the angle between the extrapolation of the product curve (AE) and a horizontal line to the point at which the extrapolation leaves the product curve (parallel to the baseline). That is, AFD is approximately DCB'. The uncertainty in the reactant area in the product peak is represented by the shaded line ADB'. A substantially complete argument for this approximation has been presented (29); a simplified version is as follows. When one-quarter of the reactant peak has been eluted, the product appears at the column outlet at a rate about the same or slightly less than three-quarters of the production rate observed at A; when half the peak has been eluted, this rate will be the same or slightly less than half the rate at A, and so on. The dashed lines of Figure 3 in the reactor chromatograms and the chromatogram baseline enclose the estimated areas of product contribution to the reactant peak.

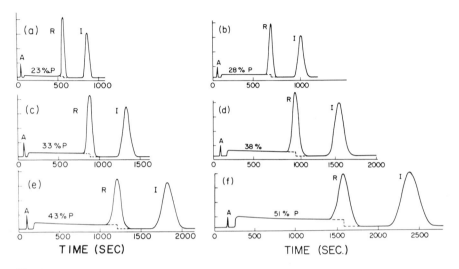

Fig. 3. Typical series of reactor chromatograms (39): dicyclopentadiene dissociation to form cyclopentadiene at 180.1°C. Column (90 × ¼ in.) contains *n*-hexatriacontane (20%) on Gas Chrom Q, 60-80 mesh: *A*, air peak; *P*, product; *R*, reactant; *I*, inert reference material. Flow rates at column temperature range from (*a*) 66.1 cm³/min to (*b*) 20 cm³/min. Areas under dashed lines approximate true product interference in reactant peak *R*.

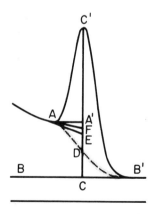

Fig. 4. Correction of reactant peak area for product interference (40). Area under *ADB'* approximated as area between *AF* and baseline (see text). Conversion is 81% here; correction is greater than normal.
By courtesy of *Journal of Physical Chemistry*.

Rate plots based on Eq. 38 appear in Figure 5; data for the reactor chromatograms in Figure 3 are included. The slope of the lines gives the apparent rate constant; thus since the gas phase rate constant for the DCDP decomposition has been reported (45, 46), the liquid phase rate constant can be calculated, and we can write

$$k_l = k_{app} - k_g(t_g/t_l)$$

$$= [5.04 - 4.29(0.0936)] \times 10^{-4} = 4.64 \times 10^{-4} \text{ sec}^{-1}$$

If k_g is unknown, and significant, the liquid phase rate constant can still be calculated with an estimate of k_g obtained with a second column of different composition or percentage of liquid phase.

The ratio of gas phase to liquid phase residence times is related to column composition by

$$t_g/t_l = f_g/f_l K$$

Since the partition coefficient K is a constant in the present context, the time ratio is proportional to f_g/f_l, the ratio of void volume to liquid phase volume; this increases with decreasing percentage of liquid phase in the column. From columns with varying levels of liquid phase, measured apparent rate constants, plotted against the time ratio for these columns, yield a straight line whose intercept is the liquid phase rate constant and whose slope is proportional to the gas phase rate constant. This is illustrated for 180 and 190°C in Figure 6. The liquid phase rate constants were 4.65 ± 0.06 and $10.31 \pm 0.21 \times 10^{-4} \text{sec}^{-1}$ at 180 and 190°C, respectively. Constants and standard deviations obtained in this manner without the gas phase data are almost identical to those calculated with the aid of literature results (45, 46) shown in Table I. The gas phase rate constants found from the slopes were 3.89 ± 0.62 and $10.92 \pm 1.82 \times 10^{-4} \text{sec}^{-1}$ at 180 and 190°C, respectively. The gas phase rate constants are adequate for our purposes but not as accurate as desired, since differences in experimental t_g/t_l ratios are small. The higher precision of the liquid phase constant in this instance is a consequence of the major portion of the conversion taking place in that phase.

The foregoing treatment is somewhat involved because the DCPD reacts in both the gas and the liquid phases. Reactions that take place in the liquid phase only are easier to analyze, for then k_g is zero and the equations yielding k_l are simplified.

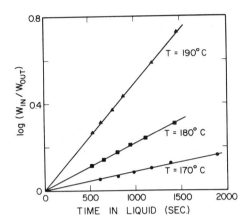

Fig. 5. First-order rate plots of ln (inert area/reactant area) at various temperatures (see Eq. 33) for DCPD dissociation in hexatriacontane (40). Points are averages of duplicate runs. Plots were arbitrarily adjusted to pass through origin.
By courtesy of *Journal of Physical Chemistry*.

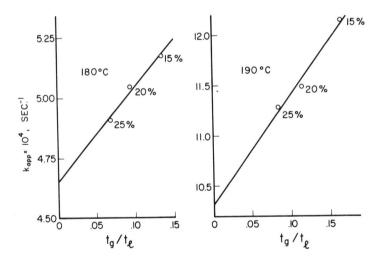

Fig. 6. Plots (40) of t_g/t_l against apparent rate constant to determine gas phase rate constant from the slope and liquid phase rate constant from the intercept (see Eq. 38). The t_g/t_l is varied by changing percentage of hexatriacontane liquid phase in the column.
By courtesy of *Journal of Physical Chemistry*.

316

TABLE I

Rate Constants for Dicyclopentadiene Dissociation
Determined in n-Hexatriacontane on Gas Chrom Q

$k_g \times 10^{-4}, \sec^{-1}$ (Ref. 45) 180.1°C	189.8°C	% $C_{36}H_{74}$ (in packing)	$k_l \times 10^4, \sec^{-1}$ (Ref. 39) 180.1°C	189.8°C
—	—	15	4.59 ± 0.12	10.66 ± 0.21
4.28	9.41	20	4.64 ± 0.07	10.40 ± 0.09
		25	4.61 ± 0.06	10.51 ± 0.08
			4.61	10.52

The GCR reactions considered to this point have been simple, with few side reactions. However, the device can be applied with some loss of accuracy to more complex first-order reactions involving a number of paths and, possibly, consecutive reactions. An example of such a reaction is the decomposition of β-pinene (47), which has been studied on a tricresyl phosphate-firebrick column (29). Reaction rate varied with aging of the column, but this is not of concern here. Chromatograms for the decomposition are shown in Figure 7. It is evident that kinetic rate constants can be determined less accurately than with a "clean" reaction. Nevertheless, a good estimate of the rate constant for decomposition of β-pinene can be made with the internal standard method, despite the apparent formation of at least four products.

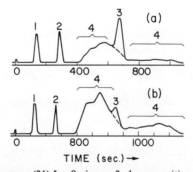

Fig. 7.   Reactor chromatograms (31) for β-pinene, 3, decomposition to several products, 4, at 100°C. The internal standards are benzene, 1, and toluene, 2. (a) Flow = 58.8 cm³/min, 70% conversion; (b) flow = 29.9 cm³/min, 91% conversion. Area of reactant peak is set off with dotted line.

By courtesy of *Industrial and Engineering Chemistry*.

2. Product Curve Method

The product curve on reactor chromatograms of the type represented in Figure 3 resembles an exponential decay. This strongly suggests the calculation of reaction rate constants from the shape of a single product curve from reactant as it moves through the column (39, 48–50). Since recorder response is proportional to concentration, it is appropriate to consider equations for the product concentration at the column outlet as a function of time.

The material balance on the product in the ideal chromatographic reactor becomes

$$f_g\left(\frac{\partial C_{g,P}}{\partial t}\right)+f_l\left(\frac{\partial C_{l,P}}{\partial t}\right)= -f_g\frac{\partial}{\partial x}[u(x)C_{g,P}]+f_g r_g+f_l r_l \tag{40}$$

where $C_{g,P}, C_{l,P} =$ molar concentration of product in the gas and liquid phase, respectively

$f_g, f_l\ =$ volume fraction of the gas and liquid phase, respectively

$u(x)\ =$ linear velocity of the gas phase

$r_g, r_l\ =$ rate at which product is formed in the gas and liquid phases, respectively

$x\ =$ position in the column measured from the inlet

$t\ =$ time measured from injection

The second subscript with the concentration is used to distinguish between reactant and product.

Since isotherms are linear in the ideal chromatographic reactor, we can write

$$K_P= C_{l,P}/C_{g,P} \quad\text{and}\quad K_R= C_{l,R}/C_{g,R} \tag{41}$$

The initial and boundary conditions are

$$C_{g,P}(x,0)=0 \quad\text{and}\quad C_{g,P}(0,t)=0 \tag{42}$$

For first-order reactions, Eqs. 40 and 41 can be combined to give

$$(f_g+K_P f_l)\left(\frac{\partial C_{g,P}}{\partial t}\right)+f_g\frac{\partial}{\partial x}[u(x)C_{g,P}]=\eta_P(f_g k_g+f_l k_l K_R)C_{g,R} \tag{43}$$

where η_P is the number of moles of product formed from each mole of reactant. This factor arises because k_g and k_l have been defined in terms of the rate of reactant depletion, rather than product formation. Substituting the expression for reactant concentration in the ICR (Eq. 10) and putting

$$\alpha_R = 1 + f_l K_R / f_g$$

$$\alpha_P = 1 + f_l K_P / f_g$$

$$\beta = k_g + k_l f_l K_R / f_g$$

into the last expression, we obtain the partial differential equation

$$\alpha_P \left(\frac{\partial C_{g,P}}{\partial t} \right) + \frac{\partial}{\partial x} [u(x) C_{g,P}] = \frac{u(0)}{u(x)} \eta_P \beta \phi [t - \alpha_R \tau(x)] \exp[-\beta \tau(x)]$$

$$(44)$$

Unlike the inert standard method, solving Eq. 44 requires specification of the form of the reactant input wave. For a square-wave input, we have

$$\phi(t) = \left\{ \begin{array}{ll} C_0 & 0 < t < t_0 \\ 0 & t > t_0 \end{array} \right\}$$

The product concentration at the outlet for $t_P + t_0 < t < t_R$ (the solution to Eq. 44) is then given by

$$C_{g,P}(L,t) = [u(0) \eta_P C_0 / u(L)][\exp(k'_{app} t_0) - 1] \exp[-k'_{app}(t - t_P)]$$

$$t_P + t_0 < t < t_R \quad (45)$$

where t_P is the total residence time of pure product $(t_{l,P} + t_g)$ in the column. (This assumes product to be retained to a lesser extent than reactant.) Hence, after t_0 sec from the initial appearance of product, a plot of ln (recorder response) versus time has a slope of $-k'_{app} = -k_{app} t_{l,R} / (t_{l,R} - t_{l,P})$ (39, 49–51). Use of a Dirac or Gaussian input, instead of the square wave, changes the form of the corresponding equations for product concentration. Both, however, decrease exponentially with time, having a slope $-k'_{app}$.

Application of this method should be limited to kinetic experiments with relatively high conversion, perhaps 50% or more. When conversion is lower, the height of the product curve changes so little with time that the slope of the semilog plot cannot be measured accurately. This can be seen in Figure 3. Therefore, the method should be avoided with slow reactions when 50% conversion cannot be attained in the chromatographic reactor. Figure 8 illustrates the application of this method.

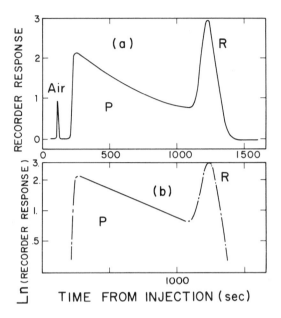

Fig. 8. Illustration of product slope method (51) for determining rate constants using the DCPD decomposition in Silicone DC 550 at 190°C. (a) Product wave P decreases into reactant R. Rate constant found by replotting on log scale (b) to find slope of product curve (solid line in b). See Eq. 45 and following text. Equations are:

$$t_{l,R} = 1110.4, \quad t_{l,P} = 105, \quad t_g = 140.9 \text{ sec}$$

$$k_{app} = -\text{slope} = 1.28 \times 10^{-3} \text{ sec}^{-1} = k_{app}[t_{l,R}/(t_{l,R} - t_{l,P})]$$

$$k_{app} = 1.16 \times 10^{-3} \text{ sec}^{-1} \quad (\text{inert substance method})$$

$$k_{app} = (1.21 \pm 0.03) \times 10^{-3} \text{ sec}$$

The product curve method can be used to advantage in two special situations. The first is for rapid estimation of the rate constant when a full kinetic study is not desired. The second is for analysis of very fast reactions when conversions of less than 80% are not attainable.

The major disadvantage of the product curve method is that the rate constant is calculated from the shape of a peak from a single kinetic run. The objections to using a single run could be minimized by repeating that run several times and averaging. More serious is the use of peak shape rather than peak area. In the inert standard method, any short-term change in the operating conditions are "averaged out" when the total reactant peak is considered. Similarly, diffusion contributes to peak broadening but does not affect peak area. Diffusion fluxes too small to affect the inert substance method would tend to smooth the product curve. The resulting decrease in slope lowers the measured value of the apparent rate constant. Although reactions involving short residence times suffer least from diffusional effects, they could involve mass transfer complications. Another potential complication not observed with the inert standard method is reverse reaction, which might occur further along the column, giving reactant that would confound the observed product curve.

3. Reactant–Product-Ratio Method

In the development of the ICR, conversion was related to the apparent rate constant by

$$\ln(W_{in}/W_{out}) = k_{app}t_l \tag{46}$$

Instead of making a series of runs with reactant area measured relative to an inert, the apparent rate constant can be calculated from a single run by utilizing the product area. The weight ratio of initial to unconverted reactant is

$$W_{in}/W_{out} = \frac{W_R + W_P}{W_R}$$

Since weight is related to the area on a chromatogram by

$$W_i = A_i S_i \tag{47}$$

where S_i is the sensitivity factor, we have

$$W_{in}/W_{out} = \frac{A_R S_R + A_P S_P}{A_R S_R}$$

$$= 1 + (A_P/A_R)(S_P/S_R) \tag{48}$$

Once the relative sensitivity factors for product and reactant have been determined, possibly through calibration, rate constants can be calculated directly from the retention time (residence time), the ratio of product to reactant area, and Eqs. 48 and 49,

$$k_{app} = \frac{1}{t_l}\ln[1 + (A_P/A_R)(S_P/S_R)] \tag{49}$$

The method is illustrated in Figure 9.

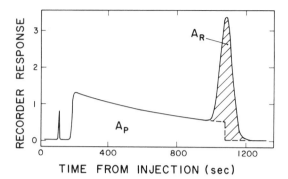

TIME FROM INJECTION (sec)

Fig. 9. Reactant–product-ratio method to determine apparent rate constant, illustrated with DCPD decomposition in hexatriacontane at 190°C. Rate constant calculated using Eq. 49, with product area $A_p = 9.07$; reactant area $A_R = 3.35$, $t_{l,R} = 976$ sec. Equations are:

$$S_R / S_p = 1.35 \pm 0.05 \quad \text{(from previous measurement)}$$

$$k_{app} = \tfrac{1}{976} \ln \left[1 + (9.07/3.35)(1/1.35) \right]$$

$$k_{app} = 1.13 \times 10^{-3} \quad \text{(inert substance method)}$$

$$k_{app} = 1.12 \times 10^{-3} \text{ sec}^{-1}$$

The major disadvantages of the reactant–product-ratio method are (a) the requirement that relative sensitivity factors be determined beforehand and (b) the sensitivity of this method to the accuracy of product interference estimation. The magnitude of any error in approximating the product interference is greater than with the inert substance method, since the correction affects reactant area and product area in opposite ways, and these areas appear as a ratio. For example, in a run with 50% conversion, +2% error in A_P and −2% error in A_R leads to 2.1% error in k_{app}. Similarly, 4% error in S_P/S_R, at 50% conversion, leads to 2% error in k_{app}. To approximate the rate constant, the sensitivity factors can be assumed to be equal. A more satisfactory approach is to determine the sensitivity ratio for product and reactant at lower temperatures, where no reaction occurs.

The reactant–product-ratio method can be used at lower conversions than even the inert substance method. Rate constants from runs having 0.1 to 2% conversion should be possible. This makes the technique attractive for measuring small rate constants.

4. Stopped-Flow Method

In the stopped-flow technique developed by Phillips et al. (49), the carrier gas flow is intermittently switched on and off after the reactant has been introduced into the chromatographic column. These periods of stopped flow produce sharp peaks superimposed on continuous chromatograms. For a first-order reaction on a catalytic surface, with no reaction in the gas phase, the conversion to product X is given by

$$\ln(1-X) = -k_s F_s t$$

$$X = 1 - (\exp - k_s F_s t)$$

where k_s is the surface rate constant, t is the total residence time, and F_s is the fraction of residence time that the reactant is adsorbed to the surface $[F_s = t_{surface}/(t_{surface} + t_g)]$. The change in conversion during a stopped-flow interval of 2ν, centered at t', is

$$X_{t'+\nu} - X_{t'-\nu} = [\exp - k_s F_s (t'-\nu)] - [\exp - k_s F_s (t'+\nu)]$$

$$= (2\exp - k_s F_s t')\sinh(k_s F_s \nu)$$

Since the area A under a superimposed peak is proportional to the product formed during the period of stopped flow ($W_{in}[X_{t'+\nu} - X_{t'-\nu}]$), rate constants can be found from two intervals of stopped flow, subscripted 1 and 2, by a rapidly converging iterative technique, as follows:

$$\ln(A_1/A_2) = \ln\frac{X_{t'_1+\nu_1} - X_{t'_1-\nu_1}}{X_{t'_2+\nu_2} - X_{t'_2-\nu_2}}$$

$$= k_s F_s (t'_2 - t'_1) + \ln\frac{\sinh(k_s F_s \nu_1)}{\sinh(k_s F_s \nu_2)} \tag{50}$$

Typically, reactions were run to 99.9% completion with stopped-flow intervals near the beginning and end of the experiment; 2 to 10% of initial reactant was converted during each interval.

Figure 10 is a typical chromatogram obtained with the stopped-flow method. This technique allows the measurement of rate constants with an apparent accuracy of about 1 to 5% (49).

D. Rate Processes Other than Chemical Reactions; Statistical Moments

The development of physicochemical applications of the gas chromatographic column frequently has involved emphasizing effects that are minimized or ignored in conventional and analytical operation. The study of the exaggerated effect then permits a more quantitative estimate for

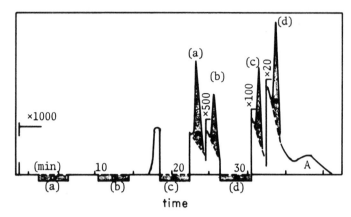

Fig. 10.   Stopped-flow chromatogram (49) for

cyclopentyl chloride (A)→cyclopentene (B) + HCl

on 10% KCl/Al$_2$O$_3$ at 120°C. Shaded peaks correspond to cyclopentene produced during the corresponding stopped-flow intervals indicated in shaded intervals on baseline. Changes in detector attenuation are indicated.

By courtesy of *Journal of Gas Chromatography*.

analytical chromatographic operation and leads to the adaptation of the chromatographic approach to the study of the physical process involved. This has happened in the development of chromatographic reactors. Space does not permit a comprehensive discussion of the adaptation of the column to the study of nonchemical rate processes. However, some review of this subject may serve as a guide to the operator of reactor columns, and as an introduction to the literature for those interested in or encountering rate-limiting physical processes.

The description of the ICR includes assumptions that other rate processes, especially transport, are not significantly affecting the rate controlling chemical reaction; and the reactor column should be operated so that this is true. Yet for heterogeneous surface reactions, particularly, such operation is often not possible or convenient. Moreover, some catalytic reactor columns are designed and run so that the rates of transport processes become important.

The use of statistical moments of elution curves for the study of physical processes in the column has been the subject of an active and informative series of investigations by Smith and co-workers (52–55) and others (56–58). These follow from the suggestions and treatments by McQuarrie (59), Kubin (60), Kučera (61), Grubner and Kučera (62, 63), and Vink (64), which apply moment analysis to chromatography and the measurement of axial dispersion, intraparticle diffusion, and adsorption. With simplifying assumptions, the theory is a nonequilibrium chromatographic treatment that considers transport processes in the column as well as chemical reactions, as in Eqs. 1 and 2.

The basic assumptions include a restriction to linear first-order processes, which are often attainable in the column. Thus a linear adsorption isotherm, a linear adsorption rate, and linearity for other mass transfer processes (as well as surface reaction rate) are all assumed in the treatment. Furthermore, negligible pressure drop is assumed.

As alluded to in the early reactor section, Kocirik (11; see also Ref. 56) extended the moment treatment to include first-order reactions. He used the first normal statistical moment and the first four central moments to characterize chromatographic curves resulting from an arbitrary input. First independent physical constants, including rate constants, were obtained from the five moment equations. However, the extraction of any single constant becomes difficult and specialized, because the moment equations are complex and sensitive to small errors and nonideality.

The statistical moments are related to the solution to the basic differential equation in the Laplace domain. The nth normal statistical moment μ_n' is defined by

$$\mu_n' = \frac{\int_0^\infty C_g t^n dt}{\int_0^\infty C_g dt} \tag{51}$$

where C_g represents a chromatographic (response) curve, $C_g(L,t)$ a function of position on the column, L the outlet here and t time of elution. The nth central moment μ_n is defined by

$$\mu_n = \frac{\int_0^\infty C_g (t - \mu_1')^n dt}{\int_0^\infty C_g dt} \tag{52}$$

Procedures for computation of the central moments are well known and are contained in articles referenced here. The normal statistical moments are related to the Laplace transforms of the function $C(L,t)$.

The zeroth reduced moment μ_0 is

$$\mu_0 = \frac{m_0}{(m_0)_{x=0}} = \frac{W_{\text{out}}}{(W_0)_{x=0}} ; \quad m_0 = \int_0^\infty C_g dt$$

where m_0 is total mass (the zeroth moment integral in this treatment) and μ_0 is equal to the fraction of unreacted reactant. Of course, this treatment contains an inherent assumption of no product interference in chromatographic elution peaks or, failing this, a means of eliminating the interference.

The first central moment is the arithmetic mean holdup time and determines a retention time; it should depend on the partition coefficient or adsorption constant (Eqs. 4 or 3). The second central statistical moment is the variance and is related to the sum of the squares of deviations from the mean. It is similar to the spreading of the familiar van Deemter HETP equation and, of course, is related to peak-spreading processes. The third moment is a measure of the symmetry of the elution curve. A positive value is indicative of tailing of the curve. The fourth statistical moment is a characterization of flattening of the distribution curve relative to the Gaussian distribution. A positive value indicates a flatter distribution than the normal distribution curve; a negative value indicates the converse.

Smith and co-workers have applied the moment method and chromatographic measurements for the following:

1.   The determination of rates of adsorption and desorption on a catalyst, as well as internal and external diffusion in catalyst pellets from the analysis of the second central moment (53, 54). The method depended on the response to a 0.5 to 1 cm^3 pulse of deuterium in a hydrogen stream. Eluted pulses were relatively broad.

2.   The determination of axial dispersion coefficients and intraparticle surface diffusion effects, as well as adsorption rates for silica gel, from analysis of the second central moment of breakthrough curves for hydrocarbons (55).

3.   The determination of surface diffusivities of ethane, propane, and butane on silica gel (65).

The application of moment equations to kinetic processes, including surface reaction, in a catalytic bed under special conditions recently has been comprehensively reviewed and discussed by Suzuki and Smith (52).

Although the elegant moment method is powerful in the hands of the skilled and informed investigator, limitations are apparent from even the sketchy assumptions described here. Experiments to measure transport limiting processes are generally performed with inefficient columns. A wealth of catalyst information is often necessary for application. The nature of moments (see Eqs. 51 and 52) emphasizes errors and unexplained effects for second and higher moments, especially at distances on the chromatogram far from the mean.

E. Experimental Measurement of Rate Constants: A Review

In recent years, the GCR has been used by a number of workers to investigate various aspects of its application to first-order and pseudo-first-order reactions. Much of the work has been introduced in Section II. D, in our development to this point; but some additional review is needed to reflect the present state of the art. The majority of workers have employed models resembling those described in Sections III. B and III. C. The common assumption of uniform pressure and velocity of the gas phase is unrealistic for most chromatographic columns, but the derived equations are the same *for first-order reactions,* with or without this assumption (39, 40). Most treatments involving reaction rate measurements, assume rates of adsorption and desorption much greater than the rates of reaction, as well as a linear distribution isotherm. Although this is justifiable in many situations, it results in some limitations on the chromatographic approach.

In their early report, Bassett and Habgood (20) assumed a linear adsorption isotherm to develop an equation relating the first-order heterogeneous rate constant for the catalytic isomerization of cyclopropane to the fractional conversion, the chromatographic retention volume of the reactant on the solid catalyst, and the carrier gas flow rate

$$kK_A = \frac{F^0}{273RW} \ln \frac{1}{1-X} \tag{53}$$

where F^0 is the corrected flow rate at $0°C$ and W is catalyst weight in grams.

The similarity of this result to the equation for a conventional flow reactor was noted (41). Rate constants, Arrhenius parameters, and the heat of adsorption were calculated for the reactant. Although the reactor chromatogram of Figure 2 was obtained, a catalytic microreactor attached to an analytical column was used for the major portion of the work.

In bimolecular reaction studies with a diene and nonvolatile liquid chloromaleic anhydride as a column stationary phase and coreactant, in a Diels–Alder reaction, Gil-Av and Herzberg-Minzly (8) considered the column concentration to be constant if small diene pulses were used. The second-order reaction thus became pseudo-first-order. Addition of an inert substance to the diene allowed the use of the inert standard method to obtain first-order rate constants. Since the nonvolatile product did not interfere with the eluted reactant peak, the ideal reactor prototype was approached. The infinite difference in retention times between product and reactant made separation fast. The true second-order rate constants and the activation energies agreed with literature values. With maleic anhydride dissolved in tricresyl phosphate as the liquid phase, Berezkin et al.

(28, 66) determined a reaction order and rate constants, with diene coreactant, by varying the maleic anhydride concentration. They claimed that the order, with respect to anhydride, differed from unity.

Joanne Yurchak's investigations (29–31) included three irreversible first-order reactions: (a) the depolymerization of trioxane to formaldehyde in a polyphenyl ether column

$$\to 3\ CH_2O$$

(b) the depolymerization of paraldehyde to acetaldehyde in tricresyl phosphate

$$\to 3\ CH_3CHO$$

and (c) the aforementioned decomposition of β-pinene in a tricresyl phosphate column.

All columns used trimethylsilylated firebrick support, and first-order rate constants were determined using the inert standard method. When the percentage of liquid phase was increased from 15 to 30%, the trioxane depolymerization rate constant decreased by a quarter, despite the absence of gas phase reaction. Since contact of reactant with the firebrick support is decreased as more liquid phase is added, these results suggest catalysis by sites on the support. This was reinforced by the observation that a Teflon support (Chromosorb T) coated with 5% polyphenyl ether, even containing 0.2% $FeCl_3$, gave a paraldehyde (more sensitive than trioxane) decomposition rate constant two orders of magnitude smaller than that found in a column using commercial silylated firebrick. Tests with untreated brick as support gave a rate so fast that it could not be measured readily.

The plate theory models of Kallen and Heilbronner (32) and others (36) are not discussed further here; nevertheless, the reports cited may be helpful to those interested in constructing theoretical responses for reactant substances.

Roginskii and Rozental (12) derived the kinetic equation for chromatographic reactions on heterogeneous surfaces. Their results cover diffusion and nonequilibrium between phases, as well as a surface adsorption rate commensurate with chemical rate. They integrated a simplified material balance for the special case of low surface coverage to relate the total reactant passing any point in the column to the adsorption and desorption constants, the diffusivity, the position, and the column length. They further suggested the possibility of investigating fluidized bed catalytic reactions by comparing chromatographic reactor results under appropriate conditions.

Considering rates of adsorption and desorption commensurate with the rate of surface reaction, a material balance (Eq. 1) with first-order reactions in both phases (but no diffusion), and a material balance on the reactant in the stationary phase (Eq. 2), Roginskii and Rozental (12) obtained an expression resembling

$$\ln(W_{in}/W_{out}) = t_g\left[k_g + \frac{k_s f_s K_s}{f_g(1 + k_s/k^\bullet)}\right]$$

$$\cong k_g t_g + \frac{k_s t_s}{1 + k_s/k^\bullet} \tag{54}$$

here $k^\bullet$ is the mass transfer coefficient in the stationary phase material balance, written as

$$\frac{\partial C_s}{\partial t} = k^\bullet(K C_g - C_s) - k_s C_s$$

Equation 54 allows direct estimation of the importance of the mass transfer coefficient (or rate of adsorption on a surface).

Gaziev, Filinovskii, and Yanovskii (38) extended the early qualitative treatment by Roginskii (37) of chemical reactions in a gas-solid chromatographic column by solving the material balance for an nth-order reaction. They made ideal reactor assumptions with incompressible flow in considering the effect of zero, first-, and second-order reactions on the shape of the exit peak for rectangular and triangular input peaks. Formulas for determining rate constants were compared with those used in conventional dynamic reactors. Under special conditions, the incompressible or constant flow assumption can be approached experimentally (52–55, 57) with wide short columns that might be inefficient for separation.

In a later work, Roginskii, Yanovskii, and Gaziev (67) enumerated the conditions essential for a reaction to proceed in the "chromatographic regime," using the dehydrogeneration of cyclohexane as an example.

Filinovskii, Gaziev, and Yanovskii (48, 68) used the same experimental apparatus but presented calculations for rate constants from the product curve of hydrogen rather than from reactant area.

Berezkin (2) reviewed chemical reactions in chromatographic reactors in his book on analytical uses of reaction gas chromatography; the greater part of this work is devoted to reactions in precolumns and auxiliary columns and major emphasis is on pre-1966 U.S.S.R. research. With co-workers, he also has reviewed gas chromatographic methods for studying the kinetics of liquid phase reactions (68, 69).

Phillips et al. (49) proposed and illustrated the stopped-flow method with studies on the elimination of HX (X = halogen) from cyclopentyl chloride, cyclopentyl bromide, and cyclohexyl chloride on alumina treated with potassium chloride. Later applications include studies of olefin isomerization (70) and catalytic dehydration of alcohols (71).

The accuracy of results obtained from the chromatographic method has been tested. Schwab and Watson (91) compared the pulse method of Gaziev et al. (38) with the more common, steady-state flow reactor, for the dehydrogenation of methanol over silver. Activation energies and reaction orders calculated by the two procedures agreed within experimental error. However, Gaziev et al. reported that the chromatographically determined activation energy for the dehydrogenation of cyclohexane was 2 to 3 kcal/mole lower than was found in a static reactor. Roginskii (72) hypothesized that a chemical reaction had a lower activation energy and fewer side reactions in a chromatographic column. Bett and Hall (73), alternatively, explained one instance of this as a consequence of nonequilibrium between the gas phase and surface adsorbed species.

Patton, Pratt, and Langer (39, 40, 42) used the decomposition of dicyclopentadiene (DCPD) (43–45) to demonstrate the validity of reaction rate constants determined in the liquid phase of a GCR. In addition to kinetic results, they demonstrated the acquisition of thermodynamic information (activity coefficients, heats, and partial molar entropies of solution for reactants and products), simultaneously from chromatographic runs, for further interpretation of the kinetic data. An Arrhenius plot for the first-order rate constants obtained from the study of the dissociation of DCPD to CPD appears in Figure 11. Additional kinetic studies in these laboratories include the dissociation of dimethyldicyclopentadiene (74) and the decomposition of di-t-butyl peroxide (39).

Patton applied the inert standard, product curve, and reactant–product-ratio methods to analyze reaction chromatograms for the DCPD dissociation (39). He found that results from the three methods were internally consistent when all factors were considered. For general application, he recommended the inert standard method. Patton attempted a

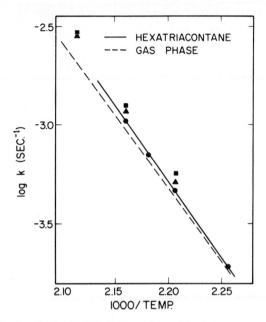

Fig. 11. An Arrhenius plot for DCPD decomposition (39, 40) in various solvents compared with gas phase results (45): circles on hexatriacontane solvent; triangles on Versamid 900; squares on polyphenyl ether. In hexatriacontane, $k_l = 10^{13.5 \pm 0.2} \exp(-35{,}000 \pm 300/RT$ sec^{-1}).
By courtesy of *Journal of Physical Chemistry*.

fourth method of analysis based on change of the total reactant plus product area with conversion due to the difference in detector sensitivity to reactant and product. He found that errors here were an order of magnitude greater than those observed with the other methods.

Some workers have used the GCR with approaches different from those just outlined. Berezkin, Kruglikova, and Shiryaeva (75, 76) applied the concept of linear variables with the chromatographic method to investigate the oxidation of benzaldehyde to benzoic acid. The benzaldehyde was dissolved in the stationary phase and air was used as the carrier gas. For a pseudo-first-order reaction in benzaldehyde, we have

$$k_l t = \ln[C_l(0)/C_l(t)] \tag{55}$$

where C_l is concentration of benzaldehyde in the liquid phase. Assuming that the retention time of 2-butene, pulsed periodically through the column, was (a) a linear function of the benzoic acid concentration in the stationary phase and (b) an indication of its concentration, they used the following equation for determining the rate of oxidation; we can write

$$k_l t = \ln \frac{(\lambda_0 - \lambda_\infty)}{(\lambda - \lambda_\infty)} \qquad (56)$$

here, λ_0 is the relative retention of 2-butene in the initial benzaldehyde stationary phase, λ_∞ is the relative retention time of 2-butene in the completely converted benzoic acid stationary phase, and λ is the relative retention time of 2-butene in a binary mixture with benzaldehyde at concentration C_l at time t. The rate constant obtained in this manner agreed with one obtained in a static reactor.

For those interested in comparing chromatographic results with data obtained in comparable unpacked reactors, there are at least two useful relevant studies (at 200–500°C). Kwart, Sarner, and Olson (46) improved a reactor system originally designed by Levy and Paul (77) which consisted of a gold coil reactor between two chromatographic columns. The first column purified reactants, which were swept into the reactor as a diluted helium stream. The second column analyzed the reactor effluent. By varying both the flow rate and the reactor volume (4–20 cm^3), first-order rate constants from 10^{-2} to 1 sec^{-1} were found, which compared favorably with earlier data for thermolysis of ethyl acetate and endo-dicyclopentadiene dissociation. Their incompressible flow model (a valid assumption for open-tube reactors) included axial dispersion, but the effect on the rate constant was less than 0.1% in a 4-cm^3 reactor, 90 cm long, of 2.3 mm i.d. They emphasized the inert character of the gold-surfaced reactor. The apparatus is now commercially available (78).

Retzloff, Coul, and Coul (79) measured first-order constants for the isomerization of vinyl-cyclopropane to cyclopentene, and for the pyrolysis of di-t-butyl peroxide, which agreed with earlier results from more conventional methods. They worked with a short, Inconel-X reactor of 0.6-cm diameter and concluded that axial diffusion was insignificant where reactor length was about ten times the diameter and where laminar flow could not take place. In both studies of unpacked columns, reactants and products flowed into a chromatographic column operating at a lower temperature, and here the reaction was stopped and the components were analyzed.

Schmiegal et al. (80) and Sethi et al. (81) used the injection port of a conventional gas chromatograph as the reactor. Combining the first-order rate law with the Arrhenius form of the rate constant, they found

$$\ln \left[\ln \left(W_{R,\text{in}} / W_{R,\text{out}} \right) \right] = (2.303A - E_a/RT) + \ln t$$

At constant flow rate, the residence time in the injector t, by Charles's law, is inversely proportional to injector temperature T. Thus, we have

$$\ln\left[\ln\left(W_{R,\text{in}}/W_{R,\text{out}}\right)\right] = (2.303A - E_a/RT) + \ln\left(t_0 T_0/T\right)$$

or

$$\ln\left[T\ln\left(W_{R,\text{in}}/W_{R,\text{out}}\right)\right] = \text{constant} - E_a/RT \qquad (57)$$

By varying the injector temperature, the activation energy can be found from a plot of $\ln[T\ln(W_{R,\text{in}}/W_{R,\text{out}})]$ against $1/T$. Residence times are on the order of one second. This technique gives only activation energies and not rate constants or frequency factors.

Chang et al. (82) studied the hydrogenation of propylene in a chromatographic reactor under conditions of both equilibrium and non-equilibrium chromatography. For a first-order, irreversible reaction, with equilibrium chromatography and a linear isotherm, they proposed an equation for obtaining a chemical rate constant from the mass center of the reaction. The mass center is defined as the time (or position) on the chromatogram at which one-half the product has been eluted. This treatment neglects diffusion, which may be significant in some systems. We have found this method difficult to apply (29, 30).

This review of GCR applications is intended to be not comprehensive but illustrative, since with increasing acceptance of the chromatographic approach, any review would soon be incomplete. The interested reader may wish to consult other sources (2, 31, 83, 84).

IV. COMPLEX REACTIONS AND PROCESSES

A. Non-First-Order Reactions

Irreversible reactions other than first-order, where reaction rate is kC^n, can be incorporated into the mass balance for the ICR (Eqs. 7 and 8) under linear–ideal (no spreading processes) conditions (38, 39) to give

$$(f_g + f_l K)\left(\frac{\partial C_g}{\partial t}\right) = -f_g\left(\frac{\partial}{\partial x}[u(x)C_g]\right) - (f_g k_g + f_l k_l K^n)C_g^{\,n} \qquad (58)$$

where the stationary phase is a liquid. This can be simplified to

$$\alpha\left(\frac{\partial C_g}{\partial t}\right) + \left(\frac{\partial}{\partial x}\right)[u(x)C_g] + \beta_n C_g^{\,n} = 0 \qquad (59)$$

where

$$\alpha = \frac{f_g + f_l K}{f_g}$$

$$\beta_n = \frac{f_g k_g + f_l k_l K^n}{f_g}$$

Substituting as before, we have

$$K f_l / f_g = t_l / t_g$$

Thus

$$\alpha = \frac{t_g + t_l}{t_g}$$

$$\beta_n = \frac{k_g t_g + k_l t_l K^{n-1}}{t_g}$$

and

$$\tau_n(x) = \int_0^x \frac{dx'}{[u(x')]^n}$$

The solution of the material balance, with the boundary and initial conditions, can be obtained with Laplace transforms (38, 39). For $n \neq 1$, it is

$$C_g(x, t) = \frac{u(0)\phi[t - \alpha\tau_1(x)]}{u(x)\sqrt[n-1]{1 + (n-1)\beta_n\tau_n(x)\{u(0)\phi[t - \alpha\tau_1(x)]\}^{n-1}}} \tag{60}$$

For a second-order reaction, for instance, the amount of reactant leaving the column of cross-sectional area a_g at the outlet L would be

$$W_{out} = a_g u(L) M \int_0^\infty C_g(L, t)\, dt$$

$$= a_g u(0) M \int_0^\infty \frac{\phi[t - \alpha\tau_1(L)]\, dt}{1 + \beta_2\tau_2(L)u(0)\phi[t - \alpha\tau_1(L)]} \tag{61}$$

which is equivalent to

$$W_{out} = a_g u(0) M \int_0^\infty \frac{\phi(t)\,dt}{1 + \beta_2 \tau_2(L) u(0)\phi(t)} \tag{62}$$

since $\phi(t)$ is 0 for t less than 0. The expression for W_{in} is still Eq. 12, namely,

$$W_{in} = a_g u(0) M \int_0^\infty \phi(t)\,dt \tag{12}$$

For a second-order reaction, it can be shown that τ_2 is

$$\tau_2(L) = \frac{t_g j_4^{\;3}}{u_0}$$

and

$$\beta_2 \tau_2(L) u(0) = j_4^{\;3}(k_g t_g + k_l t_l K)$$

where the pressure correction is (3–6, 33)

$$j_4^{\;3} = \frac{4}{3} \frac{(P_i/P_0)^3 - 1}{(P_i/P_0)^4 - 1} .$$

In Eq. 62, W_{out} is a function of the shape of the input function $\phi(t)$. This would probably be rectangular or Gaussian, since these shapes are most readily produced experimentally.

Hattori and Murakami (85) have considered the pulse reaction technique for a number of reaction models in catalytic systems, including irreversible, reversible, and consecutive reactions. They compared conversion in chromatographic and flow reactors for rectangular, triangular, and Gaussian input functions of varying width. They treated elementary reactions, assuming adsorption equilibrium, with a simplified material balance. Differential equations were converted to difference equations and solved on a digital computer.

As we have seen previously, the width of the inlet pulse is critical for several situations in which the chromatographic reactor would be advantageous relative to a flow reactor. Narrow pulses are preferable, as might be anticipated for the reactions

$$A \rightleftarrows 2R \qquad A \rightleftarrows R + S$$

Wide pulses result in decreased separation of the products, which participate in the potential reverse reaction (the column is operated with few plates, inefficiently). In the two consecutive sequences below, the buildup of intermediate concentrations is increased with wide pulses and the relative yield (compared to amount consumed) of intermediate R is decreased

$$A + B \rightarrow R; \quad R + B \rightarrow S \qquad A \rightarrow R; \quad 2R \rightarrow S$$

For nonlinear rate processes (higher-order), more R is produced with the pulse technique in the reaction schemes above than with the flow technique. For rectangular, square, and Gaussian input functions, peak width is more important than peak shape in conversion–time plots. When the cumene system (cumene$\Leftrightarrow$propylene + benzene) was studied experimentally with a silica alumina catalyst, the tendency of narrow rectangular inlet pulses to eliminate reverse reaction because of separation of propylene and benzene, was verified (85).

The Japanese workers noted, with others, the identical conversions of the pulse and flow techniques with first-order, irreversible rate processes (85). With nth-order irreversible reactions ($r_A = kK_A C^n$, where K_A is an adsorption equilibrium constant, $n \neq 1$), conversion for a rectangular pulse approaches that of a flow reactor as the inlet pulse is spread. The triangular and Gaussian input pulses, with lower average concentrations, produce lower values of C^n. For nth-order, irreversible reactions, with a narrow inlet pulse width, conversion in real columns will decrease because of the decrease in average concentration due to peak spreading.

Calculating the dependence of the rate of reaction on a concentration in an experimental column becomes a challenge, since, in real chromatography, spreading processes operate to dilute the reactant. With incompressible flow, and no spreading, for first- and second-order reactions, pulse height is reduced with rectangular and triangular input pulses (38). For a zero-order reaction, and triangular input pulse width is presumably reduced; this is suggested as a diagnostic test for a zero-order reaction.

Blanton, Byers, and Merrill (86) discussed power-law ($r_A = kC^n$) reactions in a short catalytic column, where the diffusion term of the material balance equation was not significant. They generated, with a long precolumn, a Gaussian pulse that did not broaden in the reactor. Using a finite-difference technique, they calculated the shapes of curves relating conversion or a function of conversion to a dimensionless rate constant $(Vk/Q)C^{n-1}$, where V is the reactor void volume, Q is the volumetric flow rate of reactant over catalyst, and n is order. Order could be determined through matching of curves. The shape of the exit curve, conversion versus a function of dimensionless time, enabled a power-law rate to be distinguished from a Langmuir–Hinshelwood-type mechanism. The need for equilibration of the column with the gas phase to be faster than any changes in gas phase partial pressure was explicitly stated. The system was tested with ethylene in a carrier gas stream of hydrogen. The shape of the product or exit curve in the reactor outlet did not permit a choice between alternate mechanisms because the product ethane could not be specifically distinguished from the olefin reactant at the outlet of the reactor.

B. Reversible Reactions

1. Kinetics of Interconversion of Isomeric Species

The GCR can be adapted to the study of several types of reversible reactions. The slow reversible first-order–first-order interconversion of two species (A⇌B) was a system to which the attention of our laboratory was drawn in recent years (29–31). Fortunately, we were preceded by two other groups, both becoming interested because of the complications in chromatography caused by interconverting species.

A hypothetical gas chromatogram for a mixture of A and B, where a reversible first-order–first-order interconversion

$$A \underset{k_b}{\overset{k_a}{\rightleftharpoons}} B$$

occurs is presented in Figure 12. Molecules that have traveled through the column solely as A are eluted as peak A, and molecules that spent their entire residence time as B are eluted as peak B. Between these peaks is an overlapping zone, produced by material that has been interconverted between A and B at least once in the course of passage through the column. If $k_a t$ and $k_b t$ are small, relative to unity (where t is residence time in the column), then separation tends to occur. As $k_a t$ and $k_b t$ grow larger

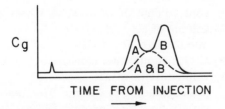

Fig. 12. Hypothetical reactor chromatogram for the reversible isomerization A⇌B (29, 31). By courtesy of *Industrial and Engineering Chemistry*.

than unity, the separate A and B peaks merge to give a single peak with a characteristic retention time between that of A and B. Those familiar with nuclear magnetic resonance should not overlook the analogy with spectra of interconverting species.

Keller and Giddings (87), in their review of kinetic complications in chromatography, recognized that reversible reactions such as A⇌B can confound conventional chromatographic analysis so that erroneous conclusions are drawn about the nature of the solute. They calculated the probability distribution for the relative times spent as A and as B for a reversible first-order–first-order system. They then generated typical chromatograms with assumed values for initial concentrations, rate constants, average diffusivity, and time.

Klinkenberg (89), independently, added a reversible reaction to the chromatographic material balance equation in describing the first-order–first-order isomerization. He used a transformation to obtain an equation for which an asymptotic solution was available. The forward and reverse rate constants were found from the peak broadening for the near-equilibrium situation (single A and B peak) and the position of the peak relative to that for the two pure isomers. The latter permits the determination of the equilibrium constant. In terms of chromatographically measurable quantities, Klinkenberg's equation (29, 89) becomes

$$
\left(\frac{1}{N}\right)_{rxn} \equiv \left(\frac{\sigma_t}{\mu_t}\right)^2_{rxn}
$$

$$(63)$$

$$
= \frac{2\left(V_R^B - V_R^A\right)^2 V_R^A K_{eq}^2 F}{\left[\left(V_R^B + V_R^A K_{eq}\right)V_g + V_R^A V_R^B (1 + K_{eq})\right]^2 \left[V_R^B + V_R^A K_{eq}\right] k_a}
$$

where V_R^A, V_R^B = true retention volumes of A and B, respectively (no dead volume)

V_g = void volume of column, as measured from retention volume of air

k_a = rate constant for A→B

K_{eq} = k_a/k_b; equilibrium constant

F = volumetric flow rate

$(\sigma_t)_{rxn}$ = peak spreading in column from reaction

μ_t = elution time for maximum of equilibrium (single) peak

N_{rxn} = plate number due to reaction

The equilibrium constant can be obtained from

$$K_{eq} = \frac{k_a}{k_b} = \frac{V_R^B(V_R^{eq} - V_R^A)}{V_R^A(V_R^B - V_R^A)} \tag{64}$$

Although Klinkenberg did not provide data, he suggested the application of the method to the interconversion of ortho- and para-hydrogen in a chromatographic column reported by Moore and Ward (88).

Yurchak and Langer (29–31) studied the reversible interconversion of *syn*- and *anti*-acetaldoxime, first described by Pratt and Purnell (90), in Carbowax 1000.

A single acetaldoxime peak, indicative of a high degree of interconversion, was observed at lower flow rate and at high temperatures as shown in Figure 13*b*.

The problem of separating $(1/N)_{rxn}$ from nonreaction broadening effects was approached by measuring N for unreactive 1-pentanol under the same chromatographic conditions (29). Then

$$(1/N)_{rxn} = (1/N)_{\text{acetaldoxime}} - (1/N)_{\text{pentanol}} \tag{65}$$

The inert standard method was used at small conversions with both acetaldoxime isomers (29). Linear, first-order plots were obtained despite the reversibility, which casts doubt about the validity of this procedure.

The values of V_R^A and V_R^B were approximated by extrapolating to infinite flow rate in order to eliminate contributions of each to the other. Actually, Yurchak (29) found it more feasible to extrapolate to zero retention volume of a reference inert substance, discarding measurements at the extreme at which mass transfer complications begin to appear. The value of V_R^{eq} was taken by extrapolating the retention volume of the single acetaldoxime peak near equilibrium to zero flow rate. The apparent gas volume in the column V_g was taken as the corrected retention volume of a small air peak. Results for the acetaldoxime interconversion are compared in Table II.

These results are in qualitative agreement. The problems and complications in studying reversible reactions were thoroughly discussed by Yurchak (29). Although both approaches provide reasonable isomerization

TABLE II

Kinetic Measurements on Acetaldoxime System at 70°C in Carbowax 1000

Method	$k_a \times 10^4, \sec^{-1}$	$k_b \times 10^4, \sec^{-1}$	K_{eq}
Klinkenberg	5.9	4.6	1.29
Inert standard method	3.9	2.8	1.38

rate constants and equilibrium constants, the difficulty in applying Eq. 63 with the necessary estimation of nonreactive peak spreading makes the simple inert standard method a method of choice at present. Further study and understanding are needed to make the Klinkenberg equation a more practical approach. It must be noted that there was difficulty in reproducing the isomerization in subsequent studies (91), and we surmised that catalytic effects influenced the isomerization rate. Nevertheless, the comparison in the study described here is valid. The rate constants simply are not characteristic of the neat Carbowax solvent.

For interconverting separable species, then, it is possible to obtain a single peak in three ways: by decreasing flow rate sufficiently to allow interconversion in the column, by increasing temperature to increase rates of interconversion, and by adding a catalyst to increase rates of interconversion. Some experimental chromatograms for the first two situations, presumably with catalyst present, are given in Figure 13.

2. Reversible Reactions Involving Multiple Reactants

More attention has been focused on reversible reactions involving multiple reactants in one direction:

$$A \rightleftharpoons P + Q$$

The early interest of Soviet workers (37) and of Hattori and Murakami (85) has been described in other sections.

The possibility of achieving yields greater than predicted by thermodynamic equilibrium, when the retention volumes of P and Q differed, was recognized by several groups, especially for catalytic situations. Dinwiddie (92), Magee (93), and Gaziev et al. (94) all obtained patents on such systems. For the dehydrogenation of cyclohexane to benzene and hydrogen, Matsen et al. (95) found that conversion depended on the ratio of pulse size to column length but was independent of residence time for instantaneous equilibrium. Matsen et al. (95) and Roginskii et al. (96, 97)

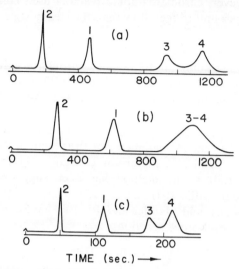

Fig. 13. Reactor chromatograms (29, 31) illustrating the resolution obtained during the reversible isomerization between *anti*- and *syn*- acetaldoxime (peaks 3 and 4, respectively); the internal standards are toluene (peak 1) and benzene (peak 2). (a) $T = 70°C$, $F = 108$ cm^3/min; (b) $T = 100°C$, $F = 30$ cm^3/min; (c) $T = 100°C$, $F = 166$ cm^3/min. The two forms are separable at 70°C; at 100°C only a single peak is observed at the same residence time. Increased flow gives separation again at 200 sec.
By courtesy of *Industrial and Engineering Chemistry*.

claimed 20 to 30% higher yields than were obtained in conventional reactors limited to equilibrium conversions. Both cyclohexane dehydrogenation and butene dehydrogenation to butadiene were studied experimentally. The problem, both experimentally and theoretically, is in achieving significant practical scale-up. As pulse size and width increase, the column reactor operates less efficiently in effecting separation and, therefore, conversion.

Magee (98) modeled a reversible reaction with instantaneous equilibrium between products and reactant using an analog computer. Numerical results indicated that a reversible reaction limited by equilibrium could be forced to completion in a chromatographic reactor when the equilibrium constant was greater than 2×10^{-7} atm. Abet et al. (99) numerically integrated the expression of Magee, using a digital computer.

Gore (100) simulated systems with impulse feeds to the chromatographic reactor on a digital computer. He considered reactions such as the dehydrogenation just mentioned with an eye to improving operation. Fast reaction rates, impulse feed, and control of input wave frequency to give separation with minimal interaction between cycles were important. However, chromatographic reactors used considerably greater catalyst weight per unit of feed flow than the steady-state reactor operating at equilibrium.

The theory for perturbation of local equilibrium has been applied to chemically reacting compounds traveling through a column for gas-solid (100) and gas-liquid (101) chromatography by Deans and his co-workers. Reacting systems containing a volatile reactant were allowed to achieve chemical and physical equilibrium before the introduction of a pulselike perturbation. The retention times of the peaks generated by the perturbation were related to the stoichiometry of the reaction, the partition coefficients, and the chemical equilibrium constant. With proper choice of experimental variables, the method has been suggested as a means of obtaining rate data (103, 104).

Harrison, Koga, and Madderom (105, 106) have considered and tested the reversible second-order reaction

$$I_2 + Br_2 \rightleftarrows 2IBr$$

using both plate theory and a material balance on each plate in a column, with the aid of a computer for numerical calculations. Both finite and infinite reaction rates were treated. The apparent equilibrium constant, calculated from the elution profile, tended to be appreciably less than the true equilibrium constant, since the reaction of I_2 and Br_2 is slowed by separation of the two elements as a pulse proceeds through the column. The behavior of the system on Kel-F grease, supported on Teflon powder, agreed with calculations. However, the system is not a conventional one, since a fast rate constant $[k > 5 \times 10^4 \ 1/(\text{mole})(\text{sec})]$ for the decomposition of IBr at 94°C was observed. This treatment of the IBr system is complex, and further work on such systems would call for strong interest and sophistication on the part of the experimentalist.

C. Catalytic Reactions and Microreactors

The use of microreactors attached directly to the inlet of gas chromatographic columns for the study of surface-catalyzed heterogeneous reactions is a well-established practice. Strictly speaking, such systems do not fall under our definition of chromatographic reactors, since most microreactors are too short to separate reactants and products. Therefore, concurrent reaction and effective separation do not take place in the reactor; the auxiliary chromatographic column effects the separation. Otherwise, processes occurring in such microreactors resemble those described for gas chromatographic reactors; in addition to operational convenience, this accounts for the popularity of the approach. Kokes, Emmett, Tobin, and Hall (107–110) pioneered this application, in which a small amount, or pulse, of reactant is introduced to the microreactor, which generally is significantly smaller than the conventional GCR. The sample pulse can be longer than the reactor and is still diluted with carrier gas.

After refinement of the microreactor technique (111), a useful design and theoretical treatment was given by Hall, MacIver, and Weber (112). Continued development of design features has assisted the ready acceptance and interest in heterogeneous catalytic microreactor systems (113–117). Advantages of the microcatalytic technique include ready control of flow rate, feed composition, temperature and pressure, and handling of radioactive isotopes on a small and relatively safe scale. Other advantages are convenient application of two detectors and determination of adsorption isotherms and rates under reaction conditions (118–120).

The concepts and chemistry involved in solid catalytic reaction studies are so extensive and yet so specialized that it is not possible to do justice to them in a review of this length and tenor. Table III, with references and reference referrals, will serve as an introduction to the literature and special topics. Later consideration of recent work by Hall and his co-workers (73, 121, 122) can give additional insight to nonspecialists into the complications in comparing microcatalytic results with those obtained with steady-state reactors.

A significant number of solid surface catalytic reactions are treated by models that take into account the kinetics of surface adsorption and desorption, as well as surface reaction. Because of the use of a Langmuir adsorption assumption, many of these are referred to as Langmuir–Hinshelwood mechanisms (123) or Hougen–Watson models (124). The characteristic feature is the assumption of Langmuir-type preadsorption, subscripted (a), of the reacting molecule A from the gas phase, subscripted (g). The sequence where A is adsorbed and desorbs, or reacts to form product, P, and desorb is represented by

$$A_{(g)} \underset{k_{-1}}{\overset{k_1}{\rightleftarrows}} A_{(a)} \overset{k_2}{\rightarrow} P_{(a)} \overset{k_3}{\rightarrow} P_{(g)} \tag{66}$$

The Langmuir adsorption concept visualizes a uniform adsorption surface with less than, or a unimolecular layer of noninteracting adsorbate molecules. In a Langmuir–Hinshelwood mechanism or a Hougen–Watson model, overall reaction rate may be governed by: (1) the rate of reaction on the surface k_2, (2) the rate of adsorption k_1, or (3) the rate of desorption k_3 for the product P. The adsorption equilibrium constant K is equal to k_1/k_{-1}.

TABLE III

A Summary of Some Microreactor Catalytic Investigations
(See Also Other Parts of This Chapter)

Topics	References
Design	
Conversion of a commercial chromatograph	113, 126
Automatic precision microreactor	117
High-pressure reactor	115, 116, 127
General	113, 114, 128–130
Reviews of catalytic research and micro reactors	2, 31, 52, 83, 102, 108, 131, 135
Catalyst comparison	136–140
Oxidation of automobile exhausts	141–145
Semiconductor catalysis	146–149
Isomerization	
Silica alumina-cyclopropane, isobutane, isopentane	150–152
Aluminum oxide–cycloalkenes	153, 154
Depolymerization of ethylene	155
Hydrogenation	5, 87, 128, 156–158
Dehydrogenation	159
Hydrodesulfurization	160, 161–163
Cracking	
Isobutane and cumene	151
Thermal degradation	164
Poisoning	165
Radioactive isotope studies	102, 109, 166
Deuterium isotope studies	53, 121, 122
Comparison of microreactor with other reactors	2, 20, 73, 91, 122, 167
Transport processes	See Ref. 52 for a summary.

Bassett and Habgood (20) showed that at low partial pressure of reactant (where there is linear adsorption) and rapid adsorption and desorption relative to surface reaction rate, identical results can be obtained from a catalytic pulse reactor and a steady-state reactor. The adsorption equilibrium constant is then related to the chromatographic retention volume of the pulse in this situation. Comparison of microcatalytic and steady-state results for the first-order dehydrogenation of methanol over silver gave good agreement; but low values of activation energy were recorded when reaction rate constants were based on disappearance of reactant (121, 125).

As indicated earlier (86), the microcatalytic reactor can be applied to the study of Hougen–Watson and power-law models by arranging the simplified material balance into dimensionless groups. We can treat the situation mathematically if we assume constant flow rate and a negligible diffusion term, since its coefficient is the inverse of the Peclet number (near unity for this system) times the ratio of the particle diameter to column diameter (a very small number). The Gaussian input essentially did not spread in the reactor. The computer solutions of the equations produced families of curves for the following rate laws:

$$r_A = kC_A^n \tag{67}$$

$$r_A = \frac{kC_A}{1 + K_A C_A} \tag{68}$$

$$r_A = \frac{kC_A}{1 + K_A C_A^2} \tag{69}$$

By comparing the outlet conversion curve shape for the hydrogenation of ethylene over alumina with the computer-generated curves (conversion versus time), Blanton, Byers, and Merrill concluded that this reaction followed the Hougen–Watson kinetics of Eq. 68, rather than power-law kinetics (86). Rate and adsorption constants were determined from the curve position.

Bett and Hall (73), who compared the steady-state flow reactor with a microcatalytic reactor for the dehydration of 2-butanol over hydroxyapatite catalyst, found the results to differ. Desorption of product is apparently rate-controlling in the pulse reactor; however, this situation is not discernible in a reactor, where inlet and outlet streams are allowed to reach a steady state over a long period of time. (Our earlier discussions of chromatographic reactors emphasized the need for equilibrium between gas and surface phases.)

With the pulse reactor, products continued to desorb for some time after the pulse had passed. Study of trapped product olefin relative to reactant addition revealed that the rate of dehydration is effectively independent of initial alcohol pressure or zero order. A plot of conversion against reciprocal space velocity obtained by varying the amount of catalyst bed was linear and gave a zero conversion intercept for zero reciprocal space velocity. However, when the space velocity was altered by varying the flow rate of helium carrier gas *over the catalyst bed*, a plot of conversion versus reciprocal space velocity extrapolated to a significant conversion intercept at zero reciprocal space velocity (infinite flow rate). This can be explained on the basis of product adsorption on passage of reactant over the bed. Since product was measured in these experiments by trapping followed by later chromatographic analysis, eluted product was not lost. The finite conversion intercept is regarded as a measure of "site monolayer" material adsorbed immediately. Thus the pulse and steady-state reactor results complement each other in this situation. Although the microcatalytic reactor complements the flow reactor study, alone it is not suited to this system.

Hightower and Hall (122) investigated the kinetics of cyclopropane and methylcyclopropane isomerization over silica-alumina in static, micro-catalytic, and steady-state reactors. They interpreted their results in terms of Langmuir–Hinshelwood kinetics, where k_2 and k_3, in a situation similar to that represented in Eq. 66, are comparable in magnitude to k_{-1}. Their experimental approach involved trapping of microreactor eluant followed by chromatographic analysis. For this situation, they proposed an equation relating the reaction rate to the partial pressure of reactant after a part of the pulse has been adsorbed.

Some of the practical problems and considerations in using a microreactor for catalytic studies are illustrated in the description of the evolution of a glass-lined reactor by Muchhala, Sanyal, and Weller (126). In their study of the decomposition of acetic acid over silica-alumina, they showed that, without the glass liner, the walls of the reactor catalyzed decomposition.

For additional discussion and survey references of catalytic microreactors, the reviews by Choudhary and Doraiswaimy (83) and Berezkin (2) are especially helpful.

V. EXTRAKINETIC APPLICATIONS

Our emphasis thus far has been on the use of the GCR to determine reaction rate constants and parameters for other kinetic processes for a variety of gas, liquid phase, and solid-catalyzed reactions. There are, however, a number of extrakinetic applications involving the use of the column as a reactor. These include: (a) the use of reaction phenomena in the column as a diagnostic probe for chacterizing the column, (b) the adaptation of the column for preparative purposes including isotope labeling, (c) the use of kinetic phenomena for chemical identification, and (d) the use of the column for obtaining solution thermodynamic data on products and reactants while measuring reaction kinetic data for eventual conciliation.

A. Reaction Kinetics as a Tool for Chromatographic Column Diagnosis

An irregularity or disturbance in an otherwise smooth product curve, in the course of elution, may give substantial information about the homogeneity of a chromatographic column. The combination of concerted reaction and separation in the absence of complications results in smooth reactor chromatograms of the type shown earlier. Reaction and separation, the fundamental phenomena in the gas chromatographic reactor, are so intimately related that a small perturbation in either distorts the product curve. Therefore, a kinetic experiment used as a diagnostic probe has the potential of characterizing the column throughout its length.

Three diagnostic tests based on reaction kinetics have been investigated. They include procedures for recognizing and locating unsilylated catalytic hydroxyl groups on the surface of a solid support, for identifying and locating temperature zones above ambient, and for recognizing the presence and location of void zones in the column packing.

Work in our laboratories (29, 30, 168) had shown that the surface of the solid support in gas-liquid chromatographic columns frequently catalyzed the depolymerization of trioxane and paraldehyde to give formaldehyde and acetaldehyde, respectively. These reactions were acid and metal catalyzed; no depolymerization occurred at 100°C in basic liquid phases, such as 4,4'-dimethylenedianiline, or hydrogen-bonding liquids, such as polyethylene glycol 1000, which quench surface hydroxyl sites to prevent surface adsorption (169). However, the surface-catalyzed depolymerization of paraldehyde was observed at temperatures as low as 60°C with liquid phases such as Silicone DC 550, squalane, and polyphenyl ether with a commercial silanized support. The rate of depolymerization of paraldehyde was about an order of magnitude greater than that of trioxane in the temperature range of 70 to 150°C.

The use of the paraldehyde depolymerization to test the surface support is illustrated in Figure 14. The curves in Figure 14a were obtained

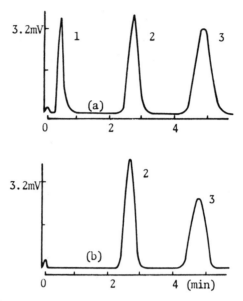

Fig. 14.   Simulated test of solid support at 175°C. Paraldehyde (0.5 μl) chromatographed on DC 550-Gas Chrom Q (15/85 w/w). Helium flow, 30 cm^3/min, $P_i/P_0 = 1.19$, column $64 \times \frac{1}{4}$ in. (*a*) Commercial brick–DC 550 (2 in.) inserted near entrance column; (*b*) After trimethylsilylation: peak 1, reaction product; peak 2, paraldehyde; peak 3, ethylbenzene.

when paraldehyde and ethylbenzene (reference material) were chromatographed on a Silicone DC 550 column. The solid support was inert except for a 2-in. insert 4 to 6 in. after the column inlet, which had been only acid washed. The characteristic retention volumes of acetaldehyde and paraldehyde would allow location of this poorly treated section. The curves of Figure 14*b* were obtained with the same column after treatment with a commercial trimethylsilylating agent recommended for silylation of the column *in situ*. Effectiveness of the treatment is evident from the disappearance of acetaldehyde in the elution chromatogram.

Adsorption and desorption behavior and associated tailing of certain compounds have been customarily used to investigate and classify surface properties of solid supports (169, 170). It is apparent that a clean-cut reaction taking place in a gas chromatographic column provides an alternate sensitive method for testing and characterizing solid supports.

For columns in which the trioxane depolymerization can take place, irregularities in the product curve from this surface-catalyzed reaction can be used to identify and locate temperature gradients in the column (162). The rate of product formation is greater in a hot spot along the column, giving rise to a small lump at a position on the product curve correspond-

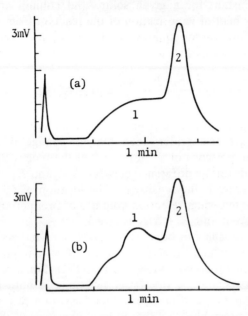

Fig. 15. Location of column section operating above ambient temperature. Chromatograms for 0.5 µl of trioxane on DC 550–silanized brick (15/85 w/w). Helium at 62 cm³/min; $P_0 = 742$, $P_i/P_0 = 1.53$. (a) Normal operation at 135°C; (b) Heated zone (4-in.) present in the column. Region 1 is reaction product, peak 2 is trioxane.

ing to the high temperature region of the column. This is illustrated in Figure 15. Figure 15a was obtained from trioxane on a $64 \times 1/4$ in. packed column containing Silicone DC 550 on 60-80 mesh commercial-acid-treated trimethylsilylated brick. The elution curve preceding the trioxane peak indicates continuous formaldehyde formation. A chromatogram from a similar column with a more thoroughly treated solid support gave a single peak with no evidence of trioxane depolymerization. Figure 15b was obtained with the column of Figure 15a except that a 4-in. heated zone 21°C above ambient was built into the column with center approximately 31 in. from the column outlet. A maximum in reaction rate, or product production, is evident.

The sensitivity of this method to column temperature can be estimated from the Arrhenius equation and the sensitivity of V_R to temperature as shown in Eqs. 70 and 71, respectively, thus,

$$k = A \exp(-E_A/RT) \tag{70}$$

$$V_R^T = B \exp(\Delta H_S/RT) \tag{71}$$

where B is a constant for a given solute and column and ΔH_S is the differential molar heat of vaporization of the reactant from solution. Since the steady-state flow reactor equation applies to a pulse situation (20) for a first-order reaction, we can write

$$\ln(1-X) = \frac{k(V_R^T)}{F} = \frac{AB}{F} \exp[(\Delta H_S - E_A)/RT] \tag{72}$$

where X is fractional conversion of the reactant and F is the corrected carrier gas flow rate. It is evident then, that the change in conversion in a column due to the presence of a "hot zone" of the type with which we are concerned depends on the difference between ΔH_S and E_A. The greater the difference, the greater is the sensitivity. The location of the "hot zone" is most easily estimated when retention volumes of product and reactant are significantly different and $E_A - \Delta H_S$ is large.

The same approach can be used to investigate void zones in chromatographic columns which sometimes occur after shipping or on large preparative columns following vibration (39, 171). For a reactant such as DCPD, which dissociates by first-order reactions in both gas and liquid phases with similar rate constants, product formation rate in a void zone does not differ appreciably from that in any other part of the column; but separation ceases there for lack of liquid phase. Material formed without immediate separation in the void zone is eluted as a small "void peak" on the production curve. The position and sometimes the volume of the zone can be calculated by using the column as a reactor, if possible.

For both a temperature inhomogeneity (rise) or a void zone (168, 171), we write

$$(V_R^T)_x = (1-x)(V_R^T)_1 \frac{\bar{u}}{\bar{u}_{1-x}} + x(V_R^T)_2 \frac{\bar{u}}{\bar{u}_x} \tag{73}$$

where
$\quad x$ = fraction of distance of inhomogeneity from column outlet (0 to 1)

$\bar{u}, \bar{u}_x, \bar{u}_{1-x}$ = average flow rate in the column from x to outlet and from inlet to x, respectively

$(V_R^T)_x, (V_R^T)_1, (V_R^T)_2$ = retention volume of "void peak" or hot spot, pure reactant, and pure product, respectively

Either corrected or uncorrected retention volumes can be used. It is readily shown (168, 171) that

$$x = \frac{(V_R^T)_1 - (V_R^T)_x j_x}{(V_R^T)_1 - (V_R^T)_2 j} \tag{74}$$

where j is the pressure correction factor (3–6, 24) for the entire column and j_x is the correction factor from x to the column outlet. An iterative procedure is necessary to solve Eq. 74, since P_x, needed to calculate j_x, also depends on x:

$$P_x = \sqrt{P_0^2 + (P_i^2 - P_0^2)x} \qquad (75)$$

The solution converges rapidly, about two or three iterations are necessary for a small change in x. Both temperature gradients and void zones can be located successfully with these equations (168, 171) under favorable conditions.

The volume of the void zone can be estimated from the area of the void peak A_P on the chromatogram. Conversion in the void zone is related to A_P, $(A_R)_{\text{void}}$, the area equivalent to reactant passing the void zone, and S_P and S_R, the respective sensitivity factors. Assuming the sensitivity factors to be equal, it can be shown (171) that the void volume V_{void} is calculated from

$$V_{\text{void}} = \frac{P_0 u_0}{P_x k_g} \frac{A_P}{(A_R)_{\text{void}}}$$

Both the DCPD dissociation reaction and the trioxane decomposition reaction respond to temperature gradients in the column; but only the former, which occurs in the gas phase, is sensitive to surface irregularities. Thus the use of the two reactions can be complementary, if the investigator is careful not to interpret any small peak from impurity in the product elution curve as a column defect (39, 171).

B. The Preparative Chromatographic Reactor

Under special, but common, circumstances, the GCR is useful for preparative purposes. For small amounts of material, the combination of a readily adaptable flow reactor and separation unit with temperature control has convenience and appeal for laboratory preparations. Of course, the reaction under consideration must be feasible in a chromatographic column. Then the chromatographic preparative reactor is capable of generating uncontaminated reagent, which is isolated by selectively trapping effluent. If the reagent produced in the preparative reactor is for use in an intermediate preparation, the chromatographic eluant can be passed directly into the next reaction mixture.

Of particular interest are reactions such as esterification (172) or decarboxylation (173), which can take place in the course of passage through the column (2). Such preparative schemes need not be quantitative, and when viable, they eliminate laboratory manipulation and apparatus, are convenient, and save time. If special equipment is not used, it is necessary only to be assured that the relatively expensive chromatographic apparatus is not chemically contaminated or damaged before subsequent analytical use. Otherwise, chemical practitioners consider only the usual factors in any chemical preparation and separation.

Problems and limitations in the scale-up of the chromatographic reactor are similar to those encountered with conventional preparative columns. These are discussed in many places, including those referenced here (4–6). Pyrolysis techniques, involving an auxiliary device before the column, are well developed at this time and have been the subject of extensive discussion (1–3).

The capability of performing reactions "beyond equilibrium" without the limitations of thermodynamic reversibility (as in the dehydrogenation of benzene), together with high separation efficiency, drew early attention to the preparative reactor and possible scale-up. Where this capability is due to the separation of two or more products, the batch nature of conventional chromatography sets limits to the amount of reactant that can be processed (95, 96, 99). Considerations include the pulsating nature of the feed, the quantitative limitations of the column, the decrease in separation efficiency for large columns, and the degree of efficiency in trapping the eluted product.

Our own investigations (29, 30, 39) led us to propose the GCR for convenient preparation of unstable reagents for immediate use in the laboratory, especially in synthetic operations (174). The preparative chromatographic reactor is particularly valuable for dissociative or depolymerization reactions. The substantial difference in volatility between reactant and product in these reactions causes an approach to the instantaneous separation feature of the ICR. Rapid and constant separation minimizes interaction between reactants and products, thereby reducing some side reactions and increasing the yield of reactions that are otherwise limited by equilibrium. The dilution of reactants in the column also increases yields from potentially reversible reactions, which proceed with an increase in the number of moles.

The processes of producing cyclopentadiene and anhydrous formaldehyde are ideally suited for illustrating the utility of the preparative chromatographic reactor. Both compounds associate to more stable materials during storage. Cyclopentadiene normally is obtained by cracking and distilling from the dimer each time it is needed, since the dimer is thermodynamically favored at room temperature. Anhydrous formaldehyde is prepared in the laboratory by the dehydrogenation of methyl alcohol over heated copper or silver, or by heating an oligomer (175). Formaldehyde is available in aqueous solution but must be anhydrous for many uses.

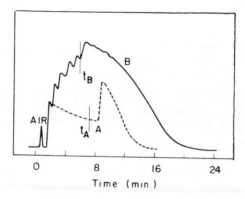

Fig. 16. Preparation of CPD reactor chromatograms for DCPD dissociation to CPD at 200°C. Column: $\frac{3}{8} \times 84$ in., flow: 60 ml/min of helium; ΔP: 323 mm, 65 g of packing containing 25% Versamid 900 on 60-80 mesh silanized brick. Curve A for one 0.3-ml injection; curve B for 5 injections at 1.5-min intervals.
By courtesy of *Chemistry and Industry, London.*

Work on solvent (stationary phase) effects showed that Versamid 900, a high-molecular-weight polyamide resin with high temperature stability, enhanced the reaction rate of DCPD depolymerization by about 20% above the rate found in hydrocarbon solvents (40). A reactor chromatogram for 0.3 ml of DCPD injected into a scaled-up preparative column over a period of 15 sec appears as curve A in Figure 16. Cyclopentadiene formed in the column is eluted initially as a broad band with a sharp front. The column here is operated close to capacity at heavy load, so that it operates outside the linear range of the sorption isotherm for both reactant and product. This also results in a sharp front and a sloping tail to the peak for unreacted DCPD (5, 7). As shown, the reactant is only about 50% converted. Since CPD was collected by trapping the portion of the chromatogram from injection to time t_A, several injections of reactant at regular intervals or a long constant injection pulse seemingly would give more efficient use of the column. Thus the first portion of the column can be used while previously injected material reacts further down the column. The chromatogram obtained from five slow, successive, 0.3-ml injections, at 90-sec intervals, is shown as curve B. Five distinct product peaks appear before t_B. Successive reactant peaks are distinguishable at later elution times.

Sample sizes of 0.1 to 0.3 ml are needed to produce practical quantities of CPD. This significantly increases concentration of reactant and product in the column. The second-order association of CPD, negligible in the small samples used for kinetic runs, must be recognized under preparative conditions, where conversion falls below 100%.

In one instance where DCPD was not eluted for 670 sec at a flow rate of 38 cm^3/min, analysis revealed 3.5% dimer present in a sample collected for the first 540 sec. Injecting 0.3-ml samples at 30-sec intervals and at the same flow rate increased reactant concentration. Longer intervals between injections diminished reactant concentration in the column, but dimer impurity did not fall below 3%. In comparison, CPD obtained with the conventional distillation procedure (176) contained only 0.5% DCPD.

Calculation showed that it would be difficult to eliminate all impurity at the concentrations necessary for appreciable production of CPD from the column. To eliminate the product impurity, a trap was devised for the reactor outlet (174) so that DCPD condensed from the outlet stream. With the addition of the trap, CPD was obtained with only 0.9% dimer impurity and 1% of an unidentified material. A CPD stream produced in the chromatographic reactor was reacted successfully with maleic anhydride at the outlet to give Diels–Alder product (174).

Trioxane depolymerizes to formaldehyde by a surface-catalyzed reaction on firebrick (29, 30). The estimated liquid phase rate constant k_l was 0.03 sec^{-1} at 140°C in a column of 20% Silicone DC 550 on untreated 60-80 mesh firebrick. Gas phase rate was negligible. Under these conditions, more than 99% of the trioxane should be converted; only formaldehyde was detected in the stream of eluted material. In an experiment, where a flask containing n-butyl magnesium bromide in anhydrous ether was fitted to receive the column effluent, formaldehyde reacted as soon as it entered the flask. After hydrolysis, the investigators isolated a 33% yield of amyl alcohol, the Grignard addition product. A 50% excess of trioxane was pulsed through the column to compensate for any reverse reaction or trapping system inefficiency. This procedure should be improved with further study (174).

From Eq. 15 it seems that the liquid phase term $k_l t_l$ would determine conversion, since liquid phase residence time usually exceeds the gas phase residence time. In addition, solvent effects and catalysis by the solid support frequently favor the liquid phase constant relative to the gas phase constant.

Examples here are illustrative only. Recycling, improved technique, automatic injection, and use of smaller samples could be advantageous.

Further exploratory work would be helpful. However, the preparative gas chromatographic reactor is established as a useful, rapid source of high-purity reagents in appropriate circumstances.

C. Isotope Labeling

Strictly speaking, the use of the GCR for isotope labeling is another application of the reactor for preparative purposes. However, the technological importance of this application and some of its features make it worthy of separate note.

The labeling exchange reaction can have the following forms:

$$AX_{n(\text{mob})} + RX^*_{(\text{stat})} \rightleftarrows AX^*_{n(\text{mob})} + RX_{(\text{stat})} \tag{76}$$

$$AX_{n(\text{mob})} + RY^*_{(\text{stat})} \rightleftarrows AY^*_{n(\text{mob})} + RX_{(\text{stat})} \tag{77}$$

where the subscripts (mob) and (stat) refer to the mobile and stationary phases, respectively. The X, X^* exchange refers to exchange for isotopes of the same element, and the X, Y^* exchange refers to exchange of one element for an isotope of another (177–180).

The labeled stationary exchange material is coated on the solid support as part of the liquid phase or exists as part of the surface of a solid phase. The compound to be labeled, AX_n, is pulsed through the column and trapped at the outlet upon emergence as exchanged labeled material.

The advantages of this labeling technique derive from our discussion of the chromatographic reactor and the special aspects of labeling and are as follows:

1. A higher degree of exchange is achieved in the dynamic system than with equilibrated static systems (181, 182). This result, which is equivalent to operating beyond equilibrium (37, 96), follows from simple plate theory if the pulsed material AX_n is considered to move through the column approaching equilibrium in each plate. If the column is completely exchanged, the pulsed material encounters a totally labeled counterreactant as it enters each new plate.

2. As with the preparative reactor, the gas chromatographic apparatus is essentially an integrated reaction-separation system in which temperature and flow rate can be adjusted readily to favor reaction.

3. Tracer and pulsed material are used efficiently. Both milligram and gram amounts of sample for labeling can be handled (7, 183).

4. The labeled product is of high purity because of concerted reaction and separation.

The stationary phase may be a treated solid support (184), a reactive solid (184–186), or a liquid (178, 187). Material adsorbed on the solid support may also be used for labeling. Elias and his collaborators have used low-melting quaternary ammonium halides with particular effectiveness (177, 178, 187). Water containing deuterium, tritium, or ^{18}O is particularly useful for labeling the stationary phase (179, 188–192). In turn, KOD (191), deuterium and oxygen-labeled phosphoric acid (192), and tritiated sorbitol (177, 179) are useful stationary phases for labeling. Tritiated hydrocarbons can be prepared by treating reactant alkyl halides with CaT_2 as a stationary phase (184, 193, 194).

Elias has written useful reviews of this area (178, 187). A number of references and applications are indicated and summarized in Table IV. Any available standard references prior to 1970 for data on static equilibrated systems can be used for comparison with the success of isotope labeling by the chromatographic reactor.

TABLE IV

Selected References on Labeling by Chromatography

Topic	References
The immobilized phase	
Treated surface	179, 180, 184, 185, 194, 195
Labeled stationary phase	177, 178, 188, 190, 191, 192, 193, 196
Labeling elements	187 (review)
Halides	184, 185, 194, 195
On $GeCl_4$, $SNCl_4$, $AsCl_3$, $FeCl_3$	180, 185
On alkyl halides	184, 193, 194
Hydrogen isotopes	
Tritium on alcohols, acids, amines, and hydrocarbons	177, 190, 193, 194
Deuterium on ketones and aldehydes	191–193
^{18}O on ketones and aldehydes	192
Adsorption chromatography	
Deuterium	197, 198
Oxygen	199, 201
Effect of chromatography on labeled compound	199, 200

D. Identification

Since the GCR can be utilized to obtain reactant retention data and reaction rates simultaneously (9, 40), it represents a potentially powerful tool for identification of reactive compounds. Unfortunately, this application, plus the popular use of auxiliary reaction and subtractive devices for identification, are often discussed under the ambiguous title of "reaction gas chromatography." The use of this term in this context probably should be discouraged since, in most instances, reaction does not occur during the chromatographic process.

The use of auxiliary devices with chromatographic columns has been described, together with the few true chromatographic reactor applications in several comprehensive reviews (1–3). Reaction sequences that have been adapted to chromatographic columns include hydrogenation, dehydrogenation, pyrolysis in varying degrees, oxidative and reductive processes for elemental analysis, derivative formation for functional groups, and miscellaneous catalytic processes described earlier.

Customarily, reactions occurring on columns have been considered complications; therefore, it is profitable to look to discussions of these complications by Hesse (202, 203) and others (5, 204, 205) for potential chromatographic reactor applications. There are several reports of derivative formation on columns (172, 206, 207) in connection with "peak shifting" (208) that involve their use as reactors. Anders and Mannering (172) prepared derivatives of hydroxylic compounds, including steroids, by following injection of these materials with injections of reactive carboxylic anhydrides and hexamethyldisilazane (208). Haaken (206) used this technique to convert alcohols to acetates, and others have employed a variation in the conversion of 2-naphthylamine to an acetamide (207).

Gil-Av and Herzberg-Minzly (9) have suggested the use of the GCR for identification in the sense of the present discussion. They chromatographed cis–trans mixtures of conjugated dienes on a gas chromatographic column of chloromaleic anhydride. The trans isomer reacts to give Diels–Alder product more rapidly than the cis compounds. Thus, during passage through the column, identifying rates or relative rates as well as retention data are determined.

E. Retention and Thermodynamic Data

Retention data for reactant and product can be collected simultaneously with the use of the chromatographic reactor for reaction kinetic studies. From this information thermodynamic data for solution processes for reactant and products can be obtained from operation at several temperatures.

Accurate reactant retention time measurement is important in both kinetic and thermodynamic investigations (24, 25, 27). The determination of the mean residence time of an unreactive material in the region of a linear adsorption isotherm is comparatively simple (19–22). However, product interference can distort the observed reactant peak somewhat, as in Figures 3 and 4. If product is more volatile than reactant, the apparent maximum of the peak occurs before the true mean residence time of the reactant if a nonselective detector is used. The converse is true for products retained longer than reactant in the column.

A quantitative treatment of the problem of determining true reactant retention volume has been described by one of us for the CPD–DCPD system (39). It is also possible to use a semiquantitative approach from a physical feeling for the reactant–product system. For instance, it is possible to reconstruct from most reactor chromatograms a product interference curve similar to that in Figure 4. By subtraction of the estimated product response from the total response, an undistorted reactant peak of the eluted material can be reconstructed for the determination of the mean residence or retention time. The contribution of product interference to a change in the reactant apparent retention time can then be estimated. The validity of the retention time or volume can be checked by using standard calculational procedures to give a "corrected retention volume" for different pressures and flow rates in the column (4–6, 24). The corrected retention volume, of course, should be independent of flow rate.

The correction for the effect of the product curve on the apparent reactant retention volume depends on the difference between characteristic retention volumes of product and reactant. The effect is least when the difference is greatest. Operating conditions for the reactor column are also important. The apparent retention volume of reactant will be least affected by product when measuring at low conversions and high column efficiencies (many plates, sharp pulse), as can be seen from Figure 3.

The product retention volume may be measured by injection of pure product. Alternatively, it can be obtained, with a liquid stationary phase, from the product wave that resembles a frontal analysis elution curve (4, 34). It is not surprising, then, that preliminary experiments indicate that the time for reaching one-half the maximum height attained in the product curve approximates the retention volume of pure product (40, 209).

The measured retention volumes of products and reactants can be converted to corrected retention volumes per gram of liquid in a column at column temperature V_g^T. In turn, V_g^T can be related to the activity coefficient of any solute (reactant or product) at infinite dilution γ_p^∞ by the well-known expression (4, 25, 26, 210)

$$V_g{}^T = \frac{RT}{M_s \gamma_p^\infty p_2{}^0} \qquad (78)$$

where M_s is the molecular weight of the stationary phase and $p_2{}^0$ is the vapor pressure of solute at the operating temperature.

Under some conditions, because of deviations from ideality caused by molecular interactions, γ_f^∞, the zero total pressure, corrected solute activity coefficient significantly differs from γ_p^∞. The equation used for this correction has been extensively treated in the literature (7, 27, 211, 212). For our work in the CPD–DCPD system (40), where helium was used as a carrier gas, the term involving solute–carrier-gas interaction gave a negligible correction. Under these conditions, the final operational expression for γ_f^∞ (infinite dilution) is

$$\ln \frac{\gamma_f^\infty}{\gamma_p^\infty} = -\frac{p_2{}^0}{RT}(B_{22} - v_2{}^0) \qquad (79)$$

where B_{22} is the second virial coefficient of pure solute, $v_2{}^0$ is the molar volume of pure solute, and γ_p^∞ is calculated from Eq. 78. Parameters used in the correction expression are discussed in the references cited. As reflected in Eq. 79, the correction increases in importance with higher vapor pressures of solute.

Some illustrative activity coefficient data and rate constants are presented in Table V, without discussion. It should be remembered that

TABLE V

Rates and Properties in Indicated Stationary Phases (40)

| Solvent | M.W. | Dicyclopentadiene | | γ_f^∞ at 180°C | γ_f^∞ at 200°C | $k_l \times 10^4$, sec^{-1} |
		V_g at 180°C, cm³/g	V_g at 200°C, cm³/g			
Silicone DC 550	1722	67.0	45.5	0.283	0.278	4.65 ± 0.15
Apiezon L	1230	97.1	66.0	0.335	0.331	4.72 ± 0.04
Hexatriacontane	507	121.4	—	0.532	—	4.61 ± 0.03
Polyphenyl ether (6 ring)	538.6	80.4	55.4	0.757	0.729	5.73 ± 0.10

according to transition-state theory, the liquid phase unimolecular rate constant k_l is related to the activity coefficient of the reactant (123, 213) by

$$k_l = k_0 \gamma_R^\infty / \gamma_\ddagger^\infty \qquad (80)$$

where γ_R^∞ and $\gamma_\ddagger^\infty$ are activity coefficients for reactant and activated complex, respectively, at infinite dilution and k_0 is the rate constant in an ideal medium. If the gas phase is assumed to be ideal, then the gas phase rate constant is k_0, and the activity coefficient of the activated complex can be calculated from

$$\gamma_\ddagger^\infty = \gamma_R^\infty k_g / k_l \qquad (81)$$

Admittedly, the gas phase may not be ideal. However, relative values of $\gamma_\ddagger^\infty$ can be calculated. The capacity for measurement of both rate constant and activity coefficients in the same liquid medium enhances the appeal of the GCR. Such data are rarely available for any particular reaction (40, 123, 213, 214). Availability of such data through the chromatographic reactor for a number of reactions should assist in the correlation of kinetic measurements with the thermodynamic properties of reactants and products in various solvents (40).

The activity coefficient may be related to excess partial molar free energy at infinite dilution, $\Delta \bar{G}_e^\infty$. From activity coefficient measurement at several temperatures, excess partial heat and entropy data also become accessible (25, 26). With these facts in mind, it can be predicted that the next few years will witness a significant amount of work on reactions in gas chromatographic columns for the purpose of obtaining kinetic and thermodynamic data concurrently.

VI. CHROMATOGRAPHIC CONDITIONS AND REQUIREMENTS

Physicochemical chromatographic work generally requires careful control of temperature and operating conditions and features that may differ considerably from those favored for analytical use. For instance, apparatus built for rapid response to temperature programming may not have the temperature stability and reproducibility necessary for obtaining kinetic and related physicochemical data (7, 25, 26, 40).

A choice can be made from standard glass, plastic, or metal tubing, which can catalyze reactions. Column materials also vary in catalytic activity depending on the reaction being considered. The stationary phase support or surface must be chosen with care. Since reactions are sensitive to impurities such as hydroxyl groups on the surface, the presence of reaction may serve as an indicator of noninert solid support (168). For preparative reactors, supports might even be chosen for their reactivity (174). The constancy of the first-order rate constant with respect to change in liquid phase-support ratio or column diameter is a useful indication of homogeneous kinetic measurements (40, 174).

The liquid phase should have enough solvating power to retain reactant for significant conversion to product. The percentage of liquid phase should be about 10 to 30% of total packing weight. The upper limit is set by the usual chromatographic consideration to avoid rate-limiting mass transfer. The possibility of solid support catalyzed reactions sets the lower limit for ordinary reactor operation well above the 1 to 3% of liquid phase used for fast chromatography. The choice within the allowable range depends on two factors—the solubility or retention volume of reactant in the stationary phase, and the desired conversion to products in the gas phase. For reactants of low solubility, the amount of liquid phase must be large enough to obtain the residence times necessary for a kinetic study. This can be done with long columns or high liquid loadings. For difficult-to-elute reactants, the converse is true, and the amount of liquid phase is kept near the lower limit. The apparent rate constant is related to column composition by

$$k_{\mathrm{app}} = k_l + (f_g/f_l K)k_g \tag{82}$$

where f_g/f_l is the ratio of gas to liquid phase volume. If both k_l and k_g are sought, two columns are necessary, one with a high percentage of liquid phase, the other with with a low percentage. However, when only the liquid phase rate constant is desired, the percentage of liquid phase should be high to minimize the effect of error in k_g when determining k_l from k_{app}.

Any reference material (inert standard method) on the chromatogram must be sufficiently distant from reactant and product that areas can be determined unambiguously. This is facilitated if the inert peak is on the side of the reactant peak opposite from the product. A separation factor of 1.5 (retention volume of inert–retention volume of reactant) ensures a 99.99% separation of areas on a 500-plate column (4, 5). But too much separation is unduly time-consuming. This establishes a maximum separation factor of 2. The retention volume ratio of inert then should be

$$1.5 < \frac{(V_g^T)_I}{(V_g^T)_R} < 2.0$$

$$1.5 < \frac{\gamma_R^\infty p_R^0}{\gamma_I^\infty p_I^0} < 2.0 \tag{83}$$

Small sample size enables kinetic runs to be conducted in the linear portion of the adsorption isotherm. For preparative work, samples can be used at the limit of the capabilities of the apparatus.

A number of assumptions were made in the mathematical treatment of the ICR regarding physical processes in the column. Since this is the prototype reactor for kinetic studies, experimental conditions should conform to these original assumptions. Since distribution isotherms should be linear, the study of reactions of polar reactants and the extension of reactor studies to many reactions on the solid surface may not always be feasible. The assumptions (a) that mass transfer is not rate limiting and (b) that diffusion is unimportant, restrict the range of flow rates to about 5 to 150 cm³/min in quarter-inch liquid-support columns. Although large residence times are needed for slow reactions, reducing the carrier gas flow rate to 1 cm³/min is not justifiable, nor is a flow rate of 300 cm³/min always acceptable for fast reactions. Chromatograms can help determine satisfactory operating conditions. Asymmetric peaks and retention volumes dependent on sample size and conversion are indicative of nonlinear distribution isotherms.

At low flow rates, diffusion can broaden peaks so that residence time distribution of reactant is a significant fraction of the average residence time; the accuracy of the measured rate constant is reduced. High flow rates may prevent mobile vapor stationary phase from equilibrating in the mass transfer regime. The effect of partition coefficients then decreased, producing erroneous rate constants. Incidentally, catalysis at the surface may serve as a means of studying solute–surface mass transfer in the presence of liquid phase (168).

The temperature of the injector (see Section IV. C) can influence column efficiency in conventional chromatography. If the injector temperature is low, the peak may be broadened because of the time required to vaporize the sample, an effect similar to that of a slow injection. Unless this result is specifically desired (80, 81), reaction in the injection chamber can result in erroneous kinetic measurements. In some instances, a spike at the maximum of the product wave with a high injector temperature relative to the column is an indicator (40). Injector temperature should be just slightly higher than column temperature; we have often found a value 5°C above column temperature to be satisfactory.

Mathematical models for reactions in a chromatographic column assume no radial concentration or velocity gradients inside the column. In practice, this means that bends in the column should be avoided. A straight column would be best but is impractical. The next choice is a U-shaped column with a large-diameter bend. Giddings (216) has proposed the following criterion for the minimum coil diameter consistent with good resolution in terms of column diameter and solid support particle diameter:

$$D_{coil} > (D_{column})^2 / D_{solid\ support} \tag{84}$$

For an 80-mesh solid support inside a 0.194-in. i.d. column, Eq. 84 gives a minimum bend diameter of 3.2 in. Basic column configurations used for our kinetic work are U-shaped or W-shaped.

VII. LIMITATIONS AND ADVANTAGES

The gas chromatographic reactor is quite unusual. Depending on circumstances, it is truly attractive for some rate studies and is useless for others.

Some disadvantages of the GCR stem from the stationary phase solvent for the reaction. Since this liquid must be nonvolatile to avoid column bleeding, the choice of solvents is limited. High temperatures, particularly, are restrictive, since only polymers and a few high-melting compounds are available for solvent use. However, commercial stationary phases include a variety of chemical types, and solvent effects on the reaction rate can still be investigated and controlled. A more serious problem is the large amount of surface area per unit volume of liquid phase. In packed columns, traces of metals and numerous silicon–hydroxyl groups offer potentially active sites for undesired reaction catalysis. Washing the support with acid and base and surface treatment can reduce these effects, although they are not always eliminated (168). Teflon solid supports, which are inert, often are difficult to use and tend to decrease column efficiency. In capillary columns, the liquid is coated on column walls, which may reduce the surface area per unit volume of liquid, although with some adverse effect; the ratio of gas phase to liquid phase retention time is large in capillary columns, whereas liquid phase rate constants are most accurate when this ratio is minimal. Also, solute may selectively adsorb on the column walls. Despite these reservations, the capillary chromatographic reactor appears to be worthy of future consideration.

Only limited classes of chemical reactions can be studied in the GCR. In the development of equations from which rate constants are calculated, an expression for reactant concentration at the column outlet has to be integrated. An analytical solution is possible only when the reaction is first-order or pseudo-first-order. Pseuso-first-order possibilities include dissolution of one reactant in the liquid phase, provided it is nonvolatile, or mixing with carrier gas at constant concentration, if it is volatile. Equations for higher-order reactions can be solved with a digital computer, and curve-fitting techniques are being developed for finding rate constants. Because of the complexity involved, the chromatographic reactor is still best suited to first-order and pseudo-first-order reactions.

Several experimental factors combine to limit the range of rate constants that may be determined. Assuming that reactant and product are separated unambiguously, the ratio of inert to reactant area can seldom be measured more accurately than $\pm 0.5\%$. This represents an error of ± 0.005 in $\ln (A_I/A_R)$. Since the rate constant is calculated from

$$\ln (A_I/A_R) = k_{app}t_I + \text{constant}$$

with the inert standard method, the difference between the maximum and minimum values in $\ln (A_I/A_R)$ should be at least ten times the expected error in each point, or 0.05. Low conversion runs can be made from 0 to 10%, and high conversion runs can be made to 90%. The range of retention times is limited to a minimum of about 10 sec by detector response, sample vaporization, and system time lag, and to a maximum of about 5 hr with the inert standard method. The range of rate constants that may be determined is approximated from the longest retention time with the smallest maximum conversion (Ca. 5% if product is studied) and the shortest possible retention time with the largest minimum conversion (90%). Hence we have

$$-\frac{\ln (0.95)}{5 \text{ hr}} < k_{app} < -\frac{\ln (0.1)}{10 \text{ sec}}$$

$$3 \times 10^{-6} < k_{app}(\text{sec}^{-1}) < 2 \times 10^{-1} \tag{85}$$

Constants from the middle of this range (10^{-4} to 10^{-2}) are most accurate. The range might be extended somewhat using other methods of analysis (e.g., reaction–product ratio).

The retention volume of the reactant is important in the determination of the apparent rate constant. As the difference between reactant and product retention volumes increases, product interference decreases and the measurement of the net reactant area is improved. Simultaneous or near-simultaneous elution of reactant and product eliminates most analytical methods other than total area or a selective detector. A separation factor greater than 1.5 to 2.0 should give adequate separation of reactant and product for measurement of reactant area and calculation of the apparent rate constant.

For a reaction to be studied in the GCR, the proper range of liquid phase retention times must be experimentally attainable. Preferably, retention times on the order of the half-life $(0.692/k_{app})$ are used. The residence time of the reactant in the liquid phase is

$$t_l = \frac{(V_g^T)_R W_l}{F} \tag{86}$$

where $(V_g^T)_R$ = specific corrected retention volume at column temperature per gram for reactant
$\quad\quad W_l$ = weight of liquid phase in the column
$\quad\quad F$ = mean volumetric flow rate in the column.

Although the retention volume is fixed by temperature, a range of residence times is attainable through variation of flow rate and the amount of liquid in the column. Usually only a single column of particular composition is used for a reaction; column length and percentage of liquid phase are chosen to match the desired residence time range. Since the percentage of liquid varies from 10 to 30%, and since column lengths from 3 to 25 ft are convenient, W_l can be varied by a factor of 25. Combining this with a flow-rate range of 5 to 150 cm³/min in a quarter-in. column, a 750-fold change in the liquid phase residence time is possible. Assuming 2.5 g of packing material per foot of column, we have

$$\frac{(3\text{ ft}\times2.5\text{ g/ft}\times0.10)(V_g^T)_R}{150\text{ cm}^3/\text{min}} < t\,(\text{min}) < \frac{(25\text{ ft}\times2.5\text{ g/ft}\times0.30)(V_g^T)_R}{5\text{ cm}^3/\text{min}}$$

$$0.30(V_g^T)_R < t\,(\text{sec}) < 225(V_g^T)_R \tag{87}$$

where the retention volume has units of cubic centimeters per gram. Despite this range of retention times, sometimes t_l cannot be made sufficiently large (or small) to achieve desired conversion. Then the alternative is to work at higher (or lower) temperature for a particular column, since reaction half-times usually decrease with temperature more rapidly than retention volumes (168). Smaller or larger columns can also be introduced at either range extreme of Eq. 87.

The chromatographic nature of the reactor column provides special advantages. First, a reactor and a means of product analysis are combined. Second, each component in the reactor is continuously purified. Trace impurities are tolerable in reactant if they are removed during passage through the column. Continuous product removal from reactant can simplify kinetics, as exemplified with the reversible DCPD dissociation. The enlarged chromatogram of Figure 17 is illustrative. Commercial

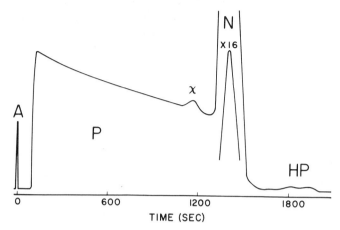

Fig. 17. Enlarged reactor chromatogram (39, 40) illustrating detection of *exo*-isomer impurity *x* and higher polymer byproduct *HP*. Column $T = 180.6°C$; flow $= 50.2$ cm^3/min at column T. Conversion 53% on 12-ft, quarter-inch column containing hexatriacontane (25%) on Gas Chrom Q.
By courtesy of *Journal of Physical Chemistry*.

DCPD contains 0.5 to 1.0% of less reactive *exo*-isomer, which can be removed by recrystallization. However, the gas chromatographic reactor removes that isomer from the reactive *endo*-DCPD peak and elutes it with the product peak. *endo*-Dicyclopentadiene does tend to react with CPD to form trimer and tetramer (45, 217). Apparently, some higher polymers or possibly a small amount of adsorbed material (73) are eluted after the reactant peak. However, separation has reduced reaction between reactant and product to negligible proportions.

The GCR is a convenient and readily controlled device for the study of chemical reactions. Since most laboratories have chromatographs available, potential reactors abound, dependent on controls and the nature of the need. A wide choice of residence times is possible from change of flow rate and column size.

Sample requirements are small, less than 0.5 μl per run. This is an important feature when reactant is of limited availability. About 0.1 g of reactant is adequate for determination of several rate constants. Small samples render heat effects negligible, and the local heating effects encountered in some reactors are eliminated.

Since retention volumes can be measured for reactant and product during the course of a kinetic run, heats of vaporization, activity coefficients, and excess partial molar quantities can be determined with rate constants and Arrhenius parameters. Thermodynamic correlations are often used to interpret solvent effects on chemical reactions (213, 218, 219). Using conventional reactors, thermodynamic data must be found in separate experiments and, therefore, are seldom available.

For special preparative purposes, the chromatographic reactor shows promise as a convenient laboratory tool. The unique advantage is elimination of laboratory manipulation and rapid conversion of small amounts of material as needed. This advantage becomes more significant in isotope labeling, where only small amounts of product are required or desired and efficient use of labeling agent may be the major consideration.

When applicable, the GCR is a valuable device for the study of chemical reactions. It has some disadvantages and limitations, but for first-order and pseudo-first-order reactions of volatile compounds with rate constants between 2×10^{-6} and 0.2 sec^{-1}, the chromatographic method is appealing. The restriction to systems involving volatile reactants and nonvolatile solvents often makes the reactor complementary to conventional reactors, which cannot be used in that situation. In general, the advantages derived from immediate separation of reactant and product, from small sample size, from the frequent availability of gas chromatographs, and from the ease with which thermodynamic data can be obtained, outweigh the limitations and disadvantages discussed here. Thus the gas chromatographic reactor is an important addition to the versatility of the chromatographer and the armamentarium of the kineticist.

Acknowledgment

We thank the Wisconsin Alumni Research Foundation, the E. I. duPont Company, the American Oil Company, and the United States government (N.D.E.A. Fellowship to J. E. P.) for grants and fellowships that helped support this work. We are especially grateful to the Petroleum Research Fund, administered by the American Chemical Society, for flexible grants over a period of years which assured the continuity of the work.

References

1. M. Beroza and R. A. Coad, *J. Gas Chromatogr.*, **4**, 199, 1966; also in *The Practice of Gas Chromatography*, L. S. Ettre and A. Zlatkis, eds., Interscience, New York, 1967, pp. 461–510.
2. V. G. Berezkin, *Analytical Reaction Gas Chromatography*, Plenum Press, New York, 1968.
3. O. E. Schupp, III, *Gas Chromatography*, Vol. XIII, *Technique of Organic Chemistry*, E. S. Perry and A. Weissberger, eds., Interscience, New York, 1968.

4. J. H. Purnell, *Gas Chromatography*, Wiley, New York, 1962.
5. A. B. Littlewood, *Gas Chromatography*, Academic Press, New York, 1962.
6. S. Dal Nogare and R. Juvet, *Gas-Liquid Chromatography*, Interscience, New York, 1962.
7. J. R. Conder, *Progress in Gas Chromatography*, in J. H. Purnell, ed., Vol. 6, *Advances in Analytical Chemistry and Instrumentation*, Interscience, New York, 1968, pp. 209–270.
8. E. Gil-Av and Y. Herzberg-Minzly, *Proc. Chem. Soc.*, 316 (1961).
9. E. Gil-Av and Y. Herzberg-Minzly, *J. Chromatogr.*, **13**, 1 (1964).
10. C. G. Collins and H. A. Deans, *AIChE J.*, **14**, 25 (1968).
11. M. Kocirik, *J. Chromatogr.*, **30**, 459 (1967).
12. S. Z. Roginskii and A. L. Rozental, *Dokl. Akad. Nauk SSSR*, **146**, 152 (1962).
13. L. Lapidus and N. Amundson, *J. Phys. Chem.*, **56**, 984 (1952).
14. H. Yamazaki, *J. Chromatogr.*, **27**, 14 (1967).
15. J. C. Giddings, *Dynamics of Chromatography*, Dekker, New York, 1965.
16. J. C. Giddings, *J. Chem. Phys.*, **32**, 1462 (1959).
17. J. C. Giddings, *J. Chromatogr.*, **3**, 433 (1960).
18. O. Grubner, A. Zikanova, and M. Ralek, *J. Chromatogr.*, **28**, 209 (1967).
19. H. W. Habgood and J. F. Hanlan, *Can. J. Chem.*, **37**, 843 (1959).
20. D. W. Bassett and H. W. Habgood, *J. Phys. Chem.*, **64**, 769 (1960).
21. R. L. Gale and R. A. Beebe, *J. Phys. Chem.*, **68**, 555 (1964).
22. J. Janak, *Collect. Czech. Chem. Commun.*, **18**, 798 (1953).
23. P. E. Porter, C. H. Deal, and F. H. Stross, *J. Am. Chem. Soc.*, **78**, 2999 (1956).
24. A. T. James and A. J. P. Martin, *Biochem. J.*, **50**, 679 (1952).
25. S. H. Langer and J. H. Purnell, *J. Phys. Chem.*, **67**, 263 (1963).
26. S. H. Langer, B. M. Johnson, and J. R. Conder, *J. Phys. Chem.*, **72**, 402 (1968).
27. A. J. B. Cruickshank, B. W. Gainey, and C. L. Young, *Trans. Faraday Soc.*, **64**, 337 (1968).
28. V. G. Berezkin, V. S. Kruglikova, and N. A. Belikova, *Dokl. Akad. Nauk SSSR*, **158**, 182 (1964).
29. J. Y. Yurchak, M. S. thesis, University of Wisconsin, Madison, 1966.
30. S. H. Langer and J. Y. Yurchak, Abstracts, 153rd National Meeting of the American Chemical Society, Miami Beach, Fla., April 1967, no. R46.
31. S. H. Langer, J. Y. Yurchak, and J. E. Patton, *Ind. Eng. Chem.*, **61** (4), 10 (1969).
32. J. Kallen and E. Heilbronner, *Helv. Him. Acta*, **43**, 489 (1960).
33. A. J. P. Martin and R. L. M. Synge, *Biochem. J.*, **35**, 1358 (1941).
34. E. Glueckauf, *Trans. Faraday Soc.*, **51**, 34 (1955).
35. A. Klinkenberg and F. Sjenitzer, *Chem. Eng. Sci.*, **5**, 258 (1956).
36. M. Nakagaki and M. Nishino, *Yakugaka Zasshi*, **85**, 305 (1965).
37. S. Z. Roginskii, M. I. Yanovskii, and G. A. Gaziev, *Dokl. Akad. Nauk SSSR*, **140**, 1125 (1961).
38. G. A. Gaziev, V. Yu. Filanovskii, and M. I. Yanovskii, *Kinet. Katal.*, **4**, 688 (1963).
39. J. E. Patton, Ph.D. thesis, Department of Chemical Engineering, University of Wisconsin, Madison, 1970.
40. S. H. Langer and J. E. Patton, *J. Phys. Chem.*, **76**, 2159 (1972).
41. O. Levenspiel, *Chemical Reaction Engineering*, Wiley, New York, 1962, pp. 107–117.
42. G. L. Pratt and S. H. Langer, *J. Phys. Chem.*, **73**, 2095 (1969).
43. J. B. Harkness, G. B. Kistiakowsky, and W. H. Mears, *J. Chem. Phys.*, **5**, 682 (1937).
44. B. S. Khambata and A. J. Wasserman, *J. Chem. Soc.*, 371 (1939); *ibid.*, 375.
45. W. C. Herndon, C. R. Grayson, and J. M. Manion, *J. Org. Chem.*, **32**, 652 (1967).
46. H. Kwart, S. F. Sarner, and J. H. Olson, *J. Phys. Chem.*, **73**, 4056 (1969).
47. M. Vilkas and N. A. Abraham, *Bull. Soc. Chim.*, 1651 (1951).

48. V. Yu. Filinovskii, G. A. Gaziev, and M. I. Yanovskii, *Dokl. Akad. Nauk SSSR*, **167**, 143 (1966).
49. C. S. G. Phillips, A. J. Hart-Davis, R. G. L. Saul, and J. Wormland, *J. Gas Chromatogr.*, **5**, 424 (1967); *Advances in Gas Chromatography*, A. Zlatkis, ed., Preston, Evanston, Ill., 1967, pp. 209–213.
50. G. L. Pratt, personal communication.
51. J. E. Patton and S. H. Langer, in preparation.
52. M. Suzuki and J. M. Smith, *Chem. Eng. Sci.*, **26**, 221 (1971).
53. G. Padberg and J. M. Smith, *J. Catal.*, **12**, 172 (1968).
54. J. C. Adrian and J. M. Smith, *J. Catal.*, **18**, 57 (1970).
55. P. Schneider and J. M. Smith, *AIChE J.*, **14**, 762 (1968).
56. S. Z. Roginskii and A. L. Rozental, *Kinet. Katal.*, **5**, 104 (1964).
57. T. A. Denisova and A. L. Rozental, *Kinet. Katal.*, **8**, 441 (1967).
58. R. Chao and H. E. Hoelscher, *AIChE J.*, **12**, 271 (1966).
59. D. A. McQuarrie, *J. Chem. Phys.*, **38**, 437 (1963).
60. M. Kubin, *Collect. Czech Chem. Commun.*, **30**, 1104 (1965); *ibid.*, **30**, 2900 (1965).
61. E. Kučera, *J. Chromatogr.*, **19**, 237 (1965).
62. O. Grubner, *Advances in Chromatography*, Vol. 6, J. C. Giddings and R. A. Keller, eds., Dekker, New York, 1968, pp. 173–209.
63. O. Grubner and E. Kučera, in *Gas Chromatographie, 1965*, H. Struppe, ed., Akademie Verlag, Berlin, 1965, p. 157.
64. H. Vink, *J. Chromatogr.*, **20**, 305 (1965).
65. P. Schneider and J. M. Smith, *AIChE J.*, **14**, 886 (1968).
66. V. G. Berezkin, V. S. Kruglikova, N. A. Belikova, and V. E. Shiryaeva, *Gaz. Khromatogr.*, 454 (1968).
67. S. Z. Roginskii, M. I. Yanovskii, and G. A. Gaziev, *Kinet. Katal.*, **3**, 529 (1962).
68. V. Yu. Filinovskii, G. A. Gaziev, and M. I. Yanovskii, *Metody Issled. Katal. Reakts., Akad. Nauk SSSR, Sib. Otd., Inst. Katal.*, **3**, 313 (1965).
69. V. G. Berezkin, *Usp. Khim*, **37**, 1347 (1968).
70. C. S. G. Phillips, M. J. Walker, C. R. McIlwrick, and P. A. Rosser, *J. Chromatogr. Sci.*, **8**, 401 (1970).
71. R. Lane, B. Lane, and C. S. G. Phillips, *J. Catal.*, **18**, 281 (1970).
72. S. Z. Roginskii, *Izv. Akad. Nauk SSSR, Ser Khim.*, 1321 (1965).
73. J. A. Bett and W. K. Hall, *J. Catal.*, **10**, 105 (1968).
74. J. Coca and S. H. Langer, unpublished data.
75. V. G. Berezkin, V. S. Kruglikova, and V. E. Shiryaeva, *Kinet. Katal.*, **6**, 758 (1965).
76. V. G. Berezkin, V. S. Kruglikova, and V. E. Shiryaeva, USSR Patent 163,008, 1964; *Chem. Abstr.*, **61**, 12688 (1964).
77. E. J. Levy and D. G. Paul, *J. Gas Chromatogr.*, **5**, 136 (1967).
78. Chemical Data Systems, Inc., Oxford, Pa.
79. D. G. Retzloff, B. M. Coul, and J. Coul, *J. Phys. Chem.*, **74**, 2455 (1970).
80. W. W. Schmiegal, F. A. Litt, and D. O. Cowan, *J. Org. Chem.*, **33**, 3334 (1968).
81. D. S. Sethi and D. Devaprabhakara, *Indian J. Chem.*, **7**, 294 (1969).
82. Su Chang, N. C. Lou, and T. Y. Chung, *K'o Hsueh T'ung Pao*, 175 (1964); *ibid.*, 548.
83. V. R. Choudhary and L. K. Doraiswamy, *Ind. Eng. Chem. Prod. Res. Develop.*, **10**, 218 (1971).
84. M. van Swaay in *Advances in Chromatography*, Vol. 8, J. C. Giddings and R. A. Keller, eds., Dekker, New York, 1968, pp. 363–385.
85. T. Hattori and Y. Murakami, *J. Catal.*, **10**, 114 (1968); Y. Murakami and T. Hattori, *ibid.*, **10**, 123 (1968).

86. W. A. Blanton, C. H. Byers, and R. P. Merrill, *Ind. Eng. Chem., Fundam.*, **7**, 611 (1968).
87. R. A. Keller and J. C. Giddings, *J. Chromatogr.*, **3**, 205 (1960).
88. A. Klinkenberg, *Chem. Eng. Sci.*, **15**, 255 (1961).
89. W. R. Moore and H. R. Ward, *J. Phys. Chem.*, **64**, 832 (1960).
90. G. L. Pratt and J. H. Purnell, *Trans. Faraday Soc.*, **60**, 371 (1964).
91. G. L. Pratt and S. H. Langer, unpublished data.
92. J. A. Dinwiddie, U.S. Patent 2,976,132, 1961.
93. E. M. Magee, Canadian Patent 631,882, 1961.
94. G. A. Gaziev, S. Z. Roginskii, and M. I. Yanovskii, USSR Patent 149,398, 1962; *Chem. Abstr.*, **58**, 5082 (1963).
95. J. M. Matsen, J. W. Harding, and E. M. Magee, *J. Phys. Chem.*, **69**, 522 (1965).
96. S. Z. Roginskii, R. A. Zimin, and M. I. Yanovskii, *Dokl. Akad. Nauk SSSR*, **164**, 144 (1965).
97. S. Z. Roginskii and A. L. Rozental, *Dokl. Akad. Nauk SSSR*, **162**, 621 (1965).
98. E. M. Magee, *Ind. Eng. Chem., Fundam.*, **2**, 35 (1963).
99. F. Abet, F. Collavo, and M. Mauri, *Quad. Ing. Chim. Ital.*, **4**, 7 (1968).
100. F. E. Gore, *Ind. Eng. Chem., Process Des. Develop.*, **6**, 10 (1967).
101. C. G. Collins and H. A. Deans, *AIChE J.*, **14**, 25 (1968).
102. C. A. Barrere and H. A. Deans, *AIChE J.*, **14**, 280 (1968).
103. R. Kobayashi, P. S. Chappelear, and H. A. Deans, *Ind. Eng. Chem.*, **59** (10), 63 (1967).
104. H. A. Deans, F. J. M. Horn, and G. Klauser, *AIChE J.*, **16** (3), 426 (1970).
105. L. G. Harrison, Y. Koga, and P. Madderom, *J. Chromatogr.*, **52**, 31 (1970).
106. L. G. Harrison and Y. Koga, *J. Chromatogr.*, **52**, 39 (1970).
107. R. J. Kokes, H. H. Tobin, and P. H. Emmett, *J. Am. Chem. Soc.*, **77**, 5860 (1955).
108. R. J. Kokes, *Phys. Chem. Solids*, **14**, 51 (1960).
109. P. H. Emmett, *Advan. Catal.*, **9**, 645 (1957).
110. W. K. Hall and P. H. Emmett, *J. Am. Chem. Soc.*, **79**, 2091 (1957).
111. W. K. Hall and P. H. Emmett, *J. Phys. Chem.*, **63**, 1102 (1959).
112. W. K. Hall, D. S. MacIver, and H. P. Weber, *Ind. Eng. Chem.*, **52**, 421 (1960).
113. L. S. Ettre and N. Brenner, *J. Chromatogr.*, **3**, 524 (1960).
114. A. I. M. Keulemans and H. H. Voge, *J. Phys. Chem.*, **63**, 476 (1959).
115. P. Steingaszner and H. Pines, *J. Catal.*, **5**, 356 (1966).
116. P. Steingaszner and H. Pines, *Magy. Kem. Lapja*, **22**, 6 (1967).
117. D. P. Harrison, J. W. Hall, and H. F. Rase, *Magy. Kem. Lapja*, **57** (1), 18 (1965).
118. K. Tamaru, *Nature*, **183**, 319 (1959).
119. K. Tamaru and J. Nakanisha, *Kagaku no Ryoiki Zokan*, **53**, 83 (1964).
120. K. Tamaru, *Advan. Catal.*, **14**, 65 (1964).
121. J. W. Hightower and K. H. Hall, *J. Am. Chem. Soc.*, **90**, 581 (1968).
122. J. W. Hightower and K. H. Hall, *J. Phys. Chem.*, **72**, 4555 (1968).
123. K. J. Laidler, *Chemical Kinetics*, McGraw-Hill, New York, 1965.
124. K. H. Yang and O. A. Hougen, *Chem. Eng. Progr.*, **46**, 146 (1950).
125. G. M. Schwab and A. M. Watson, *J. Catal.*, **4**, 570 (1965).
126. M. R. Muchhala, S. K. Sanyal, and S. W. Weller, *J. Chromatogr. Sci.*, **8**, 127 (1970).
127. E. Soloman, J. McMahon, E. Sterling, and H. Heineman, *C.R. 31st Congr. Int. Chim. Ind. , Liege, 1968*; published as *Ind. Chim. Belge. Suppl.*, **1**, 546 (1959).
128. M. Hartwig, *Brennst.-Chem.*, **45**, 234 (1964).
129. T. M. Yushchenko-Shaprinskaya, G. P. Korneichuk, V. P. Ushakova-Stasevich, and Yu. V. Semenyuk, *Kinet. Katal.*, **4**, 154 (1968).
130. J. Parasiewics-Kaczmarska and J. Ejsymont, *Zesz. Nauk Uniw. Jagiellon, Pr. Chem.*, no. 14, **257** (1969).

131. I. Ya. Gavrilina and D. A. Vyakhirev, *Usp. Khim.*, **36**, 363 (1967).
132. Y. Murakami, *Kogyo Kagaku Zasshi*, **68**, 31 (1965).
133. S. Z. Roginskii, M. I. Yanovskii, and G. A. Gaziev, *Gas Chromatogr., Akad. Nauk SSSR, Tr. Vtorio Vses. Konf., Moscow*, 1962, 27, published 1964.
134. T. Paryjezak, *Wiad. Chem.*, **22**, 481 (1968).
135. G. M. Schwab, *Chem.-Ing.-Tech.*, **39**, 1191 (1967).
136. N. Kominami and H. Nakajima, *Kogyo Kagaku Zasshi*, **69**, 233 (1966).
137. L. deMourgues, M. Fishet, and G. Chassaing, *Bull. Soc. Chim. France*, 1918 (1962).
138. C. J. Norton and T. E. Moss, *Ind. Eng. Chem., Process Des. Develop.*, **3**, 23 (1964).
139. K. V. Topchieva, E. N. Rosolovskaya, and O. L. Shakhnovskaya, *Vestn. Mosk. Univ. Ser. II*, **23**, 39 (1968).
140. V. V. Yushehenko and T. V. Antipina, *Zh. Fiz. Khim.*, **43**, 540 (1969a).
141. R. B. Anderson, K. C. Stein, J. J. Feenan, and L. J. E. Hofer, *Ind. Eng. Chem.*, **53**, 809 (1961).
142. J. J. Feenan, R. B. Anderson, H. W. Swan, and L. J. E. Hofer, *J. Air Pollut. Contr. Ass.*, **14**, 113 (1964).
143. L. J. E. Hofer, J. F. Schultz, and J. J. Feenan, *U.S. Bur. Mines Rept. Invest.*, 6243, 1963.
144. K. C. Stein, J. J. Feenan, G. P. Thompson, J. F. Schultz, L. J. E. Hofer, and R. B. Anderson, *J. Air Pollut. Contr. Ass.*, **10**, 275 (1960).
145. K. C. Stein, J. J. Feenan, J. F. Schultz, L. J. E. Hofter, and R. B. Anderson, *Ind. Eng. Chem.*, **52**, 671 (1960).
146. J. E. Germain and J. P. Beaufils, *Bull. Soc. Chim. France*, 1172 (1961).
147 J. E. Germain, J. Birourd, J. P. Beaufils, B. Gras, and L. Ponsolle, *Bull. Soc. Chim. France*, 1504 (1961).
148. Ref. 147, p. 1777.
149. J. E. Germain and L. Ponsolle, *Bull. Soc. Chim. France*, 1572 (1961).
150. J. G. Larson, H. R. Gerlack, and W. K. Hall, *J. Am. Chem. Soc.*, **87**, 1880 (1965).
151. L. deMourgues, *Chim. Anal. (Paris)*, **45**, 103 (1963).
152. J. Marechal, L. Convent, and J. van Rysselberge, *Rev. Inst. Franc. Petrol. Ann. Combust. Liq.*, **12**, 1067 (1957).
153. A. A. Kubasov, I. V. Smirnova, and K. V. Topchieva, *Kinet. Katal.*, **8**, 146 (1967).
154. A. A. Kubasov, I. V. Smirnova, and K. V. Topchieva, *Kinet. Katal.*, **8**, 351 (1967).
155. L. deMourgues and J. Capony, *Journ. Int. Etude Methodes Separation Immediate Chromatogr. (Paris)*, 1961, p. 163.
156. D. A. Cadenhead and N. G. Masse, *J. Phys. Chem.*, **70**, 3558 (1966).
157. H. J. Dutton and T. L. Mounts, *J. Catal.*, **3**, 363 (1964).
158. E. I. Semenenko, S. Z. Roginskii, and M. I. Yanovskii, *Kinet. Katal.*, **6**, 320 (1965).
159. H. Yamamoto, M. O'Hara, and T. Kwan, *Chem. Pharm. Bull. (Tokyo)*, **12**, 959 (1964).
160. P. Desikan and C. H. Amberg, *Can. J. Chem.*, **40**, 1966 (1962).
161. P. J. Owens and C. H. Amberg, *Advances in Chemistry Series*, no. 33, American Chemical Society, Washington, D. C., 1961, pp. 182–198.
162. P. J. Owens and C. H. Amberg, *Can. J. Chem.*, **40**, 941 (1962).
163. P. J. Owens and C. H. Amberg, *Can. J. Chem.*, 947 (1962).
164. M. Bartok and B. Kozma, *Acta Chim. Acad. Sci. Hung.*, **51**, 403 (1966); *ibid.*, **52**, 83 (1967).
165. A. Romanovski, *Kinet. Katal.*, **8**, 921 (1967).
166. A. Yu. Aleksandrov and M. I. Yanovskii, *Kinet. Katal.*, **2**, 794 (1961).
167. M. Barbul, Gh. Serban, I. Ghejan, and T. Filotti, *Petrol. Gaz.*, **19** (3), 181 (1968).
168. S. H. Langer, J. Y. Yurchak, and C. M. Shaughnessy, *Anal. Chem.*, **40**, 1747 (1968).
169. J. Bohemen, S. H. Langer, R. H. Perrett, and J. H. Purnell, *J. Chem. Soc.*, 2444 (1960).

170. D. T. Sawyer and J. K. Barr, *Anal. Chem.*, **34**, 1518 (1962).
171. J. E. Patton and S. H. Langer, *Anal. Chem.*, **42**, 1449 (1970).
172. M. W. Anders and G. L. Mannering, *Anal. Chem.*, **34**, 730 (1962).
173. R. J. De Pasquale and C. Tamborski, *Chem. In. (London)*, 771 (1968).
174. S. H. Langer, J. E. Patton, and J. Coca, *Chem. Ind. (London)*, 1346 (1970).
175. A. I. Vogel, *Practical Organic Chemistry*, 3rd ed., Wiley, New York, 1967, p. 318.
176. L. F. Feiser, *Organic Experiments*, 2nd ed., Raytheon Education Co., Lexington, Mass., 1968, pp. 85–88.
177. H. Elias, K. H. Lieser, and F. Sorg, *Radiochim. Acta*, **2**, 30 (1963).
178. S. Krutzik and H. Elias, *Radiochim. Acta*, **7**, 26 (1967).
179. H. Elias, *Proceedings of the Symposium on Preparation and Bio-Medical Application of Labelled Molecules, Venice, 1964*, Euratom, Brussels, 1964, pp. 467–480.
180. J. Tadmor, *J. Inorg. Nucl. Chem.*, **23**, 158 (1961).
181. F. Daniels and D. A. Alberty, *Physical Chemistry*, 2nd ed., Wiley, New York, 1963, p. 707.
182. G. Friedlander and J. W. Kennedy, *Nuclear and Radiochemistry*, Wiley, New York, 1955, p. 315.
183. R. W. A. Rijnders, in *Advances in Chromatography*, Vol. 3, J. C. Giddings and R. A. Keller, eds., Dekker, New York, 1966, pp. 215–258.
184. F. Schmidt-Bleek, G. Stöcklin, and W. Herr, *Angew. Chem.*, **72**, 778 (1960).
185. J. Tadmor, *Anal. Chem.*, **36**, 1565 (1964).
186. J. Tadmor, *Anal. Chem.*, **38**, 1624 (1966).
187. H. Elias, in *Advances in Chromatography*, Vol. 7, J. C. Giddings and R. A. Keller, eds., Dekker, New York, 1968, pp. 243–292.
188. T. Balent and L. Szepsy, *J. Chromatogr.*, **30**, 433 (1967).
189. G. J. Kallos and L. B. Westover, *Tetrahedron Lett.*, **13**, 1223 (1967).
190. H. Elias, *Proceedings of the Conference on Methods of Preparing and Storing Marked Molecules, Brussels, 1963*, Euratom, Brussels, 1964, pp. 1206–1215.
191. M. Senn, W. J. Richter, and A. L. Burlingame, *J. Am. Chem. Soc.*, **87**, 680 (1965).
192. W. J. Richter, M. Senn, and A. L. Burlingame, *Tetrahedron Lett.*, **17**, 1235 (1965).
193. G. Stöcklin, *Proceedings of the Symposium on Preparation and Bio-Medical Application of Labeled Molecules, Venice, 1964*, Euratom, Brussels, 1964, pp. 481–493.
194. G. Stöcklin, F. Schmidt-Bleek, and W. Herr, *Proceedings of Journées Internationales d'Etude des Méthodes de Séparation Immédiate et de Chromatographie, Paris, 1961*, G.A.M.S., L.N.E. 1, Paris, 1961, pp. 114–122.
195. G. Stöcklin, F. Schmidt-Bleek, and W. Herr, *Angew. Chem.*, **73**, 220 (1961).
1969 S. Krutzik and H. Elias, *Radiochim. Acta*, **7**, 33 (1967).
197. K. Mislow, M. A. Glass, H. B. Hopps, E. Simon, and G. H. Wahl, Jr., *J. Am. Chem. Soc.*, **86**, 1710 (1964).
198. P. D. Klein and E. H. Erenrich, *Anal. Chem.*, **38**, 480 (1966).
199. H. Dahn, R. Menasse, and C. Tamm, *Helv. Chim. Acta*, **42**, 2189 (1959).
200 P. D. Klein and J. C. Knight, *J. Am. Chem. Soc.*, **87**, 2657 (1965).
2019 S. M. Karpacheva and A. M. Rozen, *Dokl. Akad. Nauk SSSR*, **75**, 239 (1950).
202. G. Hesse, *Z. Anal. Chem.*, **211**, 5 (1965).
203. G. Hesse, *Z. Anal. Chem.*, **236**, 192 (1968).
204. M. J. Baldwin and R. K. Brown, *Can. J. Chem.*, **44**, 1743 (1966).
205. C. T. Bishop, F. P. Cooper, and R. K. Murray, *Can. J. Chem.*, **41**, 2245 (1963).
206. J. K. Haken, *J. Gas Chromatogr.*, **1** (10), 30 (1963).
207. D. M. Marmion, R. G. White, L. H. Bille, and K. H. Ferber, *J. Gas Chromatogr.*, **4**, 190 (1966).

208. S. H. Langer and P. Pantages, *Nature,* **191,** 141 (1961).
209. S. H. Langer, J. E. Patton, and H. Kung, in preparation.
210. E. R. Adlard, M. A. Kahn, and B. T. Whitham, *Gas Chromatography,* Butterworths, London, 1960, p. 251.
211. D. H. Everett, *Trans. Faraday Soc.,* **61,** 1637 (1965).
212. J. R. Conder and J. H. Purnell, *Trans. Faraday Soc.,* **64,** 1505 (1968).
213. A. A. Frost and R. G. Pearson, *Kinetics and Mechanism,* Wiley, New York, 1961, pp. 127–130.
214. C. A. Eckert, *Ind. Eng. Chem.,* **59** (9), 20 (1967).
215. D. E. Martire, P. A. Blasco, P. F. Carone, L. C. Chow, and H. Vicini, *J. Phys. Chem.,* **72,** 3489 (1968).
216. J. C. Giddings, *Dynamics of Chromatography,* Dekker, New York, 1965.
217. P. J. Wilson and J. H. Wells, *Chem. Rev.,* **34,** 1 (1944).
218. S. W. Benson, *Foundations of Chemical Kinetics,* McGraw-Hill, New York, 1960.
219. J. E. Leffler and E. Grunwald, *Rates and Equilibria of Organic Reactions,* Wiley, New York, 1963.

AUTHOR INDEX

Numbers in parentheses are reference numbers and show that an author's work is referred to although his name is not mentioned in the text. Numbers in *italics* indicate the pages on which the author is mentioned in the text or the full references appear.

Abcor, Inc., *139, 167, 168*(6), *171*, 173(6), *182*

Abdul-Karim, A., 94(16), *134*

Abel, R., 31(30), *81*

Abet, F., *341*, 342(99), *370*

Abraham, N. A., 317(47), *368*

Abrams, S. T., 97(23), *134*

Ackerman, D. G., 31(42), 47(42), 48(42), 64(42), *81*

Adams, M. F., 21(121), *22, 27*

Adlard, E. R., 359(210), *373*

Adrian, J. C., 324(54), 326(54), 329(54), *368*

Aho, K., 13(70), *26*

Ainsworth, C. A., 2(5), 3(5), *24*

Akagi, M., 20(111), *27*

Albert, D. K., 21(118), *27*

Alberty, D. A., 355(818), *372*

Albrecht, J., *168*, 169(102), *184*

Aleksandrov, A. Yu., 344(166), *371*

Alishoev, V. R., *215, 235*

Al Madfai, S., 140(35), *183*

Alonzo, N., 49(68), *82*

Altenau, A. G., 93(14), 94(14), 98(14), *134, 216, 235*

Althaus, I. R., 59(173), *84*

Amberg, C. H., 344(160-163), 348(162), *371*

American Chemical Society, 30(7,8), *80*

Amundson, N., *298, 368*

Anders, M. W., 352(172), 357(172), *372*

Anderson, A. H., 47(50,51), 71(50,51,207), 72(50), 73(50), 76(51), 77(50,51,214),

78(50), *81, 85, 86*

Anderson, R. B., 344(141,142,144,145), *371*

Anderson, R. E., 40(89), *82*

Andrews, L. J., 268(49), *291*

Annino, R., 124(98,99), 125-129(99), 131 (99), 132(99), *136*

Anon., 137(1-3,5), 139(1-3,5), 172-175(3), *182*

Anson, F. C., 31(30), *81*

Antipina, T. V., 344(140), *372*

Apelblat, A., *284*, 287(54), *291*

Arisawa, K., 213(41), *235*

Aslam, Kham, M., 106(72), *135*

Astil, B. D., 5(18), *25*

Ateya, K., *213, 235*

Atkinson, E. P., 167(107a), 169(107), 171 (107a), *184*

Aznavourian, W., 50(84), *82*

Bacallo, C. Z., 13(74), *26*

Bagwell, E. E., 6(25), *25*

Baker, W. J., 99(128), *134*

Baldwin, M. J., 357(204), *372*

Balent, T., 358(188), *372*

Barber, D. W., 238(188), *372*

Barbul, M., 344(167), *371*

Barker, P. E., *140, 182, 183*

Barr, J. K., 348(170), *372*

Barraclough, J. M., 57(118), 59(118), *83*

Barrall, E. cm., 100(41), 101(41), 124(41), *135*

Barrere, C. A., 344(102), *370*

SUBJECT INDEX

Advances in Analytical Chemistry and Instrumentation

CUMULATIVE INDEX, VOLUMES 1–9

Author Index

Subject Index